The International Library of Environmental, Agricultural and Food Ethics

Volume 38

The ethics of food and agriculture is confronted with enormous challenges. Scientific developments in the food sciences promise to be dramatic; the concept of life sciences, that comprises the integral connection between the biological sciences, the medical sciences and the agricultural sciences, got a broad start with the genetic revolution. In the mean time, society, i.e., consumers, producers, farmers, policymakers, etc, raised lots of intriguing questions about the implications and presuppositions of this revolution, taking into account not only scientific developments, but societal as well. If so many things with respect to food and our food diet will change, will our food still be safe? Will it be produced under animal friendly conditions of husbandry and what will our definition of animal welfare be under these conditions? Will food production be sustainable and environmentally healthy? Will production consider the interest of the worst off and the small farmers? How will globalisation and liberalization of markets influence local and regional food production and consumption patterns? How will all these developments influence the rural areas and what values and policies are ethically sound?

All these questions raise fundamental and broad ethical issues and require enormous ethical theorizing to be approached fruitfully. Ethical reflection on criteria of animal welfare, sustainability, liveability of the rural areas, biotechnology, policies and all the interconnections is inevitable.

The International Library of Environmental, Agricultural and Food Ethics contributes to a sound, pluralistic and argumentative food and agricultural ethics. It brings together the most important and relevant voices in the field; by providing a platform for theoretical and practical contributors with respect to research and education on all levels.

Marcello Di Paola
Editor

The Vegetal Turn

History, Concepts, Applications

Springer

Editor
Marcello Di Paola
Department of Humanities
University of Palermo
Palermo, Italy

ISSN 1570-3010 ISSN 2215-1737 (electronic)
The International Library of Environmental, Agricultural and Food Ethics
ISBN 978-3-031-68670-2 ISBN 978-3-031-68671-9 (eBook)
https://doi.org/10.1007/978-3-031-68671-9

This Springer imprint is published by the registered company Springer Nature Switzerland AG
The registered company address is: Gewerbestrasse 11, 6330 Cham, Switzerland

Contents

About the Editor

Marcello Di Paola is Assistant Professor of Philosophy at the Department of Humanities, University of Palermo. He writes on plants, gardens, climate change, the Anthropocene, outer space exploration, meaning in life, virtue ethics, and the thought of Baruch Spinoza. He is the author of *Ethics and Politics of the Built Environment: Gardens of the Anthropocene* (2017) and co-editor of *Plant Ethics: Concepts and Applications* (2018), *The Springer Handbook of the Philosophy of Climate Change* (2023), and *The Philosophy of Outer Space: Explorations, Controversies, Speculations* (2024).

Chapter 1
The Vegetal Turn

Marcello Di Paola

Abstract The Vegetal Turn is driven by this new, eclectic, mounting wave of theoretical and practical interest in the complexities and peculiarities of plant life, plant-human relations, plant-nonhuman relations, and plant-mediated human relations. From a mostly philosophical perspective, this volume provides a snapshot of that mounting wave in action, its kaleidoscopic interests and spreading trajectories, its less-than-fully-disciplined methods and approaches, and its central conceptual and practical aspirations. This chapter introduces the volume.

1.1 Prologue

Look up.

If you see no plants around you—no bush, no grass, no tree, no shrub, no vine, no moss, no flowers cut—you must be in a very strange place. It might be a great place, like a slick, well-lit hotel room or office, and yet in it you are being deprived of something that is otherwise copiously available all over Earth—in innumerable versions, configurations, and populations; pervasively, rambunctiously, unrelentingly, triumphantly.

That's plants. Earth's vegetal legion.

On this planet, few modes of being are as powerful as plants. They are almost everywhere, and if they are not with you now perhaps you can at least see some of them through the window, or maybe there is a painting, a print, or some drawing of a plant on your walls. If not, then the place you are in is not only strange but also built with no love, for humans or plants. And if you feel that no plant could thrive there—if you sense that there would not be enough light, air, space, good water, or fertile soil for a plant to do its thing—then you also know that place cannot be too good for you either, at least not in the long run. You are in a strange, loveless, unhealthy place. You should leave.

M. Di Paola (✉)
Department of Humanities, University of Palermo, Palermo, Italy
e-mail: marcello.dipaola01@unipa.it

M. Di Paola (ed.), *The Vegetal Turn*, The International Library of Environmental, Agricultural and Food Ethics 38, https://doi.org/10.1007/978-3-031-68671-9_1

Step outside, go to a fruit stand, get an orange, sit under a tree, peel the orange. As you do, notice the white strands and ligaments between the fruit and its external, organic armour. These are fungi, not orange, but it all comes in one package. And orange is fruit and was plant, is seed and will be plant, always enveloped with fungi past and future. The realms of fungi and plants are separated in our books but totally synergic otherwise, always, everywhere. Perhaps all that is separated in our books is totally synergic otherwise, always, everywhere.

Hard to believe, from inside a hotel room. But you are out now. There is likely very many plants around you now, seen and unseen, big and small, shooting and crawling, nascent and dying, climbing and falling, close and distant—and there is bits of plants all around you, branches and leaves above the soil and roots within it, some dead some alive, and trillions of ungerminated seeds in sidewalk cracks and under the asphalt and cement. And then the plant eating-ants, the plant-pollinating flies, the seeds eating birds—these too are all swirling around you now, just because you are around plants.

So eat the orange and feel your body recognize its juice immediately, almost as it does water. There is water, in the orange, there is water in orange trees, and there is water in the soils where these trees grow. Most of nature's "realms", and at least two of its elements, soil and water, have just come alive in your animal hands and mouth via that one orange. And the tree you are sitting under, and any other tree that is in sight or has ever been on Earth, they modulate that third element that you so desperately need—that air you breathe. And when the fourth element, fire, burns things, you know what it burns is mostly plants—either those living, or those dry, or those fossil ones that power electric lights in slick hotel rooms or offices, and your computers and fridge and heating system. We call them oil, or coal, or gas—but it's really plants, and the animals that ate them.

Plants are successful, bountiful, indispensable, mutable. They surround us silently, ubiquitously, and in infinite variations. Evolvable homages to the explosive plasticity of being, they affect and are affected by everything there is. They feed on and coagulate the elements into cellular bodies, and attract, like focal nodes, all organisms. They do a great many things seemingly without moving or talking. They disseminate, to occupy several ontological niches at once: tree is plant, leaf is plant, root is plant, fruit is plant, seed is plant. A plant that breaks into pieces often just becomes more plants. Any plant is many, and lives and dies multiple times in multiple ways.

Earth's Legion.

But how is that legion, and the non-vegetal systems both animate and inanimate that are entangled with its innumerable lives and deaths, to be distilled in and by our thoughts and words, and addressed in action? Where to begin, and how? How to approach the plant world once it becomes clearer that their world is ours too—that the human world as we know it is neither possible nor conceivable without the plant world? What concepts are now at our disposals, and what practices are currently guiding our explorations of the vegetal? And are these concepts appropriate and useful, and these practices rich, meaningful, and farsighted? Do our communities know and understand and work with plants in ways that are maximally empowering for humans, for plants, and for this planet that is filled with both humans and plants?

If you ask these questions, then you must have had your own personal Vegetal Turn, just now. You are pondering themes that, until only quite recently, very few could even be bothered with. But things have changed, and precisely these questions you now ask are propelling this new, young, eclectic, mounting wave of theoretical and practical interest in the complexities and peculiarities of plant life, plant-human relations, plant-nonhuman relations, and plant-mediated human relations. From a mostly philosophical perspective, this volume provides a snapshot of that mounting wave in action, its kaleidoscopic interests and spreading trajectories, its less-than-fully-disciplined methods and approaches, and its central conceptual and practical aspirations.

1.2 The Vegetal Turn

There are, of course, so many turns: just recently, the linguistic turn, and then the ontological turn, the posthuman turn, the digital turn and, of course, this whole Anthropocene thing, represented as nothing less than an epochal turn.[1] These are just some of the turns you have been reading about and living through, and are in various ways connected with one another. Now is the turn of the vegetal, which also connects with all these others, in often unexpected ways—because everything on Earth implicates plants, and plants, like medieval gods, complicate everything.

In the last thirty years or so, a very diverse crowd that includes botanists, philosophers, literary theorists, artists, environmental scientists, psychologists, historians, biologists, ecologists, anthropologists, health scientists and practitioners, architects and urban planners, indigenous scholars, environmental activists, engineers, and computer scientists, has contributed to bringing plants to cultural attention in and outside academia.

The Vegetal Turn (VT) consists of these increasingly sophisticated, inter- and trans-disciplinary, inter- and trans-cultural explorations of the manifold systems and networks of communication, intelligence, technical-operational capabilities, and relations articulated by and via plants. In its applicational aspects, it concerns the ethical, aesthetic, economic, technological, psychological, cultural, and political dimensions of plant-human, plant-nonhuman, and plant-mediated human relations and interactions.

[1] That of the Anthropocene is a hotly contested notion, which does not find an official place in geological taxonomies. But the conceptual core of it—that Earth's present and future course is now consitutively bound up with, if not driven by, human activities—informs the reasoning of most of the contributors to this volume. On the geological Anthropocene see, among others, Vitousek et al. (1997), Crutzen and Stoermer (2000), Crutzen (2002), Steffen et al. (2011), Zalasiewicz et al. (2011), (2016), and Finney and Edwards (2016) for a critique. On debates on the Anthropocene in the humanities and social sciences see, among others, Bonneuil and Fressoz (2015), Hamilton et al. (2015), Chakrabarty (2015), Haraway (2016), Moore (2016), and Latour (2017). For my own interventions in this debate see Di Paola (2015), (2017), Jamieson and Di Paola (2016), Di Paola and Jamieson (2018), and Di Paola and Ciccarelli (2022).

VT has at least four distinct bundles of roots. First, significant progress in understanding the complexity of plant physiology and ecology, and consequent philosophical developments in vegetal ontology and conceptions of plant agency (see, among others, Miller 2002; Trewavas 2003, 2005, 2014; Firn 2004; Baluska et al. 2006, 2009 Struik et al. 2008; Chamovitz 2012; Gagliano et al. 2014; Calvo 2016, 2017; Maher 2017; Taiz et al. 2019; Calvo et al. 2020; Hendlin 2021; Hiernaux 2021b; Calvo and Lawrence 2023; Parise and Marder 2024).

Second, the questioning of axiological hierarchies of anthropocentric (or at most zoo-centric) inspiration, now undermined by more nuanced views of what plants are and all the ways they can matter (see, among others, Stone 1972; Goodpaster 1978; Attfield 1981, 2018; Warren 1987; Arbor 1986; Taylor 1986; Pollan 1991, 2001; Sterba 1995; Plumwood 1999; Hallé 2002; Cooper 2012; Koechlin 2009; Hall 2011; Houle 2011; Ryan 2012; Marder 2013, 2014; Marris 2013; Pouteau 2014, 2018; Kallhoff 2014, 2018; Harman 2007; Coccia 2018; Odparlik 2018; Pellegrino 2018; Sandler 2018; Coeckelbergh 2018; Schörgenhumer 2018; Visak 2018).

Third, the need for normative directives for the ethics and politics of fundamental sustainability domains—including agriculture, urbanization, public and private health, ecosystem integrity, biodiversity conservation, landscape and cultural heritage protection, and climate adaptation— which all depend, directly or indirectly, on plant life and its futures (Williams 2003; Harvey 2006; Hester 2006; Beerling 2007; FECNHB 2008; Brown 2011; Mazzolai et al. 2014; Di Paola 2017; essays in Kallhoff, Di Paola and Schörgenhumer 2018; Doggett 2018; Prajapati and Singh 2019; Bibikhin 2021; Ennos 2021; Hiernaux 2021a).

Fourth, the realization that a focus on plant life—this life so often demeaned and marginalized in and by our thought and practices–can be a fertile historical and critical prism through which to look at those many other human and nonhuman lives, and forms of life, which are also systematically demeaned and marginalized by dominant conceptual schemes and political arrangements. In this respect, the marginalization of plants can be seen as one among the many examples of these self-referential, self-absolutory, often obscurantist dynamics of diminution and removal that are regularly activated by those who are ostensibly "above" against those who are ostensibly "below" in some hierarchy, and which are so easily applied also to humans, other animals, and the rest of nature as circumstances may require (see, among others, Plumwood 1993, 2000; Drayton 2000; Miller 2002; Purdy 2015; Haraway 2016; Moore 2016; Jamieson and Di Paola 2016; Puleo 2019; Kimmerer 2020; Wood 2020).

These four root bundles mingle to build something like one comprehensive research program. The plant agency that vegetal ontologies explore is considered pivotal to the construction of richer and more nuanced vegetal axiologies. These in turn are thought to disclose reasons for increased consideration for plants in our practices, and increased experimentation with the roles that plants, much like those humans and non-humans that have traditionally been conceived and treated as non-agents, may directly or indirectly play in our politics (see Jamieson and Di Paola 2016). More generally, and now put negatively, VT stands opposed to the ethical, aesthetic, cultural, and political impoverishment, as well as the social, ecological, and indeed existential risks, which are connected to delisting plants from the catalogue of things that matter most on Earth.

The way towards conceptually more nuanced understandings of plant life and more mature practical relations to it is still very long, however, and replete with obstacles. Here are two. First, reflection on plants must somehow cover an inordinate amount and variety of beings—from these potted, mass-produced houseplants sitting on our urban windowsills, to the lettuces hanging in the air in hydroponic farms, to the monumental tree specimens in jungles and forests and the mosses that grow on them, and the stubborn cacti huge and minuscule that animate deserts—and anything else in between, be those individuals or populations, plus all the seeds, and leaves, and roots, and sprouts, and flowers. In addition, all plants participate in constant exchanges with other plant as well as non-plant beings, like fungi and bacteria and insects and other animals, humans included, as well as minerals and water and all chemical compounds. So, inevitably, anyone reflecting on plants will need to make controversial choices about what will count as "plants" in any given theoretical and practical context, as well as how wide a relational perimeter around these plants they will be considering. Very few claims will be always generally true of all plants everywhere—and the same goes for plant-human relations, plant-nonhuman relations, and plant-mediated human relations. As Aristotle's disciple Theophrastus, who wrote an *Inquiry into Plants* in the fourth century BC, was never tired of repeating, generalizations about the vegetal will always be possible but will also always be local and provisional: far too mutable and unbridled is the vegetal for any one story to tell its whole story.

Second, there are issues with the quality of our appreciation of plants. Precisely because there is so many of them everywhere, always silently doing what plants have always been doing for the past half billion years (at least), it is all too easy to take them for granted. Because plants seem not to move, and to mostly react to bioecological stimuli rather than initiate actions of their own, it is easy to reduce them to a background for the more perspicuous acting of animals, particularly humans. And from these animals and humans, plants seem separated by an abyss: they seem to have no senses, no brain, no passions, no consciousness, no languages.

Yet plants are also vertiginously like us. Some similarities are obvious: like humans and other animals they feed, grow, reproduce, decay, die, feed others. Other similarities can go undetected instead, due to a lack of fit between vegetal expression and the human sensorium. For example, research suggests that plants produce airborne sounds when cut or stressed, but these sounds are ultrasonic, in the range of 20–100 kilohertz, and therefore cannot be detected by human ears (Khait et al. 2023). Movement, too, is a power that plants have—yet the ways and paces at which they move are mostly lost on us. And there seem to be strategic and communicative behaviours of plants that challenge not only the assumption that plants have no forms of mentality, but also the paradigmatic, clearly zoo- and particularly anthropo-centric picture that is typically entertained of mentality itself (Calvo and Lawrence 2023).

How to negotiate this faraway, so close ontology of plants—this tension between basic identification and radical distinction, this mirrored ambivalence both of "us" and of "them", and that familiarly estranged "thus" that together we are,—has long been a source of conceptual and practical disorientation for those who have attempted to grapple with the complexity of the vegetal. Among the most vivid expressions given to such disorientation is the ancient characterization of plants as humans in reverse, with their mouths in the soil and feet in the air—fairly diffused among early Greek philosophers (Repici 2000).[2] Another vivid expression of disorientation is the silence of the many others, in Western philosophy at least, who have not even attempted to grapple with the vegetal at all, or have seen the vegetal within us—material, earthy, unruly, growing, impulsive, and decaying—as a burden to shake off on the way to knowledge, beauty, goodness, or holiness.

Insensate, weirdly living matter; or a mutely enacted backdrop; or a stage, a part of landscapes, a thing you eat, a medicine, a symbol, a metaphor, a commodity, a nuisance without and a burden within—all that, and more, is the vegetal. What it did *not* use to be, at least not in philosophy departments until very recently, was this new and seemingly inexhaustible theme for dedicated reflection from a great many perspectives. And it was about time for that: the real-life issues calling for a far more articulated theory and practice of the vegetal are today many and important, and the need for more nuanced societal debates on vegetal matters is pressing on many fronts.

For example: How should we respond to the losses—or simply transformations—of individual plants, plant populations, plant species, plant habitats, plant-rich landscapes and ecosystems that rising global temperatures are and will increasingly be causing? From a vegetal perspective, what is bad or unfair about pollution or massive land use change, and whom or what is wronged by those exactly: plants themselves, peoples whose material and cultural heritage is tied to plants, future generations? Also to combat climate change and major land-use changes designed to accommodate animal pasturing, shouldn't we all be vegetarians—or is eating plants morally objectionable in its own rights? Are *ex situ* plant protection projects and seed banking to be seen as second-best policies compared to *in situ* conservation? What about endangered plant species whose *in situ* survival is endangered not only by poaching but also by planetary, systemic sources of local habitat loss or degradation? And is there anything peculiarly wrong with the poaching of wild plants, and does that apply to some plants more than others? How much should cities invest in plants and their maintenance when configuring the 'green infrastructures' that seem pivotal to urban climate adaptation, and what sort of civic culture should be built around the vegetal? What forms of biotechnologies, and plant manipulation more generally, and under what circumstances (ecological as much as cultural, economic, and political), can be considered normatively justified, and which ones even call for justification at all? Should synthetic biology (i.e. lab-based creation of living organisms, including plants) be prohibited, publicly funded, or left to market forces? What, if any, forms of plant-centered geo-engineering (including a range of possibilities, from re-afforestation to the genetic enhancement of plants' metabolic capacities) should be pursued or discouraged in the hyper-technologized Anthropocene, and on

[2] For further explorations in the history of philosophy of plants see, among many others, Delaporte (1982), Marder (2013), and Baldassarri and Blank (2021).

what grounds? Does data-driven and AI-directed precision agriculture fundamentally change important aspects of the relation between humans and plants? If so, what is gained and what is lost when agriculture becomes artificially intelligent? Is massive deforestation a crime against humanity, or against nature, or both, or neither? Should food and cosmetic businesses be allowed to breed and select the most marketable varieties of plants, if that accelerates the loss of less marketable varieties, thus decreasing biodiversity with accompanying eco-systemic and socio-cultural costs? What, if any, are the obligations of large businesses towards plants themselves, as well as towards members of cultures and traditions whose plants or botanical knowledge can be profitably marketed? What are the implications, in terms of global and inter-generational justice, of the emergence of an oligopolistic network of large agrochemical multinationals dictating plant research, biological structures, commercialization, and conservation? What historical and ideological facts can be read in plant bodies and landscapes, and the ways humans know and use them? What forms of epistemic and economic injustices among humans are tied to plants, or tie humans via plants, and how are they to be addressed, by whom, and in what contexts? What institutional conditions, and what forms of political mechanisms and at what scales, are required so that plants, and plant-related and/or mediated human and nonhuman interests, not be imprudently and/or unfairly discounted in public deliberation and policy making?

These questions intersect at various junctures, and all involve a multiplicity of stakeholders—human and nonhuman, vegetal and non-vegetal, local and global, individual and collective, present and future—while engaging institutions and organizations operating at different scales and often pursuing uncoordinated, in some cases even antagonistic agendas. In these and other respects our new theories and practices of the vegetal—complex, messy, weird, and embryonic as they might be—poke at much larger, multispecies, planetary issues of sustainability, governance, political legitimacy, justice, and power—and function as laboratories for the assembling of this new, far more-than-human thinking and doing that is required to address, or even to just understand, contemporary eco-anthropic changes and challenges.

1.3 Overview of the Book

This book photographs VT at a time of relative maturity, when its penetration into academia and popular culture seems irreversible and its conceptual tools are getting increasingly sharper. Nonetheless, most of its evolution is still to come. This evolution has no in-built teleology and obeys no specific schools of thought: rather, it is a multitudinous striving towards a more powerful understanding of plants—and ourselves and the planet and the future via plants—through philosophical and other styles of reflection. The chapters that follow and constitute this book are each a mode of such striving.

To an extent that is perhaps unusual, but in many ways worth cherishing, readers will encounter a multiplicity and variety of themes, concerns, approaches, and styles that will at once orient and disorient them in the germinating and branching jungles of plant philosophy and its surroundings. Nonetheless, the volume has one unitary trajectory, which goes from conceptual issues to historical reconstructions, onto current debates on vegetal powers and ontologies, and then deep into plant ethics and politics.

The first two chapters engage with two fundamental questions: how to think of plants, and how plants have historically shaped our thinking. In his contribution, **Thomas Nail** explains how plants are one of Earth's powers, the vegetal ways in which Earth is. Plants are no guests of the land but its very way, that which the land does at any given moment—never mind if it does plants that are, as we call them, native or exotic, crops or weeds. Plants are Earth's breathwork. Through, with, and because of them Earth is what it is, with all its landscapes and animals, cities, plant philosophers, space rockets, and computer scientists.

Nail's notion of the vegetal as a constitutive geological power—perennial, explosive, unrelenting, pervasive—is at the very core of VT. It may be at odds with the everyday experience that most of us, particularly in cities, have of plants: things that are there but might just as well not be, or more precisely things that we let be there but might just as well dispense with. VT insists that seeing plants as contingent in this way, as guests on Earth and in human life, falsifies not only the ontology of plants but also that of Earth and human life (van der Veen 2014).

Andrea Le Moli explores the ways in which plants have shaped Western thinking from the start, encoded in its primordial conceptual nodes as representatives of both nature and materiality at large. The ancient Greeks called the primary material of things *hyle* ("matter," also "wood, woodland")—meaning, among other things, all that which can take shape. They then proceeded to ask two questions that have set the course for much later metaphysics as well as reflection on the relations between nature (*physis*, connected to *phyton*, "plant" or "that which grows") and human art or culture (*techné*). First: is there some "raw material" from which *all* bodies are formed and of which they can therefore be considered expressions? Second: if this primary matter exists, can its internal constitution alone explain the variety and structural features of bodies—or does matter require the intervention of other, non-material principles that orient and shape it? Le Moli's contribution interprets and discusses Aristotle's complex positions on these themes, arguing that the philosopher conceived of matter generally in the way that we now conceive of plant matter specifically: not inert but always already operative, tending by itself to a form that emerges from the assemblage of its own components, and which is pursued by resorting to every possible expedient. Such natural matter, of which the vegetal is the model and most flamboyant unfolding, must therefore already have an artful, technical side to it—it must be, Le Moli argues, always already *techno-physis*.

The next four chapters deepen our historico-philosophical understanding of Western thinking about plants through a critical and somewhat revisionary lens. **Isis Brook**'s contribution begins with a brief overview of the way plants have often been downplayed in Western culture. She considers some conceptual pivots for such attitude, such as the notion of the Great Chain of Being, the mostly zoo-centric emphasis in biology, and the typically anthropocentric focus on plants' use values. The chapter then goes on to sketch four "vignettes", as Brook calls them, of historical figures who dissented and diverged from this narrative of plants as inferior, inert, and mere means: our Theophrastus (arguably the father of scientific botany), Johann Wolfgang von Goethe (eighteenth century German author whose scientific works include *The Metamorphosis of Plants*); Agnes Arber (twentieth century English botanist and philosopher who explored and tested plant agency); and Barbara McClintock (twentieth century US botanist/cytogeneticist who discovered transposons, the so-called "jumping genes"). Brook places each of these figures in their historical context, and presents something of their approaches and what they offered that can now be of use and inspiration to VT.

Adriana Craciùn explores the philosophical questions triggered during the age of the Enlightenment by new possibilities of accelerating plant movement and growth, and generally manipulating plant time in experiments and applications. The Enlightenment is often portrayed by much contemporary environmental and plant studies as the ideational cradle of an unprecedentedly exploitative approach to the natural world. Yet by looking at the work of thinkers such as Robert Boyle, Francis Bacon, and Georg Agricola—all typically thought to consider plants as just inert objects of manipulation—Craciun builds a revised and revitalized Enlightenment genealogy of VT, showing how our "new", apparently cutting-edge contemporary discourses on plant agency, sentience, and ethical standing are really a "vegetal re-turn" to themes and topics that occupied the very thinkers and scientists whom we often assume to have historically contributed to these discourses' inhibition instead.

Quentin Hiernaux's chapter explores the work of 19th French scientist René Dutrochet, one of the founders of experimental plant physiology and responsible for the discovery of osmosis and a first articulation of cell theory. Dutrochet's 1824 *Anatomical and Physiological Research on the Interior Structures of Animals and Plants and on their Motility* offers a view on the conceptual and methodological presuppositions that physiology confronted as a nascent science when dealing with nonhuman sensitivity, movement, and mind. As Hiernaux shows, Dutrochet simultaneously challenged and accepted traditional philosophical tenets on these themes. For example, he thought that plants were "irritable", not inert, and yet refrained from equating irritability to sensation because sensation was traditionally tied to consciousness, and if irritability were equated to sensation one would then have to admit that the organs and the different irritable parts of a plant would each have a consciousness. Considering this conclusion "inadmissible", Dutrochet denied the equivalence between irritability and sensation, in this way confirming the traditional view that plants "lacked" something the animals have. Yet, that sensation requires consciousness, and that irritable parts cannot be conscious, are views that Dutrochet inherited from philosophical traditions, not facts that he discovered in the lab. Ultimately, Hiernaux's chapter tells a story about the philosophical viscosity of science, even pioneering experimental science.

Claudia Rosciglione's chapter chronicles Charles Darwin's extended research on plants, with particular attention to his 1880 *The Power of Movement in Plants*, co-authored with his son Francis. Darwin's interest in plants was consistent throughout his life, and his research brought new knowledge to botany, provided grounds to reassert the importance of the plant kingdom in connection with the animal kingdom, and gave evidence for many fundamental elements of his theory of evolution. Darwin's work on plants, however, was largely overlooked or underestimated by many botanists of his time, who never recognized him as a scientist in their field and wrongly assumed that his theory of evolution would apply only to the animal kingdom. Darwin's research on plants was, instead, precisely proof that evolution was universal, and some of his most courageous theses on the vegetal—that plants possess sensibility, movement, and a complex ability to interact with their environments through their root tips—are now widely acknowledged among the bases of contemporary plant neurobiology.

That takes us to the next two chapters, which differ from one another in important ways but are both dedicated to the complex issue of ascertaining what powers plants have, and figuring out what these powers imply for our human thinking and doing. In their contribution, **Andrea Nani and Daniele Volpara** discuss plant minds from the perspectives of psychology and cognitive neuroscience. Thanks to ever more refined experimental research, plants are now known to deploy subtle and sophisticated mechanisms of action and interaction at different biological levels, in response to environmental circumstances as well as social dynamics within plant populations. Nani and Volpara provide a helpful review of these mechanisms and discuss the extent to which they license an attribution of cognition to the vegetal world. They underline that, despite the seemingly prodigious adaptations that plants can develop, they do not seem to have the representational minds that animals have (see also Maher 2017). But Nani and Volpara also argue that the complex behavioral patterns and synergistic networks that plants intelligently articulate and sustain do support the hypothesis that they might constitute "distributed minds", as well as underscore the need to reconsider the very concept of "cognition".

In their chapter, **Aaron Ellison and Eric Zeigler** ask the following disorienting question: if to understand, empathize with, and "become-with" other species we need to see the world through their eyes, what would it mean to look at the world through the eyes of a plant? The question is posed non-metaphorically. Ellison and Zeigler illustrate how they use a range of photographic techniques combined with creative visualizations of scientific data to explore plant "vision". They show that plants perceive ("see") much more of the radiant spectrum than humans do, and that they use this radiation to orient their nutrition (photosynthesis) and breathing (photorespiration). Alison and Ziegler go on to discuss how plant experience time differently from humans, their lifespans varying from minutes to millennia; and by juxtaposing their findings on plant vision with their thoughts on plant time, they provide a unified interpretation of the iterative temporality of plant growth and reproduction.

Ellison and Ziegler see reflections on plant powers and ontologies as part of a larger and more challenging project, that of finding ways for humans to learn to relate to and evolve with plants in ethically and politically novel, rich, nuanced, and meaningful ways. The rest of the volume is dedicated to the exploration of various aspect of this project.

Jenna Aarnio and co-Authors present and discuss methodological issues in plant ethics. They consider two classic and largely incompatible approaches. The first focuses on finding possible morally relevant similarities between plants and animals, including humans, and based on these similarities extend to plants familiar arguments from animal and human ethics. They dub this the "extensionist" approach. In contrast, the "otherness" approach focuses on highlighting ontological distinctions between plants and animals, humans included, and has it that morally relevant properties are ontologically indexed and thus differ from plants to humans and other animals. In this case, they argue, plant ethics must create and navigate mostly unfamiliar arguments.

The methodological indeterminacy of plant ethics is translated into, and analyzed as, the indeterminacy of plants' moral status in the next chapter by **Ethan Terrill**, who articulates a challenging analogy and comparison between plants and human patients in persistent vegetative states. He argues that while both plants and patients are obviously biologically alive, the morally pressing question is whether they display mental abilities of any kind. As Terrill shows, there are strong arguments for and against that hypothesis in both cases. This leaves us in a state of indeterminacy, which vertiginously complicates moral decision-making on behalf both of plants and persistently vegetative patients. Terrill explores this complicated notion of "indeterminate moral status", and some ways in which at least some of the complications that it carries can be tackled. In the process, he shows the depths of our practical disorientation when confronted with the "nutritive soul", as Aristotle called the vegetal powers that inhere in plants as much as ourselves, when that soul is laid bare in a human body whose other "souls"—the perceptive and the intellectual—are seemingly extinguished.

Yet plant ethics has many ways. While the moral status of plants may be indeterminate when indexed to vegetal mental powers, their legal status could nonetheless be determined by means of positive vegetal rights. In her chapter, **Amber Polk** considers this alternative route and articulates a notion of vegetal rights that needs no basis in plant minds, agency, or interests. Her chapter considers the benefits to plant ethics of making rights part of its conceptual machinery and scans potential bases for doing so. Polk proposes grounding vegetal rights on a broadly aesthetic argument, anthropocentric and of a somewhat twisted Kantian flavor, whereby certain "exceptional plants" are right holders because they trigger or contribute to the development and exercise of practical reason in humans. This enables vegetal rights to perform the task of including new individuals into the realm of moral significance, but it also means that plant ethics cannot, and should not, be thought of as co-extensive with vegetal rights: there are innumerable "unexceptional" plants too, and these—Polk argues—are better conceived of not as the subjects of vegetal rights, but as the objects of human virtues.

The notion that some plants are "exceptional" and that their exceptionality helps seeing and treating them as worthy of ethical consideration is further explored, from a different perspective and in a specific context—that of botanic gardens—by **Thomas Heyd**. His chapter begins with a brief exploration of what it means to view plants as subjects. This is followed by an account of gardens, and the role of human-plant collaborations there. The historical development of botanic gardens, and their special functions as recreational displays and sites for plant conservation are discussed next. This is followed by the proposal that botanic gardens, even more than other gardens, exhibit human-plant collaboration, and the claim that, as such, these gardens offer outstanding opportunities for viewing plants as subjects. This, Heyd argues (much in agreement with Polk), has important implications for our moral education with respect to the vegetal and more generally non-human world.

But what about "unexceptional" plants, then? These are no second-class plants, of course. They are rather plants whose relevance is less individualized but found in their belonging and participating in larger systems of normative importance—with part of that importance lying precisely in the fact that human virtues and vices also play out in these systems. **Casey Rentmeester**'s contribution thus looks at plants as constituents of that "biotic community" whose functioning, integrity, and beauty deserved, according to celebrated twentieth century naturalist Aldo Leopold, protection and promotion. In articulating his "Land Ethic", Leopold denounced the purely economic understanding of the land that results from viewing it as a mere commodity, and attempted to rectify it by revealing "land as a community to which we belong" (Leopold 1949, viii). Rentmeester argues that Leopold's Land Ethic, which mostly cuts across typical divides in environmental ethics such as anthropocentrism/non-anthropocentrism, individualism/holism, and intrinsic/extrinsic value, provides a fitting conceptual basis and quite possibly the "key log" for VT.

In her chapter, **Laura Pustarfi** draws on phenomenology to contextualize plant ethics not in an ethics for the biotic community but in an ethics of place, accounting for both the singular and the more-than-individual, relational nature of plants. Pustarfi's specific interest is with trees, and her proposal is an arboreal ethics that recognizes trees as non-human others that can be considered for some rights—in particular, a right to stand in place—while also acknowledging that our lives can often depend on taking or subjugating theirs. Pustarfi explores the asperities of this paradoxical knot, which in different versions is ubiquitous in plant ethics, and in the process interrogates various place-based approaches to environmental ethics.

In the next contribution, **Kimberly Dill** considers a very specific place and biotic community where vegetal powers are expressed, often unmediated and unconstrained—that is, forests. She explores one particular attitudinal response to the vegetal powers of forests, namely reverence, arguing on both empirical and conceptual grounds that reverence is inherently motivating and promotes efficacious conservation behaviors, partly because experiencing it is directly beneficial to human well-being. Dill's contribution is also formulated as a nudge to prioritize anti-deforestation and biodiversity conservation as top targets for philosophical inquiry, policy reform, and societal discourse.

When forests are cleared out and the vegetal individuals, populations, and species that constitute them are burnt and bulldozed, soil is often made available for new plants that we do not revere but rather manipulate and consume. This is agriculture. The next three chapters look at plants and agriculture from different perspectives. In his contribution, **Tsuyoshi Teramoto** considers ethical and political aspects of the soil-plants-food nexus. In the foreseeable future, humanity will need to both increase food production and address soil degradation. Teramoto contrasts two competing approaches to these twin challenges: on the one hand "eco-modernist" approaches, maintaining that we should make maximum use of technology to intensify agricultural production and reduce the land area used, while returning as much soil as possible to nature and encouraging soil recovery; and on the other hand "agrarianist" approaches (see Thompson 1995), maintaining that appropriate agricultural practices for soil conservation must be determined not only on grounds of quantitative assessments of environmental impacts and productive efficiency, as eco-modernist contend, but also by considering aspects such as soil values and technological uncertainties. Teramoto explores these two alternative views as a way to acquire clues for a more comprehensive articulation of the soil-plants-food nexus that, while being alert to ecological impacts and limits as well as open to technological enhancements of agricultural practices, is not ready to just renounce the ecological and anthropological values of these practices, as these are revealed by the lived human experience of working their first elemental relatum and precondition—soil.

Food and agriculture are also determinants of human health. In his chapter, **Daniel Weissglass** discusses the 'One Health' framework that is gaining traction in global health discourses and its emphasis on the interactions and feedbacks between the health of humans, nonhuman animals, and the environment more generally. One Health strategies can sometimes include significant intervention into, and manipulation of, the non-human world, so recent work has emphasized the need to protect and promote animal values in a One Health framework. Much less has been said about plant values. Yet plants are equally central to that framework, chiefly as factors in food security but also as objects of innovative technologies such as "molecular farming", which uses modified plants as living factories in the production of vaccines, pharmaceuticals, or other chemicals for use in health efforts. In his contribution, Weissglass explores a range of ethical controversies and perspectives that might be associated with a One Health practice that accepts systematic interventions into the vegetal, charting the surprisingly large scale and scope of the moral space to be navigated.

But what exactly are these technological enhancements, these novel plant or plant-related technologies that should sustain human health and nutrition in the hyper-populated, resource-depleted, hotter planet that we now inhabit? **Lotte Asveld** presents some of them and discusses their ethically controversial dimensions in her chapter. These technologies include genetic editing techniques such as Crispr Cas, increased data screening opportunities, AI and robotics that allow for smart agriculture, and generally optimized plant breeding approaches. Their promise is efficient, sustainable, and climate-adaptive agriculture, through improved crop management systems and rational trait selection in plants. But they also raise concerns about how we should relate to plants within a techno-agricultural system. These are concerns of sustainability, safety, economic justice, and naturalness. Asveld's chapter discusses each of these domains with reference to current developments in several novel plant technologies.

The next two chapters are largely critical of the manipulative approach to plants that is typical of agricultural contexts, as well as of the regimenting approach to plant knowledge and classification, and to the exploration and classification of the lands, places, and biotic communities that plants constitute, which have been characteristic of Western colonial projects, practices, and injustices.

In her contribution, **Prudence Gibson** discusses the institution of the Herbarium and its political premises and implications. As she explains, herbaria have played a vital role in the colonial enterprise of collecting, identifying, classifying, and naming plants while charting lands. Today, the Australian herbaria that she focuses on do crucially important work for understanding changes in climate, loss of habitats, and extinctions in the wild; yet they also still follow a European model of plant collection, research, and conservation, and still use the Linnaean system of naming (Latin and common names) for their cataloguing. This speaks of their colonial past and the costs it imposed onto Australia's First Nations peoples and their lands. But there are now calls to redress these past injuries and decolonise herbaria archive collections, by adding indigenous names for plants and revealing the dark colonial histories of plant collection, as well as by letting indigenous knowledge speak of plant uses and applications. Gibson's chapter discusses these decolonizing methodologies as deployed by the Sydney herbarium collection.

With her chapter, **Deepta Sateesh** takes us to India and provides a vivid rendering of just how different plants and lands appeared from an indigenous perspective compared to the perspective of sixteenth and seventeenth century European colonizers. The context is the monsoonal Southwestern Ghats, where knowledge of plants prior to colonial interventions was enmeshed in human practices and everyday encounters, and transmitted through oral stories, songs, and material making. The rigorous colonial recordings and interpretations of vegetal life in the Ghats that overwhelmed these traditional ways produced, among other things, the twelve-volume *Hortus Malabaricus*, published between 1678 and 1693, which then found its way into experimental gardens in Europe and influenced the universal norms of classifying, documenting, and codifying indigenous understandings in the vocabulary of a then emergent science of botany. Sateesh's article articulates the transformations that occurred in plant life experiences and recordings in the Ghats, as the register shifted from the fluid multi-dimensional knowledge-gathering and transmission of the locals towards the textual form and the still images that characterized colonial systematizations. Sateesh then considers contemporary possibilities for renewed understandings of plants and land, framed by a monsoonal, wet, newly traditional relational ontology.

The volume closes with and Afterword by **Michael Marder**, whose decade-long philosophical work on plants has not only been pioneering but also rigorous, erudite, generous, creative, and deep. Marder reflects on what may be ahead for VT. Will VT lead to a well-ordered intellectual greenhouse where the philosophy of plants, and plant humanities more generally, collect theories and topics and approaches and perspectives and ordain them using plants as a pretext for the show—or will VT lead to more open, uneven, dispersive intellectual landscapes where plants are supposedly learnt and told on their own terms? But then, how will these vegetal terms ever be fully represented to, and by, those humans who are learning and telling?

This very book, with its concurrent tendencies towards both unity and diversity, embodies Marder's question about VT's future. While it provides no answers, it does show some pros and cons of going either way.

1.4 Beyond and After the Vegetal Turn

Whether in greenhouses or fields, plants will always and in many ways elude the grip of our categories and conceptual schemes. As we move forward with our research programs, VT might lead to us very different places where this same elusiveness plays out quite differently. But in greenhouses as much as in the great outdoors both plants and all our newly found ways of thinking about and with them will evolve, and thrive or not, and then silently die and rot, and then they will percolate down under, into the interstices of soil and culture, to fertilize what we will one day think of as valuable and meaningful. Maybe VT needs to lead us not to a place but to a time, when plants inadvertently begin to feed more of our general reflections on living, as well as our practical experiments in living. There are many good reasons to speed up our journey to that time, as the chapters in this volume will show.

It is simply not possible to constrain all the historical, conceptual, and practical dimensions of VT in one book. What this volume aims at is only mapping some of the main terrains that present and future scholars are or will be navigating as they further reflect upon the vegetal powers of Earth. The hope is that this will also contribute to an increasingly rich and more nuanced societal debate about plants, their powers, values, meanings, innumerable relations with our lives and all other lives, and our ways of being with them.

References

Arbor, J.L. 1986. Animal Chauvinism, Plant-Regarding Ethics, and the Torture of Trees. *Australasian Journal of Philosophy* 64 (3): 335–369.

Attfield, Robin. 1981. The Good of Trees. *Journal of Value Inquiry* 15 (1): 35–54.

Attfield, Robin. 2018. Forest Ethics. In *Plant Ethics: Concepts and Applications*, ed. Angela Kallhoff, Marcello Di Paola, and Maria Schörgenhumer, 121–130. New York: Routledge.

Baldassarri, F., and A. Blank, eds. 2021. *Vegetative Powers: The Roots of Life in Ancient, Medieval, and Early Modern Natural Philosophy*. Cham: Springer.

Baluska, Frantisek, Stefano Mancuso, and Dieter Volkmann, eds. 2006. *Communication in Plants*. Berlin: Springer Verlag.

Baluška, Frantisek, S. Mancuso, D. Volkmann, and P. Barlow. 2009. The 'Root-Brain' Hypothesis of Charles and Francis Darwin: Revival After More Than 125 Years. *Plant Signaling & Behavior* 4: 14–21.

Beerling, David. 2007. *The Emerald Planet. How Plants Changed Earth's History*. Oxford: Oxford University Press.

Bibikhin, Vladimir. 2021. *The Woods*, trans, A. Tait. Cambridge: Polity Press.

Bonneuil, Christophe, and Fressoz Jean-Baptiste. 2015. *The Shock of the Anthropocene: The Earth, History and Us*. London: Verso.

Brown, Lester R. 2011. *Full Planet, Empty Plates*. New York: Norton and Company.

Calvo, Paco. 2016. The Philosophy of Plant Neurobiology. A Manifesto. *Synthese* 193 (5): 1323–1343.

Calvo, Paco. 2017. What is it Like to be a Plant? *Journal of Consciousness Studies* 24 (9–10): 205–227.

Calvo, Paco, and Natalie Lawrence. 2023. *Planta Sapiens. The New Science of Plant Intelligence.* New York: W. W. Norton and Company.
Calvo, Paco, Monica Gagliano, Gustavo M. Souza, and Anthony Trewavas. 2020. Plants are Intelligent, Here's How. *Annals of Botany* 125 (1): 11–28.
Chakrabarty, Dipesh. 2015. The Anthropocene and the Convergence of Histories. In *The Anthropocene and the Global Environmental Crisis. Rethinking Modernity in a New Epoch*, 44–56. London: Routledge.
Chamovitz, David. 2012. *What a Plant Knows. A Field Guide to the Sciences.* New York: Scientific American/Farrar, Straus & Giroux.
Coeckelbergh, Mark. 2018. What Do We Mean by a Relational Ethics? Growing a Relational Approach to the Moral Standing of Plants, Robots and Other Non-humans. In *Plant Ethics, Concepts and Applications*, ed. A. Kallhoff, M. Di Paola and M. Schörgenhumer, 98–109. London: Routledge.
Coccia, Emanuele. 2018. *La vita delle piante. Metafisica della mescolanza.* Bologna: il Mulino.
Cooper, David. 2012. *A Philosophy of Gardens.* Oxford: Oxford University Press.
Crutzen, Paul J. 2002. Geology of Mankind: The Anthropocene. *Nature* 415: 23.
Crutzen, P., and E. Stoermer. 2000. The "Anthropocene." *Global Change Newsletter* 41: 17–18.
Delaporte, F. 1982. *Nature's Second Kingdom: Explorations of Vegetality in the Eighteenth Century*, Engl. transl. Arthur Goldhammer. Cambridge (MA) and London: MIT Press.
Di Paola, Marcello. 2015. Virtues for the Anthropocene. *Environmental Values* 24: 183–207.
Di Paola, Marcello. 2017. *Ethics and Politics of the Built Environment: Gardens of the Anthropocene.* London and New York: Springer International.
Di Paola, Marcello, and Dale W. Jamieson. 2018. Climate Change and the Challenges to Democracy. *University of Miami Law Review* 72: 369–424.
Di Paola, Marcello, and Serena Ciccarelli. 2022. The Disorienting Aesthetics of Mashed-Up Anthropocene Environments. *Environmental Values* 31 (1): 85–106.
Doggett, Tyler. 2018. Moral Vegetarianism. In *The Stanford Encyclopedia of Philosophy*, ed. Edward N. Zalta. https://plato.stanford.edu/archives/fall2018/entries/vegetarianism/.
Drayton, R. 2000. *Nature's Government: Science, Imperial Britain, and the Improvement of the World*. New Haven, CT: Yale University Press.
Ennos, Roland. 2021. *The Wood Age. How One Material Shaped the Whole Human History*. London: Collins.
Federal Ethics Committee on Non-Human Biotechnology. 2008. *The Dignity of Living Beings with Regard to Plants: Moral Consideration of Plants for their Own Sake*. Berne, Switzerland: Federal Ethics Committee on Non-Human Biotechnology.
Finney, Stanley C., and Lucy E. Edwards. 2016. The "Anthropocene" Epoch: Scientific Decision or Political Statement? *Geological Society of America* 26 (3): 4–10.
Firn, R. 2004. Plant Intelligence: An Alternative Point of View. *Annals of Botany* 93: 345–351.
Gagliano, M., M. Renton, M. Depczynski, and S. Mancuso. 2014. Experience Teaches Plants to Learn Faster and Forget Slower in Environments Where It Matters. *Oecologia* 175 (1): 63–72.
Goodpaster, Kenneth E. 1978. On Being Morally Considerable. *The Journal of Philosophy* 75 (6): 308–25
Green, Harvey. 2006. *Wood*. Craft, Culture, History, London: Penguin Books.
Hall, Matthew. 2011. *Plants as Persons: A Philosophical Botany*. New York: State University of New York Press.
Hallé, Francis. 2002. *In Praise of Plants*. Portland: Timber Press.
Hamilton, Clive, Christophe Bonneuil, and Gemenne François, eds. 2015. *The Anthropocene and the Global Environmental Crisis. Rethinking Modernity in a New Epoch*. London: Routledge.
Haraway, Donna. 2016. *Staying with the Trouble: Making Kin in the Chthulucene*. Durham, NC: Duke University Press.
Harman, Elizabeth. 2007. Sacred Mountains and Beloved Fetuses: Can Loving or Worshipping Something Give It Moral Status? *Philosophical Studies* 133 (1): 55–81.

Hendlin, Yogi. 2021. Plant Philosophy and Interpretation: Making Sense of Contemporary Plant Intelligence Debates. *Environmental Values* 31: 253–276.

Hester, Randolph T. 2006. *Design for Ecological Democracy*. Cambridge (MA): MIT Press.

Hiernaux, Quentin. 2021a. The Ethics of Plant Flourishing and Agricultural Ethics: Theoretical Distinctions and Concrete Recommendations in Light of the Environmental Crisis. *Philosophies* 6: 91. https://doi.org/10.3390/philosophies6040091.

Hiernaux, Q. 2021b. Differentiating Behavior, Cognition, and Consciousness in Plants. *Journal of Consciousness Studies* 28 (1–2): 126–135.

Houle, Karen L.F. 2011. Animal, Vegetable, Mineral: Ethics as Extension or Becoming? The Case of Becoming-Plant. *Journal for Critical Animal Studies* IX (1/2): 89–116.

Jamieson, Dale, and Marcello Di Paola. 2016. Political Theory for the Anthropocene. In *Global Political Theory*, 254–280. Cambridge: Polity Press.

Kallhoff, Angela. 2014. Plants in Ethics: Why Flourishing Deserves Moral Respect. *Environmental Values* 23 (6): 685–700.

Kallhoff, Angela. 2018. The Flourishing of Plants. A Neo-Aristotelian Approach to Plant Ethics. In *Plant Ethics. Concepts and Applications*, 51–58. London: Routledge.

Kallhoff, Angela, Di Paola Marcello, Schörgenhumer Maria, eds. 2018. *Plant Ethics: Concepts and Applications*. New York: Routledge.

Khait, Itzhak, et al. 2023. Sounds Emitted by Plants Under Stress are Airborne and Informative. *Cell* 186 (7): 1328–1336.

Kimmerer, Robin Wall. 2020. *Braiding Sweetgrass: Indigenous wisdom, scientific knowledge and the teaching of plants*. London: Penguin Books.

Koechlin, Florianne. 2009. The Dignity of Plants. *Plant Signaling & Behavior* 4 (1): 78–79.

Latour, Bruno. 2017. *Facing Gaia: Eight Lectures on the New Climatic Regime*. Cambridge: Polity Press.

Leopold, Aldo. 1949. *A Sand County Almanac, and Sketches Here and There*. Oxford: Oxford University Press.

Maher, C. 2017. *Plant Minds: A Philosophical Defense*. New York: Routledge.

Marder, Michael. 2013. *Plant-Thinking: A Philosophy of Vegetal Life*. New York: Columbia U.P.

Marder, Michael. 2014. *The Philosopher's Plant. An Intellectual Herbarium*. New York: Columbia University Press.

Marris, Emma. 2013. *Rambunctious Gardens. Saving Nature in a Post-Wild World*. New York: Bloomsbury Press.

Mazzolai, Barbara, Beccai Lucia, and Mattoli Fabio. 2014. Plants as Model in Biomimetics and Biorobotics: New Perspectives. *Frontiers in Bioengineering and Biotechnology* 2: 2.

Miller, Elaine P. 2002. *The Vegetative Soul. From Philosophy of Nature to Subjectivity in the Feminine*. New York: State University of New York Press.

Moore, Jason, ed. 2016. *Anthropocene or Capitalocene? Nature, History and the Nature of Capitalism*. Oakland, CA: PM Press.

Odparlik, Sabine. 2018. The Dignity of Plants. An Overview of the Discussion in German-speaking Countries. In *Plant Ethics. Concepts and Applications*, 59–69. London: Routledge.

Parise, André Geremia, and Michael Marder. 2024. Extended Plant Cognition: A Critical Consideration of the Concept. *Theory of Experimental Plant Physiology* 36: 439–455.

Pellegrino, Gianfranco. 2018. The Value of Plants: On the Axiologies of Plants. In *Plant Ethics: Concepts and Applications*, ed. Angela Kallhoff, Marcello Di Paola, and Maria Schörgenhumer, 13–27. New York: Routledge.

Plumwood, Val. 1993. *Feminism and the Mastery of Nature*. London: Routledge.

Plumwood, Val. 2002. *Environmental Culture and the Ecological Crisis of Reason*. London: Routledge.

Plumwood, Val. 1999. Ecological Ethics from Rights to Recognition, Multiple Spheres of Justice for Humans, Animals and Nature. In *Global Ethics and Environment*, 188–212. London: Routledge.

Pollan, Michael. 1991. *Second nature: A gardener's education*. New York: Atlantic Monthly Press.

Pollan, Michael. 2001. *The Botany of Desire: A Plant's-Eye View of the World*. New York: Random House.

Pouteau, Sylvie. 2014. Beyond "Second Animals": Making Sense of Plant Ethics. *Journal of Agricultural and Environmental Ethics* 27 (1): 1–25.

Pouteau, Sylvie. 2018. Plants as Open Beings: From Aesthetic to Plant-human Ethics. In *Plant Ethics. Concepts and Applications*, 82–97. London: Routledge.

Prajapati, Aditya, and Meenesh Singh. 2019. Assessment of Artificial Photosynthetic Systems for Integrated Carbon Capture and Conversion. *ACS Sustainable Chemistry & Engineering Journal* 7 (6): 5993–6003.

Puleo, Thomas J. 2019. Incorporating Nonhuman Subjectivity into World Society: The Case of Extending Personhood to Plants. In *Modern Subjectivities in World Society: Global Structures and Local Practices*, eds. Dietrich Jung and Stephan Stetter, 211–227. Cham, Switzerland: Palgrave.

Purdy, Jedediah. 2015. *After Nature: A Politics for the Anthropocene*. Cambridge (MA): Harvard University Press.

Repici, Luciana. 2000. *Uomini capovolti. Le piante nel pensiero dei Greci*. Laterza, Roma-Bari.

Ryan, John Charles. 2012. Passive Flora? Reconsidering Nature's Agency through Human-Plant Studies (HPS). *Societies* 2: 101–121.

Sandler, Ronald. 2018. Is Considering the Interest of Plants Absurd? In *Plant Ethics. Concepts and Applications*, 40–58. London: Routledge.

Schörgenhumer, Maria. 2018. Caring for Plants: Cultivating Relational Virtues. In *Plant Ethics: Concepts and Applications*, ed. Angela Kallhoff, Marcello Di Paola, and Maria Schörgenhumer, 110–118. New York: Routledge.

Steffen, W., J. Grinewald, P. Crutzen, and J. McNeill. 2011. The Anthropocene: Conceptual and Historical Perspectives. *Philosophical Transactions of the Royal Society A* 369: 842–867.

Sterba, James. 1995. From Biocentric Individualism to Biocentric Pluralism. *Environmental Ethics* 17 (2): 191–207.

Stone, Christopher D. 1972. Should Trees Have Standing? Towards Legal Rights for Natural Objects. *Southern California Law Review* 45: 450–501.

Struik, P.C., X. Yin, and H. Meinke. 2008. Plant Neurobiology and Green Plant Intelligence: Science, Metaphors and Nonsense. *Journal of the Science of Food and Agriculture* 88: 363–370.

Taiz, L., D. Alkon, A. Draguhn, A. Murphy, M. Blatt, C. Hawes, G. Thiel, and D.G. Robinson. 2019. Plants Neither Possess nor Require Consciousness. *Trends in Plant Science* 24 (8): 677–687.

Taylor, Paul. 1986. *Respect for Nature: A Theory of Environmental Ethics*. Princeton, N.J: Princeton University Press.

Thompson, P.B. 1995. *The Spirit of the Soil: Agriculture and Environmental Ethics*. London: Routledge.

Trewavas, A. 2003. Aspects of Plant Intelligence. *Annals of Botany* 92: 1–20.

Trewavas, A. 2005. Green Plants as Intelligent Organisms. *Trends in Plant Sciences* 10: 413–419.

Trewavas, A. 2014. *Plant Behaviour and Intelligence*. Oxford: Oxford U.P.

van der Veen, Marijke. 2014. The Materiality of Plants: Plant-People Entanglements. *World Archaeology* 46 (5): 799–812.

Visak, Tatjana. 2018. Utilitarian Plant Ethics. In *Plant Ethics. Concepts and Applications*, 34–39. London: Routledge.

Vitousek, P., H. Mooney, J. Lubchenco, and J. Melillo. 1997. Human Domination of Earth's Ecosystems. *Science* 277 (5325): 494–499.

Warren, Karen. 1987. Feminism and Ecology: Making Connections. *Environmental Ethics* 9: 3–21.

Williams, Michael. 2003. *Deforesting the Earth: From Prehistory to Global Crisis*. Chicago: University of Chicago Press.

Wood, David. 2020. *Thinking Plant Animal Human: Encounters with Communities of Difference*. Minneapolis: University of Minnesota Press.

Zalasiewicz, J., M. Williams, R. Fortey, et al. 2011. Stratigraphy of the Anthropocene. *Philosophical Transactions: Mathematical, Physical and Engineering Sciences* 369 (1938): 1036–1055.

Zalasiewicz, J., C.N. Waters, M. Williams, et al. 2016. When Did the Anthropocene Begin? A Mid-Twentieth Century Boundary Level Is Stratigraphically Optimal. *Quaternary International.* http://www.sciencedirect.com/science/article/pii/S1040618214009136.

Chapter 2
On the Geology of Plants

Thomas Nail

Abstract This chapter interrogates the deep materiality of vegetal life. Plants are not living beings on Earth, they are the process of the Earth. Plants are one of the ways the Earth moves itself. And as such vegetality can be defined by the unique patterns of motion traced by the Earth. This chapter explores these patterns arguing ultimately that vegetality is a certain becoming of the entire Earth.

2.1 Introduction

Vegetal life does not emerge *on* Earth; the Earth *becomes* vegetal. The most significant error in thinking about vegetality, or plant life in general, is that plants are distinct or separate from their deep historical emergence as cosmic and *earthly* bodies. Plants are fallen stars, minerals, air, and water. They are made wholly of the stuff of the Cosmos and the Earth.

Vegetality is something the Earth *does*—not something that merely occurs *on* Earth. Vegetality is an expression of the Earth's inner *tensions* and *pressures* relative to its atmospheric and gravitational orientation. Without the memory of minerals and the fluid dynamics of the atmosphere, vegetality would have been impossible. Therefore, there is no ontological difference between plants and the rest of the Earth—only a *difference in kinetic pattern*. The question then of this chapter is: what is the distinct kinetic pattern that defines plant and vegetal life?

Just as the atmosphere is not *on* the Earth, plants are not *in* the atmosphere. Spatial prepositions inadequately describe what is ultimately dynamic and kinetic intra-penetrations. Just as the atmosphere is the continuation of minerality by other means so vegetality is the continuation of the atmosphere by other means.

By treating vegetality in isolation and outside of its deep historical evolutionary context (Meillassoux 2010), we may be tempted to look to "vital powers" to explain its inner movements (Marder and Roussel 2015). In my view, this is a mistake. A theory of vegetality requires deep material history.

T. Nail (✉)
University of Denver, Denver, USA
e-mail: Thomas.nail@du.edu

M. Di Paola (ed.), *The Vegetal Turn*, The International Library of Environmental, Agricultural and Food Ethics 38, https://doi.org/10.1007/978-3-031-68671-9_2

The German philosopher Immanuel Kant once said that there could be no Newton of life because life is so radically creative and unpredictable that it resists all the tools and formulas of classical mechanical physics.[1] Complex life, it is still often said, is entirely unlike the randomness or law-like structure of physical particles (Schrödinger 2017). Therefore, the story goes: there must be some hidden *creative force* inside life that dead matter does not have.

These explanations, I believe, are just another way for humans to tell a story about nature that makes a tiny fraction of the universe (i.e., themselves as living) more special or consequential than the rest. We think our life is so extraordinary that only a physical or spiritual miracle could have produced it.

The assumption in this way of thinking is that non-living matter is *not* creative. But if non-living matter is so uncreative, how could it have produced life in the first place? Matter, as I have argued elsewhere, is neither random *nor* fully determined. Matter is not the passive stuff Newton and others thought it was. Therefore, not only is there no Newton of life, there is no Newton of matter, either. Chaos and quantum theory have sufficiently attenuated Newtonian mechanics. Today, we need a different kind of theory of life that is consistent with the indeterminate mobility of matter. This chapter is a contribution to this project.

2.2 A Materialist History of Plants

The first step in this direction is to take the deep history of matter seriously. This means, first, that nature has no taxonomic "kingdoms" or substances but rather emergent patterns of motion. So instead of defining vegetality by what it *is,* this chapter tries to define what it *does* and how it *moves.*

There are two problems with the idea of biological kingdoms that distinguish animals from plants. The first is that it fetishizes life by conceptually distinguishing it from the non-kingdoms of inorganic matter, viruses, and prebiotic proteins. The second and related problem is that thinking of life in terms of "kingdoms" tries to purify and hierarchize various forms of life—where there are no natural hierarchies. It is a form of imperial thinking.

Philosophically speaking, the idea of kingdoms comes from the substance-based metaphysics of types, kinds, or categories that goes back to Aristotle. The assumption was that matter is made up of ontologically divisible stuff with essential properties and forms. Darwin showed that Aristotle was wrong about essences in biology, and yet we continue to use the idea of kingdoms even if we must admit they are just notations. It is thus important to remember that organisms are patterns in motion and that kingdoms are abstractions of real historical processes.

In contrast to the ideas of kingdoms, the history of geokinetics I have tried to work out in *Theory of the Earth* argues that minerals, gases, liquids, and life are

[1] Kant doubted that biology would ever find a Newton to explain even a "single blade of grass."

not kinds of substances but metastable processes and patterns. Each pattern emerged from flows that have folded up into fields.

For example, symbiosis shows us that all lifeforms historically emerged from the coordinated activity of microbes. Biological complexity emerged from diverse symbiosis—not from kinds or types. Every life thus contains the history of all life and matter. *Ontogeny recapitulates cosmogony.* Without this ongoing process, we would not be alive today.

But even the idea of sym*biosis* does not go far enough. It stops at life and goes no further. We need another more encompassing idea that gets at the immanent contribution of non-living matter as well. This is what I call *"kinopoiesis:"* the co-emergence of matter through and as motion.

Life, in other words, is thus mineral and atmospheric. Plants arrange minerals that came from deep space and deep earth. They contain gases and liquids made possible by volcanism, rain clouds, and lightning. Plants are not a type of *substance* but a new *pattern* of material motion. The concept of "kingdom" privileges static forms of life over processes and thus tends to lose sight of the bigger geokinetic picture. Just as it is absurd to think about plants without their fungal symbionts, it is ridiculous to think about either without their "*kinopoetic*" relationship with the cosmic, mineral, and atmospheric patterns of motion that sustain them. This is the idea I will try to express in this chapter on vegetal motion.

A deep historical perspective allows us to see a kinetic relationship between beings that we often treat as separate *substances* or *kinds* such as protists, plants, fungi, and animals.

At one philosophical extreme, some believe all of being is just one continuous substance; at the other extreme, every being is singular. However, I think that monism, pluralism, and even the combination of monism and pluralism are metaphysical and idealist positions that turn away from matter, history, and motion. Being is neither one substance nor many singularities. It is neither identity nor radical difference—but rather a multiplicity of *moving flows* and historically emergent patterns. Vegetality, I argue, is the historical name of a distinctly *tensional* pattern of the Earth.

Approaching the history of life from a kinetic perspective allows us to see more clearly how specific differences emerged in the first place. The theory of biokinetic patterns is more primary than the theory of kingdoms. In particular, what I am calling "biokinetics" shows that bacteria, protists, plants, and fungi are all defined predominantly by what I am a "tensional pattern of motion." Their differences are not in *kind* but in *degrees* of tensional motion. The historical rise of this specific tensional biokinetic pattern does not emerge out of nowhere, of course. It is a way of extending and transforming the centripetal, and centrifugal patterns of planetary respiration, which I discussed in *Theory of the Earth.*

This deep historical perspective also overturns the retroactive *fallacy of extension* that humans frequently impose on their study of nature—as if animals had "human-like" qualities and plants had "animal-like" qualities, and so on.[2] *Deep* history teaches

[2] An example from McCourt (2005), "A project of engaging philosophy on its classical terms and subjecting 'the plant' *to* those terms—terms of resemblance, difference as degrees from similarity

us that the process is precisely the *inverse*. Later historical formations emerge from *earlier* ones. This is why a *cultural* history of plants that is limited to humans and human history is insufficient (Marder 2016).

Most natural histories of the Earth have been plagued by the "great chain of being" with form on top and matter on bottom. If one believes that history is progressive, then the present is by default at the top of a hierarchical pyramid. The great chain of being is thus an effect of an arbitrary philosophical assumption. The chain of being also assumes that the links are discrete beings and not processes. However, the deep historical approach this kinetic approach differs from the analogical structure of the "chain of being" because it treats each "type" of being not as a substantial rung or link but as a regional pattern of kinetic flux.

It is therefore not enough to show that the plant "link" of the chain is much more animal-like than we thought before and thus deserves of our moral consideration. Or to move to a metaphysics of vital forces and vitalism (Nail 2023). Neither is it enough to invert the chain of being such that plants are superior to animals because they were historically before them.

The whole structure of the ladder of being and all its rungs has got to go. Nature is not analogical, teleological, vital, or merely composed of singularities. It is radically kinetic and patterned. There are no types, rungs, or substances, only emergent and dissipative patterns here and there, now and then without origin or end (Nail 2018). Each of these patterns, including plants, is not like something else but is the continuation of the cosmos and its patterns by other means. There are no substances–only kinetic processes. This is what the theory of vegetality or phytokinetics in this chapter aims to show.

Therefore, by the term "vegetal," I mean not only plants or the kingdom *Plantae*, but a historical structure of geokinetic tension produced by all forms of life with *cell walls*, including bacteria, protists, plants, and fungi. These walled and pressurized cells in tension with themselves and one another saturated the Proterozoic Earth—*effectively cellularizing the Earth itself*.

2.3 Tensional Motion

The vegetal field has three kinds of kinetic tension or pressure: an inner-cellular tension within cells, an inter-cellular tension among cells, and a trans-cellular tension through the environment.

The historical emergence of tensional motion did not replace centripetal and centrifugal motions. It emerged immanently from them. Centripetal mineralization continued as the Earth gathered and distributed rocks and minerals, producing *terrestriality*. Before plants or animals came on land, the land came onto land centripetally.

---

of function, relevant functions and their relative value anchored by 'the human' *and* of hoping, as was the case with the animal, to find a common ground and a 'common logic between these two kingdoms' so that plants, now, too, can be *taken seriously*."

Centrifugal atmospherics also continued to circulate and respire volatiles into the air and water during this time. This produced the vortical patterns through which life emerged and gathered together again and again.

It is precisely the continuous operation of these two prior patterns of motion that gave birth to a third new motion. The living matter continued to fold over itself through centripetal and centrifugal convective respiration patterns. However, this process slowly introduced a new relation of kinetic *tension* between distinct bacteria that were vortically thrown together but also held apart by their cellular walls. Single-celled bacterial life, for example, was increasingly thrust together by specific mineral topologies (holes, cracks, shorelines, and energy-rich mineral sources) as well as atmospheric vortices (current patterns and cycles in air, water, convection cycles in volcanic vents, and weather). This is not just symbiosis but a whole *kinopoiesis* that allows bacteria to gather together.

Terrestrial flows threw organisms together. The cellular tension within and between organisms was like the birth of a new ocean composed of a multitude of tiny oceans. The emergence of life was like the birth of a new sky consisting of a variety of atmospheres; or the birth of a new Earth composed of a multitude of new Earths. Each cell contained a liquid ocean, mineral earth, and metabolic respiration. Accordingly, this made possible a new kinetic *tension* between cells and the larger ocean, sky, and earth. It also introduced a new kinetic tension *between cells and other cells.*

Ultimately, as bacteria saturated the ground, ocean, and sky, the entire Earth developed a kind of kinetic tension with itself. The membranes of *cell walls,* filled with hydrostatic pressure, were like little Earths that make the Earth manifold with and against itself. Vegetality is the Earth in tension with itself using *walled membranes.*

During the Proterozoic, the Earth became membraned and enfolded like a single fluid skin or surface folded up into little compartments, rooms, or labyrinthine vacuoles—to which the Latin word *memebrum* refers. It is, therefore, not just plants, in the restricted sense, that are tensional, but the vast proliferation of *walled cellular life* that continuously "membranized" the Earth into a "vegetal Earth."

There is no opposition between centripetal, centrifugal, and tensional patterns of motion. Life is the complex cellularization and "membranification" of the Earth itself, holding itself together and apart *kinopoietically.* Vegetality is the Earth turned inside out and folded up like a cellular foam or a vortical labyrinth woven from the flux of matter. Each cell was centripetally gathered from the cosmos, centrifugally breathed out from the Earth and ocean, and manifolded with and against itself. Vegetality holds the Earth together and apart from itself through using pressurized metastable membranes: a pomegranate Earth.

The historical rise of this new pattern of motion followed three main kinds of tensional motions: *within, between,* and *through* cellular life.

2.4 Inner Cellular Tension

The first tensional motion of the Proterozoic occurred *within* living cells in the form of cell wall pressure, cytoplasmic strands, and organelles.

Living cells are nothing other than the flows of earthly matter folded back over themselves. They are the product of a continuous vortical movement of biomolecules swirling around a shared region. Most scientists today believe that the first cell membranes began as chains of self-organized lipids and proteins that folded over themselves to form a sphere. They speculate that at some point, RNA must have been trapped inside the sphere and was able to flourish under protection from the walled-in area. Eventually, this RNA produced some DNA that could reproduce the protective membrane—and thus ensure continued security.

This story raises an important question: why did lipids and proteins self-organize this way, and how did RNA get inside them? The hydrophobic theory, as we discussed in the previous chapter, is a step in the right direction but still insufficient. The hydrophobic theory still assumes the *random* encounter between lipids. It thus assumes that lipids were the sole agents of order—and therefore leaves unanswered the real kinetic conditions of this order in the first place. Instead of looking only at the molecular structure of water, we need to look also at the *kinetic* structure of water. For example, the movement of water is easily capable of organizing into vortices and eddies that could have brought the hydrophobic lipids together in the first place. The molecular polarity of water can also explain why the lipids stayed together.

Just as bubbles emerge from the turbulent waters of eddies, lipids likely gathered in much the same way. Just as a bubble of water can trap air, so RNA was probably swirled up into a vortical lipid "bubble" structure. The so-called mysterious origins of life are, therefore, only mysterious if we ignore the agency and self-organization of non-living matter. If we focus only on the biomolecular building blocks (lipids, amino acids, nucleotides, etc.), we miss the kinetic patterns that underlie them.

Over enough iterations of self-organizing structures in nature, new formations can emerge without leaving a *genetic* or linear biological trace behind. The most significant "leaps" in evolution are precisely where genetic and biocentric explanations break down and where, perhaps, a kinetic account may fill in the gaps. We may also be able to explain the leap from lipids to membrane-enclosed RNA or from simple cells to Eukaryotic organelles in this way.

From the perspective of a linear "*bio-substantialism,*" however, everything looks random and miraculous.

2.5 Symbiosis and Organelles

This leads us to a leading source of internal tension within the cell: The introduction of other cells that eventually became organelles such as mitochondria and chloroplasts, which define eukaryotic cells. The jump from prokaryotic cells with no internally distinct organelles and eukaryotic cells with distinct organelles is one of the most profound leaps in evolutionary biology after the emergence of life itself. The origins of this leap have also remained shrouded in mystery because there is no linear genetic trail showing the precise evolution of organelles from something prior. Physicists love to point out that the odds of organelles emerging randomly from particles were astronomically low.

However, the theory of endosymbiosis, developed by Lynn Margulis, shows how bacteria could have been captured inside the cell membranes of other bacteria and continued to live together. Those relationships with an advantage tended to survive longer, transfer their genes horizontally without sexual reproduction, and eventually reproduce as a single whole organism.

For example, about 700 million years ago, a blue-green alga (*glaucophyte*) swallowed a single cyanobacterium (*Cyanophora paradoxa*) and began extracting energy from its photosynthetic process. The two organisms shared pieces of their DNA, and the cyanobacteria eventually became a part of the algal cell called a chloroplast. Other ancient bacteria that breathed oxygen and produced energy from it allied with other bacteria, providing energy, disposing of cell waste, and eventually becoming cellular mitochondria (Margulis and Sagan 1997).

The theory of evolution is incomplete without the story of symbiotic encounters, trans-species cooperation, and the biokinetic tensions inside the cell. For example, the fastest and most superior method for a cell to change itself and adapt to its environment is not through DNA mutation and centrifugal cellular reproduction. Instead, the quickest way for cells to change is by increasing their inner *tension* with other cells. These other cells are vortically gathered in, swirled around, and recombined through horizontal gene transfer within the original cell. Outside cells can be sucked in and given a chance to cooperate or not. If they do, then they are reproduced as part of the cell much faster than random mutation and evolution ever could. However, such a "tensional symbiosis" leaves little or no vertical genetic trace of its process or origins (Margulis and Sagan 1997). This is why it looks like a "leap" has occurred in the genetic record—where kinetic symbiosis has done all the work.

Symbionts and organelles thus introduced a biokinetic tension inside the cell. They did this by holding themselves together and apart from one another in the same cell. Each directly fed off others in a linked tension. Evolution is vertical and centrifugal, but symbiosis is horizontal and tensional. Both harness the pedetic flows of the environment that continuously support them.

We can see now that the cell has at least three kinetic patterns of motion. First, there is the centripetal accumulation of energy from the periphery. Then there is the centrifugal emanation of DNA from the center. Then there is the vast tensional and manifold cellularity of walls, strands, filaments, organelles, and symbionts that structure the whole balance between centripetal and centrifugal motion.

2.6 Inter-Cellular Tension

Cells are also in external tension with other cells. So, we can see a second tensional motion *between living cells* in the form of mucilage, filaments, and the thallus.

2.6.1 Mucilage

As we have seen, some of the earliest many-celled structures were gathered together by the vortical patterns of mineral and atmospheric flows. But why did they stick together? Cells are held together and apart in tension with one another because of *mucilage*—a sticky, sugary fluid secreted by certain protists and bacteria.

When these cells whirled together, they formed what biologists called "colonial" structures. Some had shapes like plates and others like spheres. Some developed in dendritic patterns others in rhizoid patterns. Some colonial structures contained moving cells whose flagella began to move together in collective mobility. This movement also produced internal vortices that would eventually become the sub-cellular interior.

Each of these colonic cells is appropriately called *coenobium*, the Latin word meaning cell, cloister, convent, or monastery. Just as the individual cells of the monastery surround the empty center garden (garth), held together and apart in linked tension with one another, so the *coenobia* surrounds the hollow center of the biological cell.

In the case of the spherical-shaped colonies (*volvox*) coenobitic cells eventually began to exude not only mucilage but cytoplasmic strands that connected them more firmly. These strands bound the inside of each cell with the exterior of other cells without dissolving the cell walls that kept them apart and without reproducing the whole assemblage as a single multicellular organism.

2.6.2 Filaments

When this process of mucilage and strand connection was extended linearly between cells, it formed a long tensional filament that could extend vertically, supported by the water, toward the sunlight to photosynthesize. This is the basic structure of many algae that form thread-like chains of cells held together and apart in linked tension with one another. Importantly, each cell still functioned and reproduced relatively independently from the others in these "pluricellular" filaments.

2.6.3 *Thallus*

The emergence of the thallus or single body, however, combined the inner tensions of multicellular bodies with the external tensions of filamented and branched bodies. This produced a single multicellular organism without any central organization. The structure of the thallus is fascinating because it is still not divided into specialized tissues or organs. Even without specialized organs, though, different regions of the thallus can always take on different morphologies. Kelp, for example, is a thallus and still has at least three distinct areas: the holdfast (anchor), stipe (supports the blades), and the blades (for photosynthesis).

The differentiated structure of the thallus allowed for a new type of tension. For example, thalloid plants, fungus, and lichen can grip onto a surface with part of their body and use another portion to increase photosynthesis. This allowed for a new macroscopic tension between bottom and top, rock and water, and eventually rock, water, and air.

The differentiation of the thallus into select regions, thus puts the plant in tension with itself. The locations of energy accumulation, conversion, storage, and reproduction are different yet coordinated. Even without organs, regions of the plant acted in linked tension with one another and had to share and exchange energy between other regions in tension.

Furthermore, this was not just a kinetic tension in plant and fungal bodies. It was a way that the Earth and cosmos were put in tension with themselves as light, air, water, and minerals were gathered, held together, and kept apart inside the manifold body of the vegetal earth. Sunlight and solar rays, for example, touched and mixed with water, air, and dirt without entirely dissolving into one another inside the thallus. The thallus kept them together and apart in the same body.

Kelp, for instance, fill parts of their bodies with air and float to the surface. Fungi take solar energy from the sky and bury it under the ground. The kinetic cellularity of vegetal life, therefore, is designed precisely to gather the Earth together into pressurized cells, store it away, and release it in a linked tension with the world.

2.6.4 *Trans-Cellular Tension*

The third and final tensional motion occurred *through* vegetal cells in the form of capillary, communicational, and cosmic tensions. Vegetal motion is in tension with itself, other plants, and the material cosmos more broadly.

2.6.5 Capillary Tension

Trans-cellular vegetal motion is a tensional movement of matter *through* cells via osmotic pressure. After vegetal cells began adhering to one another and forming composite multicellular bodies (*thalli*), they produced a new kind of kinetic tension that ran through their cells, connecting them. Capillary tension or surface tension is the flow of liquid through narrow spaces without the assistance of, or in opposition to, gravity.

In plants this occurs in two complementary ways: (1) the movement of water evaporating out of leaf pores creates a *transpiration* in the water tubes that pulls water up through the cell walls using surface tension, and (2) the *osmosis* of water and nutrients moves from the roots upward. This is how many plants and vegetal bodies move fluid from outside their bodies throughout their cells. Water adheres to the inside of the narrow cell tubes, coheres with other water molecules, and is pulled up by the surface tension of water pressure.

First, however, the cell wells become connected as new cells grow out of old ones. Cells are no longer held together and apart by mucilage but by a sticky inner membrane of their cell wall made of *pectin*. As new cells grow, they adhere to one another tightly except for a series of pores or passages between the cells (plasmodesmata). These plasmodesmata, or intercellular junctions between plant and fungal cells, enable the transportation of hormones, enzymes, sugars, amino acids, ions, and proteins between cells.

This tensional flow between pressurized cells is regulated by a cytoplasmic sleeve that can open or close to adjust the size of the sieve, and thus control what can flow through. Fluids can travel between cells inside the cell wall or through these cellular junctions. Furthermore, these junctions internally connect cells through cytoplasmic strands adhering to the cytoskeletal filaments within each cell.

These junctions and their tensional linages make possible a trans-cellular motion *through* the cells of the vegetal body. As water pressure increases at the roots, it is moved upward in a continuous cellular tube toward the sky. As fluids move upward and *through* the chain of cellular junctions, they can also produce vortical eddies around the edges of the junction (*septum*), creating kinetic sub-compartments. In fungal cells, these eddies capture cell nuclei and accumulate them to stabilize flow-stressed pores (Pieuchot et al. 2015).

In short, capillary tension introduced an enormous trans-cellular movement of fluids through huge numbers of cells.

2.6.6 Communication

Trans-cellular vegetal motion also occurs *through* or *across* other vegetal bodies via *signaling* and *communication*. Tensional vegetality is a dramatic way of being “together-apart”. Communication between plants is kinetically tensional because it allows them to act together at a distance via chemical, electrical, and mechanical tensions above and underground.

However, vegetal communication is not a linear, representational, code-based language. Vegetal communication is kinetic, haptic, performative, and dynamically netted in a vast mineral and atmospheric network of intersecting material flows. Vegetality is not reducible to a single object or body but is the whole tangled and *in-tensional* web of fluctuating signals and responses (Willemse 2009).

Vegetal communication is not merely a linear, mechanistic response to stimuli. It is a change across or through a whole network. For example, when one cuts sagebrush, it gives off a volatile chemical that helps wild tobacco resist grasshoppers and cutworms (Heidel et al. 2009; Karban et al. 2000). Each change is a change in the whole. The chemical language of plants does not have a fixed semiotic structure. Plants can use old chemical words in new ways or new contexts with different meanings.[3] The language of plants is a collective affair where plants can eavesdrop on one another and use this shared knowledge to prepare themselves for an attack beforehand (Paré and Tumlinson 1999; Karban et al. 2000; Heil and Ton 2008). Underground communications travel through vast fungal mycorrhizal networks affecting numerous bodies in the process (Dicke et al. 2003).

Recent discoveries have even shown that plant relatives across a wide range of taxa can recognize one another using specific leaf gestures and light signals (Crespi and Casal 2015) and use this to work together cooperatively (Dudley and File 2007; Murphy and Dudley 2009). Plant language thus always emerges in an embedded context and is not a fixed property of the organism or the chemical compound. Plant language is a material language. It is a performative kinetic act that changes in tension with the world and with or across other plants.[4]

Plants can even use biochemical cues to summon the predators of herbivore insects that endanger them. They can regulate their root volumes in a tensional response to their neighbors. They can share and distribute nutrients among different plants. They can release chemicals into their leaves to ward off predators. Root apices even have brain-like sensitivities to all kinds of minerals and dangers that can affect the whole plant (Baluška et al. 2006).

All of this trans-plant signaling and communication take place *through* a material kinetic movement of volatile gases, chemicals, light, and electrical pulses that touch and move *through* a tensional network. Material signaling thus requires a tension where different bodies hold together and apart in a network across which material signals flow.

[3] "We now know that plant chemical language, for example, is endowed with true semantic flexibility, so that new meanings may be assigned to old chemical words and used in novel interactions and new contexts," Gagliano (2017). See also Holopainen (2004).

[4] For discussion, see Kravchenko (2011).

2.6.7 Cosmos

The trans-cellular vegetal motion also follows a tensional movement of *cosmic* matter *through* vegetal bodies. The Earth is composed of meteoric minerals accreted from the cosmos, vaporized into the atmosphere and ocean, and then folded up inside-out into billions of tiny walled cells in tension with one another. Vegetality takes matter from the cosmos and celluarlizes it into compartments in kinetic tension with one another. Photosynthesizing plants thus not only dig up the dust of fallen stars they also live off the luminous dust of our dying sun. Plants are star-eaters. They live on the death and decay of cosmic bodies.

Vegetal bodies are connective tissue that brings the cosmos to Earth in a series of pressurized pockets. Through them, mineral, atmospheric, oceanic, and cosmic matter is brought together and held apart in a network of energy-rich tissue. Terrestrial plants thus have their own little worlds—with their own minerality, atmosphere, and cosmos. They are not just "hyperseas," they are "hypercosmoi." They are not just pockets of water on land; they are pockets of sun, pockets of minerals, pockets of air. They are "hyperearths," "hypersuns," "hyperatmopheres." Vegetal life remakes the entire Earth, sky, and ocean, cell by cell—tensionally—link by link.

Plants are, therefore, the ultimate material-kinetic tension on Earth. They live in constant tension between the sky and Earth. They grow up to the sky to pull the sun into the Earth. At the same time, they dig deep into the Earth to push it up to the sky. Plants are the Earth inside out, upside down, and *com-pleatly* (un)folded. We live on a vegetal-origami Earth. Since their entropic heat eventually radiates into space, plants are the cosmos-made Earth and the Earth-made cosmic. In their way, they continue the deep Earth's task of transforming the Earth into a kind of sun, radiating heat into space. They are, therefore, the cosmos transformed, remade, and returned to itself.

Below the ground, the dead bodies of plants store liquid sunlight as carbon. In this way, their tensional movement is inverted. Instead of turning the Earth into the sun, they transform the sun into the Earth. Vegetality is thus a becoming Earth of the Sun and a becoming Sun of the Earth in the same tensional movement that materially courses through their pressurized bodies. Sugars move down from the sun and the leaves to feed the roots and the fungus while minerals move up to feed the leaves. Sun and Earth are thus held together and apart in the material body of the plant.

There is no alternation or final mixture of Earth, air, ocean, and sun that erases their differences. The plant retains the differences in the mix through the tension of its cells. This is why the rooted plant is a heliotrope, geotrope, amostrope, and aqueotrope all at the same time without contradiction or synthesis (Miller and Findlay 2007). Through plants, the Earth becomes an astrological body in tension with all other astrological bodies. Plants are heaven on Earth—the sky-made Earth and the sky-made Earth. They are the ocean-made Earth and Earth-made ocean again. They are the ultimate shape-shifters and metamorphic bodies that convert everything into everything else.

2.7 Conclusion

These are the three main tensional patterns of motion that define the vegetal kinetics of the Proterozoic Eon. Without adopting the imperial metaphysics of kingdoms or vital forces, this chapter has tried to show a different way of thinking about what plants are and how they work as cosmic-terrestrial patterns of motion. During the Proterozoic Eon, the entire earth took a vegetal turn by becoming tensional. With all the latest discoveries in so-called plant "neurobiology," we are not discovering how much more plants are like animals or humans, but how plant-like humans and animals are.

In short, *the Earth began to cellularize itself*, folding itself inside itself, to work on and transform itself into something new. The Earth became a producer and releaser of tensions between light, water, and air. With the help of the sun, the Earth began to weave a new fabric by unraveling itself. *Phytality* is the weaving of sunlight into the folded surface of the Earth. It is the sun-made Earth: the textile-like tension between distinct threads held together and part in the new vegetal body of the Earth.

References

Baluška, František. 2006. Introduction. In *Communication in Plants: Neuronal Aspects of Plant Life*, eds. František Baluška, Stefano Mancuso and Dieter Volkmann. Berlin: Springer-Verlag.

Crespi, M.A., and J.J. Casal. 2015. Photoreceptor-Mediated Kin Recognition in Plants. *New Phytologist* 205 (1): 329–338.

Dicke, M., A.A. Agrawal, and J. Bruin. 2003. Plants Talk, but are They Deaf? *Trends in Plant Science* 8 (9): 403–405.

Dudley, Susan A., and Amanda L. File. 2007. Kin Recognition in an Annual Plant. *Biology Letters* 3: 435–438.

Gagliano, Monica. 2017. Breaking the Silence: Green Mudras and the Faculty of Language in Plants. In *The Language of Plants: Science, Philosophy, Literature*, 84–102. Minneapolis: University of Minnesota Press.

Hegel, Georg W. F., A.V. Miller, and J.N. Findlay. 2007. *Hegel's Philosophy of Nature: Being Part Two of the Encyclopaedia of the Philosophical Sciences*. Oxford: Clarendon.

Heidel, Andrew J., Oz Barazani, and Ian T. Baldwin. 2009. Interaction Between Herbivore Defense and Microbial Signaling: Bacterial Quorum Sensing Compounds Weaken JA-Mediated Herbivore Resistance in Nicotiana Attenuate. *Chemoecology* 20 (2): 149–154.

Heil, M., and J. Ton. 2008. Long-Distance Signalling in Plant Defence. *Trends in Plant Science* 13 (6): 264–272.

Holopainen, J.K. 2004. Multiple Functions of Inducible Plant Volatiles. *Trends in Plant Science* 9 (11): 529–533.

Karban, R., I. T. Baldwin, K. J. Baxter, G. Laue, and G. W. Felton. 2000. Communication Between Plants: Induced Resistance in Wild Tobacco Plants Following Clipping of Neighboring Sagebrush. *Oecologia* 125 (1).

Kravchenko, Alexander. 2011. How Humberto Maturana's Biology of Congition Can Revive the Language Sciences. *Constructivist Foundations* 6 (3): 352–362.

Marder, Michael. 2016. *Grafts: Writings on Plants*. Minneapolis: University of Minnesota Press.

Marder, Michael, and Mathilde Roussel. 2015. *The Philosopher's Plant: An Intellectual Herbarium*. New York: Columbia University Press.

Margulis, Lynn, and Dorion Sagan. 1997. *Microcosmos.* University of California Press.
McCourt, Peter. 2005. Hormones and Developmental Timing Boundaries in *Arabidopsis.* Abstract circulated from University of Toronto Department of Botany.
Meillassoux, Quentin. 2010. *After Finitude: An Essay on the Necessity of Contingency*, trans. Ray Brassier. London: Bloomsbury.
Murphy, Guillermo P., and S.A. Dudleys. 2009. Kin Recognition: Competition and Cooperation in Impatiens (Balsaminacreae). *American Journal of Botany* 96 (11): 1990–1996.
Nail, Thomas. 2018. *Being and Motion.* Oxford: Oxford University Press.
Nail, Thomas. 2023. What's the Matter with Life. In *Life in the Posthuman Condition: Critical Responses to the Anthropocene*, eds. S. E. Wilmer and Audrone Zukauskaite, 241–260. Edinburgh: Edinburgh University Press.
Paré, Paul W., and James H. Tumlinson. 1999. Plant Volatiles as a Defense against Insect Herbivores. *Plant Physiology* 121 (2): 325–331.
Pieuchot, Laurent, Julian Lai, Rachel Ann Loh, Keng-Hwee. Chiam, Jason Stajich, and Gregory Jedd. 2015. Cellular Subcompartments through Cytoplasmic. *Developmental Cell* 34 (4): 410–420.
Schrödinger, Erwin. 2017. *What Is Life?: The Physical Aspect of the Living Cell with Mind and Matter; Autobiographical Sketches.* Cambridge: Cambridge University Press.
Willemse, Michiel T.M. 2009. Evolution of Plant Reproduction: From Fusion and Dispersal to Interaction and Communication. *Chinese Science Bulletin* 54 (14).

Chapter 3
PRIMARY MATTER: The Technical Nature of Wood in Ancient Thought

Andrea Le Moli

Abstract The essay reconstructs the origins of the concept of matter in the Greek and Latin traditions from its archaic use to its use in literary and philosophical works to show how this concept was born linked to the experience of the technical manipulation of plant materials, in particular wood. The aim of the essay is to evaluate, on this basis, the hypothesis of a technical-vegetal derivation of the primary ontological structures and cognitive repertoire of the human being. From the archaic Greek philosophical tradition through the thought of the great systematizers Plato and Aristotle to arrive at the most important representatives of the Latin tradition (Cicero, Lucretius, Seneca, Calcidius, Isidore of Seville) up to the threshold of contemporary anthropology, the essay reconstructs the more important stages in the evolution of humanity's first true "companion species".

Keywords Matter · Wood · Material Ontology · Technical Ontology · Technophysis

3.1 Introduction: *Materia* and *Silva* in the Latin World

The word "matter", in Italian as in the major European languages (Eng. *Matter*, ted. *Materie*, fr. *Matière*, sp. *Material*) comes from Latin. It ranks among those words in the Latin language that are intrinsically endowed with a reference to the simple needs of an essentially peasant civilization—but which, as the civilization is transformed, prove capable of adapting to changing conditions and meet new linguistic and conceptual needs (Malaspina 1991).

Derived from *mater* ("mother"), the word has originally a rather circumscribed meaning. The meaning of *materia* is in fact vegetal, precisely that of tree wood (Ernout and Meillet 2001, 390) and in the peasant language it indicates the trunk considered as that which produces shoots, so by extension the hard part of the plant as opposed to the bark (*cortex*) or leaves (*frondes*). Because it is this part of the

A. Le Moli (✉)
University of Palermo, Palermo, Italy
e-mail: andrea.lemoli@unipa.it

M. Di Paola (ed.), *The Vegetal Turn*, The International Library of Environmental, Agricultural and Food Ethics 38, https://doi.org/10.1007/978-3-031-68671-9_3

tree that provides the material to the art of carpentry, that meaning is distinct from firewood (*lignum*), and also in the plural *materies,* by that of timber as a building material (*tamquam substantia ad fabricandum apta*, Ernout and Meillet 2001, 358; Malaspina 1991, 42 n. 3)—so capable of generating adjectives such as *materarius* ("relating to the art of carpentry") and verbs such as *materior* ("to procure lumber"). Except for Plautus' strongly metaphorical uses, this semantic field of the term remains unchanged until the first century B.C., when, following independent lines of development, the word *materia* begins to appear as translation of "origin," "motive," "cause," "occasion," "pretext," and "disposition" (Malaspina 1991, 42 n. 3).

The semantic association of matter and wood still persists today, for example, in the words for wood used in Spanish (*madera*) and Portuguese (*madeira*). The word taking on the sense of "matter in general," whereby one can at some point speak of bronze matter, golden matter, or even "wooden matter" (Ernout and Meillet 2001, 390), is the result of the encounter of peasant language with the translation into the literary and philosophical lexicon of more abstract terms from the Greek world. Thus it is that, in the works of some precursors, but certainly at the hands of the great canonizers Lucretius and Cicero, *materia* turns out to be suitable for translating the Greek expression that, beginning with Aristotle and then the Epicureans, Academics and Stoics, had come to indicate the generality of material being.

That expression, again from the plant world, is *hyle*: a collective noun that already in Homer shares the three basic meanings of (1) forest, woodland; (2) thicket, undergrowth; (3) wood cut and to be used, for construction and combustion mainly (Happ 1971, 274 n. 996). Subject to a possible derivation from the Akkadian *sullulu* ("dark" adj.; verb "to make shade") or *sillulu* (shade of tree, shaded place—Semerano 1994, 299), *hyle* may turn out to be related to an ancient Indo-European root, **sul*—from which comes the Greek word for "wood" (*xylon*)—and **sula*—which indicates a collection of wooden stems such as the towering logs of a forest. Another etymology has it close to the root **sil* (to stand out)—indicating a set of poles driven into the ground or, in an even broader sense, a forest of columns (in German *Saeule*) in a temple (Happ 1971, 274 n. 996). This range of meanings is still present today in the English partition between Wood and Woods, in the German overlap between *Holz*-wood and *Holz*-woods, in the French overlap between *Bois*-wood and *Bois*-woods, and again in the Russian *Les*, at once forest and wooden material (Bibikhin 2021).

The first author to extend the sense of *hyle* to mean "matter in general" is, as is well known, Aristotle. His uses of *hyle* and his innovative conception of matter have, since at least the mid-nineteenth century, been the subject of a debate among scholars that does not yet seem to have come to a conclusion (Happ 1971; Leszl 1973; Berti 2011).

In addition to the identification of *hyle* as a technical term to denote "matter" in general, from both a physical and biological perspective, Aristotle is also credited with the first figurative extensions of the term *hyle* to the matter of a discourse or argument in the sense of its subject. The medium that enables this transition is the equivalence Aristotle draws between *hyle* and *hypokeimenon* ("subject-substratum")—also destined to be canonized by Cicero in *De Inventione* and *Rhetorica ad Herennium*, to indicate both the subject of a discourse and more generally the subject of the art of rhetoric (Malaspina 1991).

But it is with the sense of *hyle* as "physical matter" that Cicero and Lucretius mainly engage, as they attempt to account, respectively, for the Platonic-Stoic and the Democrito-Epicurean traditions. Lucretius uses *materia* and the plural *materies* to render in one all the fundamental vocabulary of Epicurean materialism. In Lucretius' *De rerum natura* I, 54–61, *materies* appears along with *genitalia corpora* and *semina rerum* as equivalent to the first material principles and constituents of nature (the so-called *primordia*), from which nature creates things and in which things are resolved after their dissolution. Cicero, on the other hand, makes widespread use of *materia* in his philosophical works, from the *Lucullus* to the *Academica* (excluding the *Timaeus,* in which he uses *materia* as a translation of *ousia*), to translate *hyle* or refer to "matter" in pre-Aristotelian authors who, however, had no such use of the term. For example, in *Ac.* II, 37, 118 Cicero states: "Anaxagoras said matter to be infinite… Plato attributes the world to the activity of an aternal god out of a matter which contained everything in itself" (Anaxagoras [dixit esse] materiam infinitam… Plato ex materia in se omnia recipiente mundum factum esse censet a deo sempiternum).

However, when the nascent Latin literary-philosophical culture was faced with the task of rendering the meaning of the Greek *hyle,* it found on its hands not one, but two terms endowed with a similar origin and richness: *materia* and *silva,* another word drawn perhaps from the same root as *hyle* but more related to forest as a non-man-made space, a realm of impenetrability and entanglement.

Silva recurs as a translation of *hyle* in the *Commentary on Plato's Timaeus* by the fourth-century author Calcidius, who ties it to primordial chaos by stating:

> For after Chaos, which the Greeks call *hyle* and we call *silva,* [he explains] that the earth remained firmly at the centre of the circle of the universe, fixed to its foundations and immobile. (Calcidius 1975, CXXIII)

In his treatise, Calcidius takes up the doctrine of a chaotic and recalcitrant matter as the origin of evil from the Middle Platonic and neo-Pythagorean Syrian Numenius of Apamea (second century CE.), thus inaugurating a tradition in Latin that, from Isidore of Seville (560–636) onward, would so define *silva* (in the *De Arboribus* section of his *Etymologiae sive Origines* XVII, VI-33): "A thicket (*silva*) is a dense and small grove (*spissum nemus et breve*). *Silva* is so called as if the word were *xylva,* because there wood is felled, for the Greeks call wood *xylon.*" A place that, because of its difficult penetrability, appears apt to express the sense of something to be disciplined. Again, in the *Etymologiae sive Origines* (XIII, III-1) Isidore states:

> The Greeks call the primary material of things *hyle* ("matter," also "wood, woodland"), which is not formed in any way, but is capable of underlying all bodily forms; from this material the visible elements (*elementum*) are formed, whence they took their name from this derivation. 2 Latin speakers have named this *hyle* 'matter' (*materia,* also meaning "wood") because every unformed substance, of which something is made, is always called matter. Whence the poets have named it *silva* (lit. "woodland"), not inappropriately, because *materia* is connected with woods. (*Isidore* 1911, 520, eng. transl. Isidore 2006, 272)

We find *silva* used in this way in other medieval sources, such as the *Cosmographia* of Bernard of Tours (1085–1178), for whom:

> through unyielding, formless chaos, combative mixture, colorless face of substance, inharmonious mass, it impatiently tends in its disarray to harmony, in its informality to form, in its coarseness to elegance and, yearning to come out of its primitive disorder, it seeks a harmony that composes it, as it were, as in musical bonds. (Bernard of Tours 1978, I, 1)

This bond is already announced in earlier Latin literature. As indistinct mass *silva* is found in connection with *materia* in the *Aetna*, a poem in 644 hexameters about volcanic eruptions handed down in the *Appendix Vergiliana*—attributed either to Lucilius, the recipient of Seneca's epistles, or to the younger Virgil. Here, (vv. 444 ff.) *silva* stands also for "lava," "volcanic material": *atque haec ipsa [scil. Insula Vulcani] tamen iam quondam extincta fuisset / ni furtin adgereret Siculi vicinia montis / materiam silvamque suam.* But it also appears in Plautus and Pliny the Elder in the sense of an undifferentiated whole, and in Varro, *De lingua latina*, V, 6, regarding the distinction between *campus* and *saltus*, i.e., those expanses of land that could not be cultivated or could only be used to graze sheep (Malaspina 1991)—thus indicating resistance to discipline. The latter nuance evokes the use of the word attested in the first philosopher-botanist in history, Aristotle"s student and friend Theophrastus, of the verb *hybrizo* to indicate the arrogance (*hybris*) and "intractability" of those plants that tend to grow unrestrainedly in every direction (Buccheri 2020, 151). This use, in turn, evoked a more general problem, concerning the origin of the variety and difference of forms from the manipulation of a material base.

3.2 Woods and Matters in Plato

In oscillating between the sense of *hyle* as *silva*, a dark and impenetrable ground, and that of *materia*, a set of resources to be exploited, Latin culture reproposed two fundamental problems that animated the debate on matter from its origins up to the Stoics and Epicureans, via Plato and Aristotle. First: is there some "raw material" of which bodies are formed and of which they can therefore be considered an expression? Second: if this matter exists, can its internal constitution alone explain the variety and structural nature of physical bodies? Or does matter require the intervention of other non-material principles that orient and shape it?

This debate culminates in internal discussions within the Platonic Academy. Opposing the materialism of his day, Plato wanted to show that the material constitution of bodies alone was unable to explain their existence as intelligible and structured entities. He started by admitting the assumption that all bodies (at least those in the so-called "sublunar realm") would be formed from the four elements of tradition (water, air, earth and fire), pointing out, however, that these elements (in Greek *stoicheia*: also the letters of the alphabet), considered by other traditions to be self-generating, or at least un-generated, would instead themselves have been generated as modifications of a primal matter, which he qualifies from time to time as *ekmageion* (matrix) *hypodoche* (support) *chora* (receptacle, empty space), the indefinite *dyas* (dyad) and sometimes as a *triton genos* (*third genus*).

Indeed, if the elements have the characteristic of being converted into one other as the proportions of the four fundamental principles (hot/cold/dry/wet) change in them, they cannot, strictly speaking, originate *from* each other. Just as the letters of the alphabet whose name they bear (*stoicheia*), they can generate themselves only by virtue of a common operation that lies at their foundation (e. g. *writing*). So the elements, according to Plato, must generate themselves from some sort of common operative substratum. Because of its lack of determination and the fact that it coincides with motionless space, this substratum cannot explain by itself the origin, structure, and intelligibility of bodies. For that is required the intervention of a form (as the principle of structure), a soul (as the principle of movement) and a *logos* (as the principle of intelligibility).

This primary and fundamental matter of which the elements are composed, and which is opposed to form/soul/*logos* is not identified by Plato with a single term, nor does he ever employ the word *hyle* to refer to it. The latter word does appear in his later (*Timaeus*, *Critias*, *Laws*) and mature dialogues (*Philebus*), but in a number of particular meanings, of which *Timaeus* 68D-69 E is exemplary:

> It is god who possesses both the knowledge and power required to mix a plurality into a unity and, conversely, to dissolve a unity into a plurality, while no human being could possess either of these, whether at the present time or at any time in the future. And so all these things were taken in hand, their natures being determined then by necessity in the way we've described, by the craftsman of the most perfect and excellent among things that come to be, at the time when he brought forth that self-sufficient, most perfect god. Although he did make use of the relevant auxiliary causes, it was he himself who gave their fair design to all that comes to be. That is why we must distinguish two forms of cause, the divine and the necessary. First, the divine, for which we must search in all things if we are to gain a life of happiness to the extent that our nature allows, and second, the necessary, for which we must search for the sake of the divine. Our reason is that without the necessary, those other objects, about which we are serious, cannot on their own be discerned, and hence cannot be comprehended or partaken of in any other way. We have now sorted out the different kinds of cause, which lie ready for us like lumber (*hyle*) for carpenters. (Plato 1997, 1270)

The use of *hyle* continues in the later dialogues, particularly in the *Critias*: in 107 C, still related to a *techne*, the pictorial one representing "rivers and woods" (*potamous kai hylen*); 114 D about "All that the forests offer for the work of carpenters" (*hosa hyle pros ta ton tektonon diaponemata parechetai*); 118 DE: "and it was by this way that they brought down timber from the mountains to the city" (*ek ton oron hylen kategon eis to asty*). And in the *Laws*, where at IV, 705 C we are asked, "In what way is the place in our region supplied with timber for shipbuilding?" (*naupegesimes hyles*); at VI, 761 CD it is said, "The young men should erect in every quarter gymnasia for themselves and senior citizens, construct warm baths for the old folk, and lay up a large stock of thoroughly dry wood (*hylen auen*)." In VIII, 843 E, it is prescribed that "If anyone burns his own wood (*hylen*) without taking sufficient precautions to protect his neighbor's, he must be fined a sum decided by the officials", and in VIII, 849 D: "Any foreigner who wishes may buy any kind of firewood (*hylen kausimon*) wholesale any day from the district agents and sell it to other foreigners whenever he likes and in whatever quantity he pleases."

Examination of these passages shows that *hyle* is always in Plato a collective noun and that it almost never appears in naturalistic settings, but rather in anthropically and technically characterized contexts. One partial exception is *Politicus* 272 A where, with reference to the myth of the Golden Age, it is recounted how "Everyone came back to life from the earth, and there was no remembrance of the previous situation: these goods then were lacking, however they had abundance of fruits from trees and much other vegetation *(apo te dendron kai polles hyles alles*), without being generated through agriculture, but spontaneously offered by the earth."

The lesson we could learn from Plato is thus that even if the necessity of conceiving of a primary matter as common substratum and operational basis of the elements' transformations can be deduced from many parts of his works, this matter seems to remain—also according to the variety of terms used—a sort of abstract notion, a para-ideal principle whose role is to represent—each time in different ways and contexts—the opposition to the principles of determination (*logos*, soul, unity, identity). But, more important, we don't seem supposed to identify this primary matter within the semantic—vegetal—spectrum of *hyle*.

3.3 Wood as *Technophysis*: Aristotle

Having no other decisive indications from Plato's immediate successors such as Speusippus and Xenocrates, one must undoubtedly attribute to Aristotle the extension of the term *hyle* first to materials other than wood (especially bronze and stone for statues) and then to the meaning of "matter" in general. Such extensions and concretizations also served to answer the fundamental question of the debate: does a *prote hyle*—a first, primary matter—exist? And if so, does it alone succeed in providing things with the forms, structures and purposes with which they appear to be endowed?

Aristotle's answer is, as is well known, especially articulate (to the point of being aporetic, according to some interpreters). What really interests us is why Aristotle, when confronted with the problem of the existence or non-existence of a primary matter that explains the existence of the whole, would turn to the ancient Greek word for wooden matters in their natural state (trees and plants in the forest) and for wood as a technical resource. What is it, in this word that holds together the forest and the timber, the living and the no longer living, that prompts Aristotle's choice? What is the secret of *hyle*?

A first clue is contained in the only passage in Plato's *Philebus* where *hyle* appears in a sense reminiscent of "material" in general (54 BC):

PROTARCHUS: By heavens, what a question to ask me! You might as well ask, "Tell me, Protarchus, whether shipbuilding goes on for the sake of ships or whether ships are for the sake of shipbuilding," or some such thing.

SOCRATES: That is precisely what I am talking about, Protarchus.

PROTARCHUS: What keeps you from answering your questions yourself, Socrates?

SOCRATES: Nothing, provided you take your share in the argument.

PROTARCHUS: I am quite determined to.

SOCRATES: I hold that all ingredients, as well as all tools, and quite generally all materials (*hylen*), are always provided for the sake of some process of generation. I further hold that every process of generation in turn always takes place for the sake of some particular being, and that all generation taken together takes place for the sake of being as a whole (Plato 1997, 433–434).

The context is here still shipbuilding, and thus the apparently general sense of "material" (*hylen*) must be understood as that of all the wooden components that pertain to the ship. However, this is a first step in the direction of a generalization of wood as a "material *par excellence*," which may have served as a cue for Aristotle to connect with another Platonic insight: that of considering the creation of the world from the human point of view as an artisanal operation by an artisan God (the Demiurge). In short: as a technical operation.

Putting these elements together, it is possible to endorse the hypothesis that Aristotle picked up a particular way of viewing nature from the time of his affiliation with Plato's Academy. Which seems evidenced not only by the acquisition of this idea already in the early works (particularly in the early dialogue *Protrepticus*, see especially fr. 13 Berti), but also by the fact that in his *Corpus*, apart from a number of examples circumscribed to the dimension of wood and living plants (almost all in *De Historia Animalium*, as noted by Happ 1971), the use of *hyle* appears without variation from the earliest to the latest works as a picture of natural things as analogous to *technical artifacts* (the original exploration of this systematic parallel, though developed in an anti-Aristotelian vein, is in Heidegger 1998).

Simplifying the complexity of the Aristotelian argument, we can say that nature (*physis*) appears as a system of ordered and oriented relations between organisms only through a technical scanner- that is, thanks to the guiding idea that natural and technical entities share a tendency toward an end that constitutes their good—understood as the best possible execution of a function by a complex of parts organized according to an intelligible structure. In natural entities the end is internal, that is, it emerges from or in the composition of the parts; while in technical entities it is imposed from outside, that is, present in the mind/soul (*psyche*) of a maker who assembles the parts according to a design that pursues the performance of a function.

The two dimensions do not, according to Aristotle, stand in a simple relationship of analogy, that is, in a parallel that only relates them extrinsically. They are rather constitutively intertwined, in a new kind of *physis* (for which I have proposed the name *technophysis*, Le Moli 2022). Indeed, *physis* has the characteristic of tending from itself to its own end, but this tension does not guarantee its success: the proof is the failure to which all natural entities are exposed in the realization of their own *ergon*, their own constitutive work. And the more complex the structure and growth process of an organism, the more threatened by powers foreign to its natural purpose (fate and chance—*tyche* and *automaton*) will be its existence.

Now, because of its constitutive entanglement with nature, technique can reproduce and re-make nature (this is the sense of *mimeitai,* with which Aristotle sometimes characterizes the entanglement)—that is, technical capacities intrinsic to the natural entity can intervene, from within, to correct the course, redirect the path, remove an obstacle, or heal a wound. A *technique of nature* (*technophysis*) is thus structurally predisposed to come to the rescue of natural organisms when they are in danger of missing their purpose. And again, the more complex an organism, the more articulate the repertoire of technical solutions to remedy an interruption or threat on the way to the realization of its *ergon*. The main strategy to accomplish this task even in the smallest organisms is to modify the constitution of their body, and this can be done by purposedly reorganizing the internal disposition of their parts or by making use of external material bodies, attaching to them or inglobating them. In both cases the original nature (*physis*) of the organic body is replaced by *a new physis*, a second nature originated by the modifications produced in the body after an environmental change or solicitation.

The presence of this original technique inscribed in the form-material constitution of *physis*—the *technophysis*—shows that the matter for which Aristotle is seeking the term is already understood as an operative one, tending by itself to a form that emerges from the assemblage of its own components, and which is pursued by resorting to every possible expedient, including attaching to or inglobating another body. Such a natural matter must therefore already possess a technical side. That is, it must be capable of assembling and organizing itself in the most diverse forms, and of continuing to sustain the possibility of forms even after its natural course is exhausted or interrupted. Far beyond the example Plato used to qualify his original matter-matrix (the mass of wax as absolute absence of form), Aristotle turns to pre-structured materials such as stone or, in an early *Vegetal Turn*, wood—understanding those as naturally apt to hold a repertoire of forms somehow already prefigured in their material constitution. Like the marble block, the log and the wooden board turn out to be configurable and re-configurable according to a repertoire of forms offered and oriented by their very composition. Unlike inert matter, however, the living matter that *hyle* stands for contains a greater openness (the *physis* that resonates in all that is *phyton*, vegetal) and a heightened ability to assume ever new forms as a way, demanded and foreseen by nature itself, to address, correct or repair itself.

If, therefore, matter does not oppose form, but rather offers it a "resistance" that also evokes it, it does not merely passively undergo its action but rather legitimizes this imposition of form according to precise affordances. As primary matter, forest/wood bodies this forth in the broadest way. Unlike the four elements of tradition, devoid of form (for what is the form of fire, water, air, earth?), Aristotelian matter is apt to sustain a form because it is potentially capable of accommodating a purpose by assuming an aspect (*eidos*) that progressively comes forth through it. In this sense, Aristotle's matter is generative—it is "mother".

The connection between *hyle* and *mater*, "mother", that Aristotle establishes several times throughout his works was, in the eyes of the Latins, evidence of the appropriateness of translating *hyle* with *materia* (also evidently connected with *mater*). Matter is mother because it is able to welcome, sustain and bring to fruition new forms (Soardi 2017), and thus be the ultimate support (*hypokeimenon*) of a life that perpetuates and diversifies itself technically. Live matter generates living inventions.

Aristotle, then, while not giving explicit prominence to wood (*xylon*) as a privileged material over, say, bronze or stone, makes *hyle* the contact point of generation and invention, spontaneity and artifice. Unlike bronze and stone, the vegetal side of *hyle* is not simply that of a natural entity that can be manipulated plastically, but rather that of a living complex of organized parts that expresses the permanence of a form, in that the characteristics that held the organism together are not lost when it disintegrates but are led to new functions. This is what is meant, perhaps, even today when wood-related practices are interpreted as a way of giving a "second life" to a material that would otherwise remain destined for mere re-entry into the natural cycle (Nakashima 1981; Green 2006).

This can, of course, be debated. But if there is a *prote hyle*, a primary matter in the Aristotelian sense, it must be understood, rather than in the sense of an original and indeterminable substratum, in the direction of the *vegetal matter* which is the first (*prote*) to stimulate us both from the technical and the conceptual point of view. A matter/material destined to remain the closest to us because it never ceases to accompany the evolution of the human world.

It is worth remembering the fact that the greek word we translated with "nature", *physis*, is related to the vegetal world because of its derivation from the verb *phyo*, which in its active form means "to generate" and, in a metaphorical sense, "to make appear what is not evident or what is hidden", whereas "the mediopassive form *phyomai* indicated being generated, being born (including the sprouting of a plant), but also, by extension, the continuation of generation, i.e. growth" (Bondì and La Vergata 2014, 13). And that the substantive *phyton* indicates plants, suckers and vegetables in general as "designates [...] everything that sprouts or sprouts [...] trees, shrubs, herbs, vegetables, grains" (Repici 2020, 7).

But, and perhaps most important, also the apparent counterpole of *physis* in the ordinary natural/technical opposition, *techne*, derives from the vegetal world because it can be linked—together with the related term *tekton* which indicates all craftsmen who work with wood—to an Indo-European root **tek* which denotes the activities of arranging, putting in order or making straight (Semerano 1994, 290) with particular reference to the art of manufacturing a wooden framework (Pellegrin 1997, 1191–1192). And that can also be extended to the general practices of repairing or "curing/regenerating" something damaged (Semerano 1994, 290).

3.4 Conclusion

Primary matter (*prote hyle*)—can thus be considered the wood as the first and most abundant thing that, because of its vegetal/natural origins, we humans perceived as able to sustain different forms and to offer chances of self-regenerating. This might have been the reason why, to an extent larger than other organic materials such as bone, shells and skin, wood and fibers in general have been intuitively presented to us as natural bodies apt to be used for purposes that overcome their initial condition.

Although it is debated, among anthropologists and paleoanthropologists, whether the birth of human technology should be backdated to before the Paleolithic, following the (precarious) remnants of a "wood age" (Ennos 2021) that preceded that of stone, bronze, and iron (Leroi-Gourhan 1943; Oakley 1961), surely much of human civilization coincides with an increased capacity for combustion that begins with wood as fuel and progressively develops a whole series of other products (charcoal or coke), eventually enabling furnaces to be brought to higher and higher temperatures and bring more and more complex materials to fusion, from copper to bronze and from iron to steel. And until very recently machines, structures but especially the connective parts and main joints of our most important technical assemblies were still made of wood.

It is thus that the wooden material is "primary", as that which triggers and never ceases to accompany human technical evolution—at least until the beginning of the industrial age. To then return, surprisingly, today—to challenge the hegemony of metals and plastics in sustainable technology projects (Natterer et al. 1996).

In a new twist of the *Vegetal Turn*, then, we should look at wood today with the same twofold gaze that the ancients turned to *hyle*: each time full of gratitude for the possibilities offered by a felled forest, as well as by respectful awe for the unfelled wilderness that at every step can close in around us.

References

Bernard of Tours. 1978. *Cosmographia*, ed. P. Dronke. Leiden: Brill.

Berti, Enrico. 2011. Hyle nei testi aristotelici. In *Materia. XIII Colloquio Internazionale del Lessico Intellettuale Europeo*, eds. D. Giovannozzi and M. Veneziani, 41–52. Firenze: Olschki.

Bibikhin, Vladimir. 2021. *The Woods*, trans, A. Tait. Cambridge: Polity Press.

Bondì, Roberto, and Antonello La Vergata. 2014. *Natura.* Bologna: Il Mulino.

Buccheri, Alessandro. 2020. Crescite troppo rigogliose: modelli botanici del (superamento del) limite. In *Crescere/svilupparsi. Teorie e rappresentazioni fra mondo antico e scienze della vita contemporanee*, eds. F. Giorgianni, P. Li Causi, M.C. Maggio and R.R. Marchese, 147–184. Palermo: Palermo University Press.

Calcidius. 1975. Timaeus a Calcidius translatus commentarioque instructus. In *societatem operis coniuncto*, eds. P.J. Jensen and J.H. Waszink. London-Leiden: Brill.

Ennos, Roland. 2021. *The Wood Age. How One Material Shaped the Whole Human History.* London: Collins.

Ernout, Alfred, and Alfred Meillet. 2001. *Dictionnaire Étimologique de la langue latine. Histoire des Mots.* Paris: Klincksieck.

Green, Harvey. 2006. *Wood*. Craft, Culture, History, London: Penguin Books.
Happ, Heinz. 1971. *Hyle. Studien zum Aristotelischen Materie-Begriff*. Berlin-New York: De Gruyter.
Heidegger, Martin. 1998. *On the Essence and Concept of* Physis *in Aristotle's* Physics *B I*. In *Pathmarks*, ed. W. McNeill, 183–230. Cambridge-New York-Melbourne: Cambridge University Press.
Isidore of Seville. 1911. *Isidori Hispalensis Episcopi Etymologiarum sive Originum Libri XX*, recognovit brevique adnotatione critica instruxit W. M Lindsay. Oxford: Clarendon Press.
Isidore of Seville. 2006. *The Etymologies of Isidore of Seville*, eds. Stephen A. Barney, W.J. Lewis, J.A. Beach and O. Berghof. Cambridge: Cambridge University Press.
Leroi-Gourhan, André. 1943. *L'homme et la matière*. Paris: Albin Michel.
Le Moli, Andrea. 2022. *Technophysis. Le tecniche della natura*. Palermo: Palermo University Press.
Leszl, Walter. 1973. Matter in Aristotle. *Critical Review of the History of Philosophy* 28 (3): 243–270.
Malaspina, Ermanno. 1991. L'introduzione di materia nel vocabolario retorico e filosofico a Roma: Cicerone e Lucrezio. *Atti Della Accademia Delle Scienze Di Torino* 125 (2): 41–64.
Nakashima, George. 1981. *The Soul of a Tree. A Woodworker's Reflections*. Tokyo-New York-San Francisco: Kodansha.
Natterer, Julius, Thomas Herzog, and Michael Volz. 1996. *Holzbau Atlas Zwei*. Munich: Institut für internationale Architektur-Dokumentation.
Oakley, Kenneth P. 1961. *Man the Tool-Maker*. London: Jarrold and Sons.
Pellegrin, Pierre. 1997. Techne ed episteme, in *I Greci*. In *Storia Cultura Arte Società*, ed. S. Settis, 1189–1203. Torino: Einaudi.
Plato. 1997. *Complete Works*, ed. J. M. Cooper. Indianapolis-Cambridge: Hackett.
Repici, Luciana. 2020. *Uomini capovolti. Le piante nel pensiero dei Greci*. Pisa: Edizioni della Normale.
Semerano, Giovanni. 1994. Le origini della cultura europea, vol. II. *Dizionari etimologici. Le basi semitiche delle lingue indoeuropee. Dizionario della lingua greca*. Firenze: Olschki.
Soardi, Marzia. 2017. As a Mother: A Reading of the Relationship Between the Aristotelian Theory of Reproduction and the Concept of hypokeimenon, in *Prote Hyle*. In *Notions of Matter in the Platonic and Aristotelian Traditions*, ed. A. Le Moli and L. Alexidze, 55–66. Palermo: Palermo University Press.

Chapter 4
Heroes of Botany: Voices of Dissent from the Western Tradition's Devaluation of Plants

Isis Brook

Abstract This chapter begins with a brief overview of the way plants have often been downplayed in Western culture so that their attributes and their beingness have been overlooked. To illustrate this, reference is made to ideas such as: the great chain of being; the zoocentric emphasis in biology, which encourages a misreading of plants as essentially passive (virtually inanimate); and the anthropocentric focus on use value. The chapter goes on to sketch four vignettes of historical figures in botany who explored and wrote about the plant realm in ways that dissented from the Western orthodoxy. These figures are: Theophrastus (pupil of Aristotle and arguably the father of scientific botany); Johann Wolfgang von Goethe (eighteenth century German poet whose scientific works include *The Metamorphosis of Plants* and whose approach to science required a human/plant interaction); Agnes Arber (twentieth century English botanist and philosopher who understood the useful vestiges in earlier approaches to the appreciation of plant agency); and Barbara McClintock (twentieth century US botanist/cytogeneticist who discovered transposons, the so-called "jumping genes"). Each of these figures is placed in their time, and the chapter presents something of what they offered to the vegetal turn. The picture of plants that emerges is that of dynamically engaged beings open to relationships—a picture which, in turn, can connect us to a broader/deeper conception of the world.

4.1 Passive Useful Material

How has it come about that the predominant Western view of plants is that they are just passive useful material (Calvo 2022; Trewavas 2015)? This backgrounding of plant life is so pervasive that, like the plants themselves, it typically goes unnoticed. While we might be moved to wonder by plants, or feel unconsciously calmed by their presence, in the main plants are just background stuff. The theory of 'plant

I. Brook (✉)
Bath Spa University, Bath, UK
e-mail: isisbrook.stroud@gmail.com

M. Di Paola (ed.), *The Vegetal Turn*, The International Library of Environmental, Agricultural and Food Ethics 38, https://doi.org/10.1007/978-3-031-68671-9_4

blindness', proposed by James Wandersee and Elisabeth Schussler, who also coined the term (1999a), described the phenomenon thus:

> the inability to see or notice plants in one's environment—leading to: (a) the inability to recognize the importance of plants in the biosphere and in human affairs; (b) the inability to appreciate the aesthetic and unique biological features of the life forms that belong to the Plant Kingdom; and (c) the misguided, anthropocentric ranking of plants as inferior to animals and thus, as unworthy of human consideration. (Wandersee and Schussler 2001, 3)

In exploring possible reasons for plant blindness, Wandersee and Schussler focus mainly on the way the eye provides data from which the brain then sifts the important information. As they point out: "psychologists would say that plants have a *low signal value* for many US citizens today" (2001, 5). The predominance of green, with its density of overlapping leaf shapes, seems to invite backgrounding. Plants do not seem to stand forward to present themselves to the casual observer, except perhaps when they are unusual or in flower. Even then, once noticed, they are often passed over as a background mass or item in the landscape. For obvious evolutionary reasons, the human eye/brain is attuned to movement, as that may be a signal for immediate danger. The sessile nature of plants, the fact that they (apparently) do not move, means that more than casual attention is needed to foreground them.

The use of time lapse photography has done much to bring the plant-time dimension to our understanding of the movement of plants. Footage like that in the January 2022 BBC television series *The Green Planet* helps to shift our typical picture of plants, from static entities to agentic mobile beings visibly engaging with their environment. But these movements were available to patient observers before. For example, a patient observer can see that the stems of ivy-leaved toadflax (*Cymbalaria muralis*) after flowering, over a period of days, will turn away from the light and enter crevices to plant their seeds (Fig. 4.1).

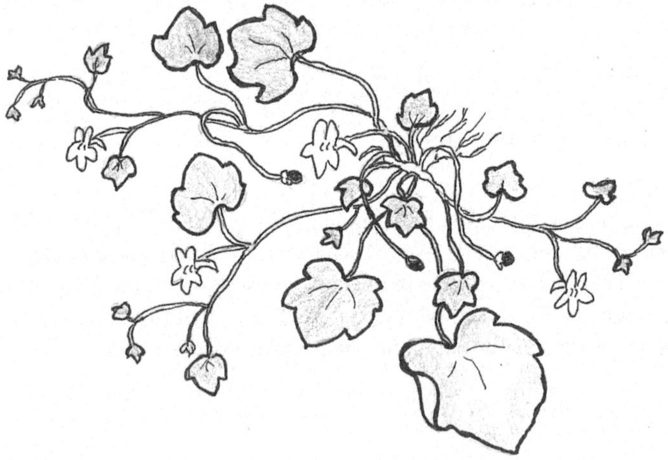

Fig. 4.1 Ivy-leaved toadflax

And there is, of course, Darwin's discovery of the circumnutating of seedlings with just the naked eye and a notebook (Darwin and Darwin 1880). Without time lapse photography or patience, our appreciation of plant movements will be dependent on showy peculiarities, such as the Venus flytrap (*Dionaea muscipula*) closing its 'jaws', or the Sensitive plant (*Mimosa pudica*) collapsing its leaves. A telling anecdote in a paper in a special issue of *Plants People Planet* dedicated to plant blindness has a researcher who touched a sensitive plant exclaim, upon seeing the immediate collapse of its leaves, "It's alive!" (Jose et al. 2019, 169). The expectation that living things move, and do so in a way that is perceptible to us, is at the very heart of the backgrounding of plants.

Historically, the lowly position given to plants came from the perception of them—from Plato onwards—as *passive*. With Aristotle comes the division of faculties of the soul into the nutritive, perceptive, desiderative, locomotive and intellectual—and plants, he thought, had only the nutritive (Hardy and Totalin 2016, 66). They can feed and reproduce themselves, but cannot move or sense their environments, and they certainly can't think. Their vegetative processes thus came to be understood as "passive mechanical processes" (Hall 2011, 23), barely ensouled at all. This is evident in Aristotle and later in Pliny (Hardy and Totelin 2016, 20), and shaped the focus of much ancient botany, which was then transmitted through to the mediaeval period and thus became part of the Western heritage that we still inhabit. Integral to that heritage is the picture of plants as very low on a hierarchical ladder that traditionally goes, in descending order: God, angels, humans, animals, plants (Lovejoy 1976; Hiernaux this volume). Even the usefulness of plants as foods is downplayed in the traditional valuing of meat and fish over vegetables and grain (Chiles and Fitzgerald 2018).

From Aristotle onward, it is evident that a certain zoocentrism also played a part in the cultural marginalization of plants. The focus of science was, for the most part, animals, and this helped make animals the norm of science too. The study of plants then takes on a deficit-based structure, whereby what gets noticed about plants becomes their lack of obvious animal-like features. They become a passive biological background for the important stuff. This zoocentrism is still evoked in the contemporary debate about plant-blindness, with biology teachers complaining of their students' lack of interest in botany compared to zoology, and researchers complaining about the lack of funding that goes into plant conservation compared to animal conservation (Jose et al. 2019, 169).

Seeing plants as, primarily, food or medicines, is another shaping presupposition for their everyday marginalization in the West; a way of backgrounding their being anything beyond our use. This focus on the practical application of plants is demonstrated in the history of botany, which for centuries was taken up with agriculture and herbals; and generally plants' immediate usefulness (Morton 1981). In other cultures the being of plants as requiring respect and thanks and, e.g., their link to sacred rituals is more evident (Kimmerer 2020).

Even in a rudimentary scientific understanding we are aware that the usefulness of plants extends way beyond the immediate allure of food or the practical benefits of building materials, their utility it total: without photosynthesis there would be no breathable atmosphere no animals no us. However, their absolute essentialness to everything else does not define the totality of their being.

Nonetheless there were, throughout Western history, those who took exception to the marginalization of plants. In what follows I present some of these dissenting voices, and what they were telling us.

4.2 Theophrastus of Eresus (C. 372–287 BCE)

It seems odd to think of Theophrastus as a dissenting voice when he was also the originator of botanical science as we know it—in Linnaeus' words, "the father of botany" (Grene & Depew 2004, 11). Theophrastus was a student and collaborator of Aristotle at the Lyceum, and was his designated successor, taking over the school at the master's death and then leading it for thirty-five years. Theophrastus wrote widely, including on physics, metaphysics and ethics; but he is best known today for his extensive work on plants. *Enquiry Into Plants* (originally ten books) and *Causes of Plants* (originally eight books), were major works, and most of the material they contained has survived.

4.2.1 Studying Plants as Plants

Reading Theophrastus, one is immediately struck by the sense of new intellectual territory being carved out. Of course there was Aristotle in the background, and Theophrastus did cite other commentators on plants, such as his contemporary Diocles of Carystos (Morton 1981, 66). However, the focus for Theophrastus was the plant first and foremost. He began *Enquiry Into Plants* thus:

> In considering the distinctive characters of plants and their nature generally one must take into account their parts, their qualities [behaviours], the ways in which their life originates, and the course which it follows in each case: (conduct and activities we do not find in them, as we do in animals). (Theophrastus 1916, 1.1.1)

This comparison with animals and the pointing out of what plants lack compared to animals would have been standard practice at the time, and Theophrastus goes on to criticise it on the very next page. He begins to paint the peculiarity of plants by discussing what, for a plant, would constitute a part. For example, should this include only permanent parts, or also those that appear seasonally? If one includes all, then the number of parts "will be indeterminate and constantly changing" (Theophrastus 1916, 1.1.2–1.1.3). And yet if one doesn't include things like new shoots or fruit, then one is no longer describing/defining a plant. By the second page, Theophrastus can exclaim that, "we should not expect to find in plants a complete correspondence with animals" (1916, 1.1.2–1.1.3). Most originally, and unlike Aristotle and later botanists, Theophrastus did not see that lack of correspondence as a lack on the part of plants. For him this indeterminacy of parts is one of the strengths of plants. As he went on to say:

> And that is why the number also of parts is indeterminate; for a plant has the power of growth in all its parts, inasmuch as it has life in all its parts. Wherefore we should assume the truth to be as I have said, not only in regard to matters now before us, but in view also of those that will come presently; for it is a waste of time to take great pains to make comparisons where that is impossible, and in so doing we may lose sight also of our proper subject of enquiry. (Theophrastus 1916, 1.1.3–1.1.4)

In that short quotation we have something of the essence of Theophrastus' work. He made a break with past assumptions about how to proceed, relished what is special about the subject of enquiry, and kept the subject in view. Such an approach freed up his thinking to be able to observe both the differences between plants and certain common principles. He brought the multiplicity of the plant kingdom into view and began grouping plants by their natures rather than their uses. The quality of Theophrastus' botanical observation and theorising is explained in Morton's *History of Botanical Science* (1981), where he notes points such as: Theophrastus's feel for fundamental taxonomic groupings, including separating plants from fungi; his grouping related species in a way similar to the genus, species and variety categories used today; his developing a logical system for naming plants, and coining new technical terms to explain his observations (Morton 1981, 37). He also spotted what centuries later became the significant distinction between dicotoledons and monocotoledons (Arber 1950, 22).

4.2.2 Rejecting a Solely Use Value

Although his *Enquiry Into Plants* and *Causes of Plants* contain detailed and perceptive descriptions of individual plants, this was not their main purpose. Theophrastus was not writing a herbal book, which describes each plant and its effects on human beings. His aim was more broadly scientific—he wanted to understand plants as plants. He focused on phenomena such as the way plants respond differently in different environments, and how moving plants to different environments can bring about mutations (Theophrastus 1976, 2.16.6). In Theophrastus' writings, plants are agentic: "each tree seeks out its appropriate locality, that locality being appropriate in which it is at its ease" (1976, 2.16.8). He also gave examples, such as deciduous trees providing nutrients to the roots of evergreens, to explain that plants "collaborate to preserve and propagate others" (1976, 2.18.1).

As Michael Hall has pointed out, Theophrastus' presentation of plant faculties goes beyond what is possible for the merely 'nutritive' soul Aristotle assigned them. Theophrastus' plants sense things and respond to the world: they are volitional, minded, intentional creatures that show autonomy and purpose (Hall 2011, 8). This comes through in Theophrastus' lectures on agriculture as well, where he imparts useful information to advance agriculture as a practical science, but is also keen to point out that some of the manipulations we undertake are for us and not always in the plant's own interest. He made a clear distinction between what might be good for us as consumers of plants, and what the plant would prefer for itself. Indeed, he ultimately rejected the Aristotelian notions of a chain of being and of an external *telos* for the plant, namely serving animals including humans (Hardy and Totelin 2016, 67).

After Theophrastus, botany rather stagnated. There was what Morton (1981, 82) describes in a chapter heading as "The Dark Ages of European Botany: 200–1483". Plants lost the agency aspect that was described in Theophrastus' work. Botany became largely about pharmacology, and scholarship declined into the copying and rigid application of 'authoritative' texts such as those of Discorides and Pliny (Morton 1981, 85), who present plants more as "useful subservient objects" (Hall 2011, 41). There were other voices even in the depths of the Dark Age, for example, there was St Hildehard von Bingen (1098–1179) writing the herbal texts *Physica* and *Causes and Cures* and characterising plant life as both life and spirit at its most vital (Clendenen 2012; Marder 2021). However, for a bridge back to botany, as usually understood, we can move to the eighteenth century and turn to Goethe. In his work we can find a clear picture of someone who was tuning in to the aliveness of plants, but in something closer to what would have been seen as a scientific way of working.

4.3 Johann Wolfgang Von Goethe (1749–1832)

Goethe, the great German poet, believed his scientific endeavours were more important than his literary legacy (Seamon and Zajonc 1998, 1). His scientific work included studies in geology, meteorology, osteology, botany, and the study of colour. The reason he felt his contribution to science was so important was that his approach (a unique fusion of older and newer ideas) was fresh. To describe the object of his studies he coined the term morphology (Morton 1981, 346), which he defined as: "principles of structured form and the formation and transformation of organic bodies" (Goethe 1995, 57). Goethe's peculiar ways of inspecting reality is clearly delineated in and by his botanical work. Moreover, there is a sense in which engagement with the plant realm may have led the way, refined, and perhaps confirmed the efficacy of his methods.

4.3.1 Journeying with the Plant

Goethe's interest in plants began in earnest when he moved to Weimar to take up an administrative post. This brought him into the "pure atmosphere of country, forest and garden" (Goethe 1952, 150). In 1776, he was given a garden by Archduke Karl August, which he used for early observation and experimentation. His studies included Linnaeus' classifications, as would those of any educated Westerner at the time, but he was ultimately dissatisfied with the emphasis on cataloguing (Morton

1981, 263), which seemed to be biology's master project in the eighteenth and early nineteenth centuries. Goethe's passion was to understand nature as a whole, and its generative powers. To investigate what a plant is, Goethe did not study dry preserved specimens in herbarium samples, as most others did, but living plants, focusing on annuals primarily (Goethe 1995). He could thus observe whole life cycles, and experiment with changing their growing conditions. He was also free of some weighty conceptual assumptions: in the eighteenth century, the hierarchical ladder of creation concept dictated that trees—as the most perfect plants—were the closest to humans. Linnaeus was burdened with both a focus on trees, as the model plant, and a belief that plant species were fixed for all time, because they were created individually by God (Tantillo 2002, 76). Goethe had no such conceits: if he had an overarching model at all, it was that Nature was creative. For him, understanding plants, or rather how nature does 'plant', was a way of seeing into the generative process of the specific plant, plants in general, and nature as a whole.

Goethe's observations and ideas coalesced on his extensive and life-transforming trip to Italy (1989). There he saw that the way a plant grows reflects aspects of its environment, such that there can be no stable sample of a plant—its size and development are shaped by its interaction with where it grows. Understanding of plants, like plants themselves, cannot be static: it must morph with them: plants' aliveness is constantly in flux and thus not something that can be pinned down. Goethe also criticised the scientific use of the term *Gestalt* for the structured form of organisms, because it excluded what was in a state of flux: "If we wish to arrive at some living perception of nature, we ourselves must remain as quick and flexible as nature, and follow the example she gives" (Goethe 1995, 63–64). Following the path laid down by the plant itself, Goethe was drawn into its nature and thus felt he could report his observations as authentically from the plant realm, and not as predetermined by any tightly held theoretical lens (Brook 2021). Two experiences on his plant journey will suffice to give a taste of what was to inform his subsequent publication in 1790 of *The Metamorphosis of Plants.*

4.3.2 Discontinuous Metamorphosis

In Italy Goethe saw familiar plants growing more abundantly and exciting new plants, but it is in the botanical gardens in Padua in 1786 that he first noticed the discontinuous metamorphosis of the shape of the leaves of a particular plant: the dwarf palm (*Chamaerops humilis L.*) (Goethe 1989, 15). This is a common feature of plant development, but at the time its relevance still needed to be identified. With his attention to flux, Goethe was able to read the way a plant produces at first simple leaves on petioles, then more elaborated leaves, then back to simple shapes that hug the central stem, as a process of metamorphosis, not of the individual leaves, but of the plant whole. By engaging his imagination enough to follow that story, he then saw that the flower parts are all also a continuation of that same process. The plant is engaged in a process of transforming its material—a "striving" (Tantillo 2002, 69) through time by shaping and reshaping itself. It does not unfold according to a strictly determined plan, nor is its form dictated by some mysterious external force: the plant itself expresses a vitality and a drive to develop and reach a point of fruition, when circumstances allow. Goethe saw that the plant substance, expressed as leaf, undergoes a transformation to become the other plant parts: sepal, petal, carpal, stamen, etc. The plant's creative process is visible and laid out in front of us, once we shift our attention from static form to fluid process.

4.3.3 *Imaginative Excursion*

Goethe gave an account of his imaginative gift being fructified by the plant realm.

> When I closed my eyes and lowered my head, I could imagine a flower at the centre of my visual sense. Its original form never stayed the same for a moment; it unfolded, and from within it new flowers continually developed with coloured petals and green leaves. … Here the appearance of an afterimage, memory, creative imagination, concept, and idea all work simultaneously, revealing themselves through the unique vitality of the visual organ in complete freedom and without intention or direction (Goethe 1995, xxi)

The observation and identification of key aspects of the plant realm, plus an open-minded contemplation of them, helped to reveal the activity of the plant. This mode of engagement, Goethe thought, was of crucial importance for the progress of botany. His use of the term *Anschauung* captures this well. Although it is not directly translatable, Agnes Arber rendered it as: 'intuitive knowledge gained directly through contemplation of the visible aspect' (Arber 1954, 122). Goethe was resistant to the notion of a free-floating idea or concept that is not tied to, or indeed given by, the phenomenon being studied. The idea or concept for Goethe needed to come to the human mind through a kind of shared contemplation with the thing studied (Stephenson 1995).

One of the examples discussed in *Approach to the Metamorphosis of Plants* is the proliferous rose (where a rose flower produces a new stem instead of the generative parts). He said of this abnormality:

> All that we have been seeking to grasp by powers of imagination and thought is shown most clearly in the instance of the proliferous rose. The calyx and corolla are arranged and developed around the axis; but instead of the seed-vessel contracted in the centre of the blossom, […] The stem goes on growing, thorns appear on it again, the coloured petals which now appear singly become smaller and at last transform into variegated stem-leaves, half red and half green; a series of regular nodes is formed and from their eyes small, though imperfect, rosebuds appear once more (Goethe 1995, 93)

Goethe did not, unlike most of his contemporaries, discard what were seen as monstrosities in nature, but rather relished them as clues to how normal development takes place. The proliferous rose demonstrates the accuracy of Goethe's insight about plant development as a process in constant flux. His insight was borne out by plants themselves, facilitated by Goethe's dissatisfaction with the sciences of his time. He

Fig. 4.2 Proliferous rose

understood that neither a mechanistic nor an idealistic/mystical picture of plants would capture what they are. Both these approaches deny the plant its individual agency—and it is that agency that Goethe was able to see by holding those approaches in abeyance, while opening himself to the plant realm to learn from the plant itself (Fig. 4.2).

Morphology, as an approach, was later taken up as a component of biology. However, the necessary 'lively apprehending'—the *Anschauung*—did not entirely survive the effects of an increasingly mechanistic approach in the following centuries. To find a proponent of plant agency embedded in botanical science, we now move to the first half of the twentieth century, and the work of English botanist Agnes Arber.

4.4 Agnes Arber (Nee Robinson) (1879–1960)

Arber authored more than eighty scientific papers and several books of original botanical research. The quality of her work was recognised with a Fellowship of the Royal Society in 1946 (the third woman to be so recognised), and in 1948 with the Linnean gold medal by the Linnean Society (the first woman to receive it). However,

Arber was not constrained by the norms of scientific botany. Her talents extended to art, which comes across in her extensive illustrations; historical study, such as her 1912 seminal book *Herbals: their origin and evolution*; as well as philosophical scholarship, as particularly evidenced in her last two books (mentioned below). What might seem like additional interests that happen to occasionally overlap or serve her scientific work would be better interpreted as all part of a seamless whole. This whole scientific endeavour—her calling—was to understand plants. These fields of learning are woven together most explicitly in her 1950 book *The Natural Philosophy of Plant Form.* Here she sets out her own research in the context of botanical morphology, informed by a sophisticated philosophical questioning of scientific objectivism.

4.4.1 Partial-shoot Beholding

To get a glimpse of her work I will juxtapose it with the more mainstream botany of her time through an examination of her partial-shoot theory. Focus on the partial-shoot theory will give a sense of how, for Arber, the plant is an active entity with an urge toward its own development. (It is worth mentioning that her interpretation of the word 'theory' goes back to its Greek roots and is better understood as 'beholding' (Kirchoff 2001, 1205) so rather than bringing a theory to the plant she is seeing it in or drawing it from the plant).

The contribution of morphology to botanical science was, in Arber's times, rather fixed and limited. The plant was seen—morphologically—as having three distinct parts: root, stem and leaf (the flower parts accepted as leaves). Thus identified, these three parts or organs are seen as non-homologous: they cannot overlap or stand in for each other, in either position or function. This is the classical approach to morphology. It is a tidy perspective, useful as a "rule of thumb" (Rutishauser and Isler 2001, 1179). However, it has its drawbacks, not least of which is just how often the rule is violated by a diversity of plant species.

For Arber, this kind of hierarchical separation of a living being, where the parts make a whole but the whole is not reflected in the parts, fails to capture the dynamic qualities of an organism. At the beginning of *The Natural Philosophy of Plant Form* she explains that morphology is the study of form, yet the term 'form' has become degraded. What should be meant by "form", what morphology proper studies, is "the whole intrinsic nature of which any individual was a manifestation" (Arber 1950, 1–2). To study morphology is to see through the changing forms and the parts, to the living processes and the whole. This requires a way of observing and of thinking that does not make hard and fast distinctions. Taking such an approach has come to be called Fuzzy Arberian Morphology (Rutishauser and Isler 2001, 1179), in recognition of Arber's work.

Now, the partial-shoot theory resists the hard and fast categories of classical morphology—the root-stem-leaf scheme. This enabled Arber to provide a new perspective on leaves. If the shoot (everything above ground, e.g., a stem cutting, or sprouting potato eye) is taken as a significant unit, then the leaf can be seen in a way that is more explanatory both of its multiple forms and of the process of leafing in the plant. With our attention on the shoot, it becomes plausible to think of leaves as partial shoots. Their expression as stems (radial/ vertical) is held back, and their expression as leaves (dorsiventral / spreading horizontal) is given free reign. However, their vestigial whole shoot qualities are still there as potential. Arber expressed this potential, controversially, as "the urge to whole-shoot-hood in the leaf" (1950, 93). As she explained:

> The thesis that the leaf has an urge towards self-completion as a whole shoot—cannot be considered in isolation; we must visualise the leaf and its behaviour on the background of the life of the plant as a whole. The likelihood will be increased that this postulated 'urge' has a real existence, if it can be shown that there is, throughout the shoot-complex and root-complex, a tendency for the members of each generation to compete with, and attempt to dominate, those of the previous generation. The unconscious effort of the leaf to reach the status of the shoot, which is its parent, and even to assume its place, would then no longer be a solitary phenomenon, but as one link in a continuous chain of tendency. (1950, 93)

She then presented, with multiple examples, evidence of how this urge to shoot-hood was expressed by different plants. As recent botanical science, using more extensive genetic data than was available to Arber, has gone on to show, the radiality of most stems and the dorsiventrality of most leaves are indeed just extremes on a continuous spectrum, and compound leaves express both leaf and shoot properties (James 2009; Hofer et al. 2001; Rutishauser and Sattler 1997).

4.4.2 Mining History and Philosophy to Enrich Botany

Arber's historical explorations would have given her examples of previous botanists suggesting the partial-shoot theory. She weaves these figures into her discussion. Particularly relevant were Nehemiah Grew (1641–1712), who saw leaves as ramifications of the branch; Casimir de Candolle (1836–1918) who saw the leaf as a branch altered by atrophy; and paleobotanists Henry Potonie (1857–1913) and Arthur Tansley (1871–1955), whose work on ferns established that the fern frond is better understood as a branch-system. On the plant as having a particularly active engagement in its life, its endeavour to "preserve its own being", Arber called on Spinoza's notion of *conatus* (Arber 1950, 77). And of course, her lifelong appreciation of Goethe's botanical work allowed her to read his 'leaf' not as a fixed unit separated from stem and root, as it had become in classical morphology, but as an indication of active agentic mutability. She noted that forty years after publishing *Approaches to the Metamorphosis of Plants,* Goethe wrote that "when leaves sub-divide, or rather they produce a multiplicity out of themselves, this is a striving to become more complete, in the sense that each leaf bethinks itself [*gedenkt*] to become a branch, as well as each branch to become a tree" (Goethe, quoted in Arber 1950, 93).

4.4.3 A Dialectical Approach to Re-Enliven a Process Morphology

It would not be Arberian to set up an Arberian approach to plant morphology as an antagonist to its classical counterpart. She pointed out that "the shoot hypothesis undoubtedly fits better with the broad facts, but it disregards the usual sharpness of distinction between leaf and stem" (1950, 71). Thus both approaches have virtues, and "we should aim at including and transcending both in a synthesis, which, while treating the shoot as a primary unit, will yet have absorbed into itself such truth as is to be found in the antithesis of stem and leaf" (1950, 72). Arber's skill in, and commitment to, dialectical thinking comes across in all her work, but is explored more extensively in her philosophical books *The Mind and the Eye: a study of the biologist's standpoint* (1954) and *The Manifold and the One* (1957).

Arber was not universally appreciated in her field. Her use of terms such as 'urges' did not fit the mechanistic direction that biology was taking (Boulter 2017, 81), her insistence on the possibility of parallel development in evolution did not fit with the overemphasis on descent from common ancestors at the time (Claßen-Bockhoff 2001, 1170), and she was a woman (Packer 1997). The sense of a plant's striving that runs through Arber's work helps to redefine morphology or perhaps return it to the path that Goethe hinted was possible. Fuzzy Arberian Morphology paved the way for process morphology, whose point is to analyse and describe the dynamic continuum of plant form, no longer describing the structure of a plant and its processes as separate (Sattler 1992).

In conclusion, it is interesting to note that Arber's first published article, in 1894, was in her school magazine (she was aged fifteen). It was an illustrated study of the ivy-leaved toadflax, the plant mentioned earlier (in Sect. 4.1) whose agency is particularly evident when it turns inward to deposit its seeds in wall crevices. In the young Agnes Robertson's summer holiday project notebook this phenomenon was accurately described (Robertson 1894, 15). It is rather tantalising to think that recording this very agentic movement was part of what inspired Arber to become the style of botanist that she was.

In the year Arber died (1960) another scientist with a feeling for plant agency was making huge strides in botanical understanding. This was the North American cytogenetisist Barbara McClintock.

4.5 Barbara Mcclintock (1902–1992)

McClintock began her extremely long and productive career with a breakthrough: while on a graduate training programme in the botany department of Cornell University, she developed cytological techniques that allowed her to identify maize chromosomes (Federoff 1994). This was just the first in a long series of breakthroughs—including, most famously, the discovery of transposons (jumping genes). McClintock's scientific publications are dauntingly complex, and I will not be exploring them here. I will rather give an account of her lifelong relationship with one plant species: maize (*Zea mays*). That relationship, McClintock's commitment to it, and her determination to share her discoveries with the (largely unreceptive) scientific community, places her among the dissenting heroes of botany. For an account of her relationship with maize, it will be helpful to rely on the biographical account given by Evelyn Fox Keller in *A Feeling For The Organism* (1983), with its generous use of direct quotation of McClintock from interview material.

4.5.1 Rejection of Simple Answers

When McClintock began her scientific career, it was common to study maize following genetic traits through the different coloured kernels on each cob. The differences come about because each kernel is an embryo produced from a separate double fertilization. Although the work involved microscopic examination and manipulation, it also required growing plants. McClintock was mostly out in the field amongst these plants, with their complex demands over a long growing season. She wrote:

> No two plants are exactly alike. They're all different, and as a consequence, you have to know that difference ... I start with a seedling, and I don't want to leave it. I don't feel I really know the story if I don't watch that plant all the way along. So I know every plant in the field. I know them intimately, and I find it a great pleasure to know them. (Quoted in Keller 1983, 198)

Work in genetics at that time also focused on the fly *Drosophila*, whose generations could be bred in days, and later the experimental organism of choice became the bacterium *Escherichia coli,* whose generational turnover was down to hours. As Keller makes clear (1983: 6), the dogma operating behind this shift to ever simpler organisms was the mistaken belief that what was true for one organism was true for all. Another dogma of biology at the time was that information in an organism originated in the DNA and only travelled one way: from DNA to RNA to proteins, and was not subject to modification. These fashions and dogmas in the science of the day did not sway McClintock. By 1960 she was able to explain many of the ways the maize plant could regulate the expression of its genes to produce what it needed when it needed it (Keller 1983, 180): this meant that the plant, far from unfolding to a DNA dictated order, had what she termed "transposable" elements, which allowed for greater flexibility to respond to environmental conditions. This was so far from the established science that she was ignored. As she later said:

> Organisms can do all types of things; they do fantastic things. They do everything that we do, and they do it better, more efficiently, more marvellously ...So if the material tells you 'it may be this,' allow that. Don't turn it aside and call it an exception, an aberration, a contaminant ...That's what's happened all the way along the line with so many good clues. (quoted in Keller 1983, 179)

4.5.2 *Working With, Not On, the Organism*

McClintock's determination to stick to a complex organism and listen to what it could tell her was finally vindicated when her work became widely acknowledged as a precursor of much later research in genetics that 're-discovered' that genes were no longer a fixed blueprint that completely dictates the unfolding of a developing organism. Perhaps we can see the delay in recognition for McClintock's work as an example of plant blindness. Because it is only when "analogous research carried out on animals confirmed the existence of genome lability in other species" (Mancuso and Viola 2015, 25) that in 1983 she was awarded the Nobel Prize for her work on gene transposition.

Her relationship with maize drove a way of working that involved emotional engagement (Keller 1983, 198) and intuitive flashes, seemingly prompted by the plants themselves, as with Goethe (Brook 2021). Through such engagement her reasoning itself becomes "plantlike" (Hendlin 2021, 13). But I will leave the last words on her understanding of plants to McClintock herself:

> Animals can walk around, but plants have to stay still to do the same things, with ingenious mechanisms … Plants are extraordinary. For instance, … if you pinch a leaf of a plant, you set off electric impulses. You can't touch a plant without setting off an electric pulse … there is no question that plants have [all] kinds of sensitivities. They do a lot of responding to their environment. They can do almost anything you can think of. But just because they sit there, anybody walking down the road considers them just a plastic area to look at, [as if] they're not really alive. These organisms are fantastically beyond our wildest expectations. (quoted in Keller 1983, 199–200)

4.6 Conclusion

The work of these four figures across the centuries continues to reach out to us to rethink our relationship with plants. There are common themes and personal qualities that resonate across the vignettes I sketched in this chapter. All four scientists were ready to submit themselves to plants, rather than submitting plants to their own human preformed concepts. They rejected a use value as the sole purpose of plants, and approached them with curiosity, a kind of fellow feeling, and wonderment. They engaged in personal conversations with plants as agentic beings, perfectly willing to disclose their secrets. They experienced plants as beings subversive of all attempts to impose a simple (or even a complex) mechanistic picture upon reality, and allowed plants to lead them into a rich understanding of their real nature.

Bibliography

Arber, Agnes. 1950. *The Natural Philosophy of Plant Form*. Cambridge: Cambridge University Press.

Arber, Agnes. 1954. *The Mind and the Eye: A study of the biologist's standpoint*. Cambridge: Cambridge University Press.

Arber, Agnes. 1957. *The Manifold and the One*. London: John Murray.

BBC Studios Natural History Unit. 2022. *The Green Planet* (a four episode television series presented by David Attenborough).

Boulter, Michael. 2017. *Bloomsbury Scientists: Science and Art in the Wake of Darwin*. London: UCL Press.

Brook, Isis. 2021. A New Science from a Historical Figure: Goethe as Holistic Scientist. In *Subtle Agroecologies: Farming with the Hidden Half of Nature*, ed. Julia Wright, 71–80. London: CRC Press.

Calvo, Paco. 2022. *Planta Sapiens: Unmasking Plant Intelligence*. London: The Bridge Street Press.

Chiles, Robert, and Amy Fitzgerald. 2018. Why is Meat So Important in Western History and Culture? A Genealogical Critique of Biophysical and Political-economic Explanations. *Agriculture and Human Values* 35: 1–17.

Claßen-Bockhoff, Regine. 2001. Plant Morphology: The Historic Concepts of Willhelm Troll, Walter Zimmermann and Agnes Arber. *Annals of Botany* 88: 153–172.

Clendenen, Avis. 2012. *Experiencing Hildegard: Jungian Perspective*. Wilmette, Illinois: Chiron Publishing.

Darwin, Charles, and Francis Darwin. 1880. *The Power of Movement in Plants*. London: John Murray.

Federoff, N.V. 1994. Barbara McClintock. *Genetics* 136: 1–10. https://doi.org/10.1093/genetics/136.1.1.

Goethe, J.W. von. 1952. *Goethe's Botanical Writings*, trans. B. Mueller. Hawaii: University Press of Hawaii.

Goethe, J.W. von. 1989. *Goethe: Italian Journey*, eds. T. Saine, and J. Sammons, trans. R. Heitner. New York: Suhrkamp Publishers.

Goethe, J.W. von. 1995. *Goethe: Scientific Studies*, ed. and trans. D. Miller. New York: Suhrkamp Publishers.

Grene, Marjorie, and David Depew. 2004. *The Philosophy of Biology: An Episodic History*. Cambridge University Press.

Hall, Matthew. 2011. *Plants as Persons: A Philosophical Botany*. New York: SUNY Press.

Hardy, Gavin, and Laurence Totelin. 2016. *Ancient Botany*. London: Routledge.

Hendlin, Yogi. 2021. Plant Philosophy and Interpretation: Making Sense of Contemporary Plant Intelligence Debates. *Environmental Values* 31: 253–276.

Hiernaux, Quentin. 2004. *The Vegetal Turn*, ed. Marcello Di Paola. New York: Springer.

Hofer, Julie, et al. 2001. Genetic Control of Leaf Morphology: A Partial View. *Annals of Botany* 88: 1129–1139.

James, P. 2009. Tree and Leaf: A Different Angle. *The Linnean* 25: 13–19.

Jose, Sarah, Chih-Hang Wu, and Sophien Kamoun. 2019. Overcoming Plant Blindness in Science, Education, and Society. *Plants People Planet* 00: 1–4.

Keller, Evelyn Fox. 1983. *A Feeling for the Organism: The Life and Work of Barbara McClintock*. New York: Freeman and Company.

Kimmerer, Robin Wall. 2020. *Braiding Sweetgrass: Indigenous Wisdom, Scientific Knowledge and the Teaching of Plants*. London: Penguin Books.

Kirchoff, Bruce. 2001. Character Description in Phylogenetic Anlaysis: Insights from Agnes Arber's Concept of the Plant. *Annals of Botany* 88: 1203–1214.

Lovejoy, Arthur. 1976. *The Great Chain of Being*. Cambridge, MA.: Harvard University Press.

Mancuso, Stefano, and Alessandra Viola. 2015. *Brilliant Green: The Surprising History and Science of Plant Intelligence*. Washington: Island Press.

Marder, Michael. 2021. *Green Mass: The Ecological Theology of St. Hildegard of Bingen.* Stanford: Stanford University Press.

Morton, A.G. 1981. *History of Botanical Science.* London: Academic Press.

Packer, Kathryn. 1997. A Laboratory of One's Own: The Life and Work of Agnes Arber, F.R.S. (1879–1960). *Notes and Records of the Royal Society of London* 51 (1): 87–105.

Robertson, Agnes. 1894. 'The Ivy-leaved Toadflax' *Our Magazine.* The North London Collegiate School for Girls. (The notebook and the published paper are held in the Hunt Institute for Botanical Documentation, Carnegie Mellon University, Pittsburgh).

Rolf, Rutishauser, and Rolf Sattler. 1997. Expression of Shoot Processes in Leaf Development of *Polemonium Caeruleum* as Compared to Other Dicotyledons. *Botanische Jahrbiicher fur Systematik* 119: 563–582.

Rutishauser, Rolf, and Brigitte Isler. 2001. Developmental Genetics and Morphological Evolution of Flowering Plants Especially Bladderworts (*Utricularia*): Fuzzy Arberian Morphology Complements Classical Morphology. *Annals of Botany* 88: 1173–1202.

Sattler, Rolf. 1992. Process Morphology: Structural Dynamics in Development and Evolution. *Canadian Journal of Botany* 70 (4): 708–714.

Seamon, David, and Arthur Zajonc. 1998. *Goethe's Way of Science: A Phenomenology of Nature.* New York: SUNY Press.

Stephenson, R.H. 1995. *Goethe's Conception of Knowledge and Science.* Edinburgh: Edinburgh University Press.

Tantillo, Astrida Orle. 2002. *The Will to Create: Goethe's Philosophy of Nature.* Pittsburgh: University of Pittsburgh Press.

Theophrastus. 1916. *Enquiry into Plants 1*, trans. A. Hort. Cambridge, MA: Harvard University Press.

Theophrastus. 1976. *De Causis Plantarium 1–2*, trans. and ed. B. Einarson and G. Link. Cambridge, MA: Harvard University Press.

Trewavas, Anthony. 2015. *Plant Behaviour and Intelligence.* Oxford: Oxford University Press.

Wandersee, James, and Elizabeth Schussler. 1999. Preventing Plant Blindness. *The American Biology Teacher* 61: 84–86.

Wandersee, James, and Elizabeth Schussler. 2001. Toward a Theory of Plant Blindness. *Plant Science Bulletin* 47 (1): 2–9.

Chapter 5
The Vegetal Return: Plant Time and the Curious Engines of Enlightened Knowledge

Adriana Craciun

Abstract Enlightenment sciences turned to plants and seeds as animate vessels of knowledge about the limits of life and the nature of time. A wide range of natural philosophers and imaginative thinkers, drawing from a continuum of mechanism, vitalism, materialism, neoPlatonism, alchemy, converged on seeds as "curious engines" (Robert Boyle's term), worthy of the highest scrutiny. Seeds were collected, preserved, dissected, imaged, artificially constructed, and imagined as inhabiting a little-known world of vegetal life that was thought to extend from deep inside the earth to the reaches of the atmosphere. This essay explores plant sensibility, longevity, and immortality that this early flowering of vegetal thinking pursued through experiment and speculation. The essay focuses on innovative seed longevity experimental and philosophical inquiries, and how these investigations of plant time shaped contemporary thinking on the scales of human time. Enlightenment efforts to extend plant viability across centuries or millennia by slowing plant time—or conversely, to accelerate plant time to make it perceptible to human senses—often included an eschatological, sacral component. This sacred dimension of plant time existed within mechanistic natural science, requiring a nuanced re-evaluation today of the underappreciated value of Enlightenment sciences for critical plant studies today.

Plants are the longest-lived forms of life that we can perceive unaided. Individual trees like the Fortingall Yew in England are approximately 5000 years old, and the clonal grove of quaking aspen known as Pando in Utah, U.S.A., is over 10,000 years old. Plant time is characterized by such extreme longevity, as well as other properties alien to animal life—a seasonal existence, seed dormancy, and a propensity for infinite growth (Marder 2015b). In our daily interactions with plants, however, all these "pulsations of vegetal temporality are often imperceptible to a conscious human observer," argues Michael Marder, "because even when they share a physical space, the two beings do not live in the same homogenous time but are non-contemporaneous

A. Craciun (✉)
Boston University, Boston, US
e-mail: acraciun@bu.edu

M. Di Paola (ed.), *The Vegetal Turn*, The International Library of Environmental, Agricultural and Food Ethics 38, https://doi.org/10.1007/978-3-031-68671-9_5

with one another" (Marder 2013, 187). This alterity of plant temporality, or "onto-logical time-lag" compared to human perceptions of time, is central to why Western traditions, traced back to Aristotle and Plato, have consistently denigrated vegetal life as mechanistic, un-minded, inanimate, and lacking in soul and even life (Marder 2015a).

Given this asynchronicity between plant time and human time, some of the most remarkable speculations on the lives of plants involve the attempts to align plant time with human time. The focus of such efforts is typically the study of plant movement, for the most part imperceptible to the human eye. Current researchers in plant behavior and intelligence emphasize the importance of overcoming the anthropocentrism stemming from our inability to perceive most plant movement—as Anthony Trewavas notes, "If it does not appear to move, then it is not behaving" (Trewavas 2014, 14). In fact, as Dov Koller explains, despite being "rooted in place most of their life," plants are "constantly in motion, restless" (Koller 2011).

This essay explores the radical temporal and philosophical possibilities opened up to the human imagination by accelerated plant movement and growth, i.e., the larger significance of manipulating plant time. Focusing on the Enlightenment—marked by much contemporary environmental and plant studies as the origin of our destructive approach to the natural world—the essay argues for a more nuanced vegetal return to the complexity of Enlightenment work on the extraordinary properties of plant time.

5.1 The Meanders of Botany

The most curious possibilities inspired by the non-contemporanaiety of plants and people emerged during the Enlightenment, from the so-called Scientific Revolution of the late seventeenth century onwards. Talking cabbages, walking trees, lusty flowers, human-plant hybrids all stalked the pages of philosophical, scientific, and fictional writing, pushing the limits of representations of intelligent life on earth (and beyond) to extraordinary imaginative levels, often assumed to originate with twentieth century science fiction. Natania Meeker and Antónia Szabari locate in the seventeenth century "the emergence of a vegetal modernity" in the sciences and letters, which began to reject the Aristotelean Scholastic tradition seeing plants as passive and inanimate, and instead "affirm[ed] an enlivening, fiery libido that pervades and animates plants" as ensouled beings with intention, intelligence, and desire, perceivable by careful observers (Meeker and Szabari 2019, 33). These enlivened and enlightened plants are central to the Norwegian/Danish imaginary voyage *Niels Klim's Journey Underground* (1741), for example, where the hero Klim discovers a world of such ensouled, speaking, walking trees inside the hollow Earth, a world distinguished by the shocking acceleration of plant time. The trees of Potu (a reversal of utop/ia) see Klim as inferior, because as they tell him, "That very Quickness of Parts permits you only to see the Surface of Things, and not the Substance" (Holberg 1742, 41). The trees, meanwhile, would repeat their statements slowly three times, with long pauses for solemn deliberation. Klim prided himself on his mental and physical speed above ground, but in the land of intelligent trees, he is merely "the animal," only fit for relaying their messages quickly across long distances. Klim experienced what Francis Hallé asks us to imagine: that "In their own time," plants "never stop moving, and we disappear, erased by our own mobility!" (Hallé 2002, 104).

Such walking, intelligent plants inhabited a mysterious world to which natural philosophers turned their attention and imaginations, at a time when we often assume plant studies to have been reduced to mere classification and extraction of economic value. Economic botany, imperial botany, and taxonomy have dominated the last few decades of scholarship on seventeenth and eighteenth-century European plant studies, and in turn these practices often feature in today's critical plant studies as foils to our more "liberatory" and (ironically) enlightened approaches to both plants and people. In *The Order of Things,* Foucault popularized such a vision of Enlightenment sciences—as dominated by the table and the grid, and intent on rendering all of nature inert and orderly, a method exemplified by Linnaean botany as taxonomy. But as Sheehan and Wahrman argue in *Invisible Hands*, "these claims are wrong," because a nuanced look at Enlightenment sciences reveals that, in many ways, they "did not form anything like coherent systems," and neither did they impose "coherence, uniformity, and symmetry" onto nature (Sheehan and Wahrman 2015, 180).

When we overemphasize the cohesion of Enlightenment taxonomies, especially in botany (a science often reduced to taxonomy), we lose sight of the often overwhelming sense of bewilderment—somewhat akin to delirium—that Enlightenment savants experienced. Linnaeus himself, the Swedish savant whose ambitious taxonomic system is usually used as synecdoche for "a totalizing order of European making" (Pratt 2008, 37) met his match before tiny Arctic flowers that he saw for the first time in 1732:

> All the rare plants that I had previously met with and which had from time to time afforded me so much pleasure, were here as in miniature, and new ones in such profusion, that I was overcome with astonishment, thinking I had found more that I should know what to do with. (Linnaeus 1811, 1–284)

Linnaeus thought of his systematic taxonomy as "Ariadne's thread," because he said, "all those that have been without it have lost their way and gone into the meanders of botany" (Linnaeus 2003, 213). Enlightenment botanical collections and taxonomies were technologies of control (biological, colonial, epistemological), but they were more than that. Linnaeus's astonishment reminds us of the excess and disorder such technologies were drawn to, into "the meanders of botany."

Beyond taxonomy and imperial botany, one experimental tradition that enlivened the strange possibilities of plants and seeds rather than deadening them was the mechanistic natural philosophy of Robert Hooke, Robert Boyle, and Stephen Hales. This strand of the New Science is often seen in critical plant studies and environmental humanities as the primary source of the hyper-rationalism, mechanization, and instrumentalism of the Enlightenment "ecological crisis of reason" (Plumwood 2001).[1] As

[1] Val Plumwood, *Environmental Culture: The Ecological Crisis of Reason* (2001). Matthew Hall reiterates Plumwood's reduction of Enlightenment plant thinking to "Baconic hierarchies" and

Carolyn Merchant summarizes, the New Science "saw nature not as an organism but as a machine—dead, inert, and insensitive to human action," culminating in the "death of nature" and the disenchantment of the cosmos persisting today in both modern sciences and late capitalism (Merchant 2005, 45–46). This longstanding association of the New Science's mechanization of matter with a soulless and essentially lifeless natural world remains out of step with scholarship emphasizing the central roles that the soul and spiritual agency still played in the natural world according to natural philosophers like Boyle, and the important continuities between mechanism and various strands of vitalism. As Simon Schaffer argues, the soulless mechanization of the universe was seen as dangerously atheistic by natural philosophers, and the much lamented "divorce between mechanical and occult philosophy" did not happen (Schaffer 1987, 65). Instead, "an active cosmology was being deliberately constructed in England during the 1670s", one in which "experimental instruments were used to show spirits working" throughout natural forces (Schaffer 1987, 70–73). Boyle, the leading figure in the corpuscular theory of mechanized matter, conducted extensive experiments on seeds and plants, and was fascinated by the potential immortality of seeds and their meaning for the afterlife. His experimental pursuit of palingenesis (the "resurrection" of plant life from lifeless ashes or seeds), voiced a curiosity about the seminal principle theory of active matter, which he supposedly abandoned as a mechanist.

"I must admire the strange power of the formative power of the seeds of things," Boyle wrote (1954, 167), drawing upon Van Helmont's "seeds of things" theory derived from alchemy (holding that the "seeds" of all matter are animated by spiritual forces)—a tradition that the mechanization of nature should have overturned. Boyle pondered extensively about the powers of those "curious engines"—seeds—and other "seminall principles" of living plants and animals that might prove the Christian resurrection of the decayed body to be not a supernatural violation of reason, but rather "above reason" (Boyle 2000, 13–286). Like Newton, Boyle remained open to the occult (hidden) active powers at work in nature's mechanisms, drawing alchemical, neo-Platonic, and hermetic traditions of vital agency into the new science's emphasis on experimental knowledge, in an effort to maintain a central role for divinity in the natural world (Henry 1986).

Stephen Hales, a follower of Newton and a friend of Boyle, focused more intense experimental attention on plant and seed physiology, and is seen as a leading figure in the reduction of plants to hydraulic machines. But by confirming that plants and seeds respire and perspire, and using novel instruments to measure such activities, Hales also reaffirmed a new dimension of liveliness in plants, after centuries of

---

entrenched mechanistic approaches that render plants passive and lifeless (Hall 2011, 37–52). Along with Carolyn Merchant's *The Death of Nature: Women, Ecology, and the Scientific Revolution* (1990) and *Radical Ecology* (1992; 2005), and Evelyn Fox Keller's *Reflections on Gender and Science* (1985), Plumwood's foundational feminist critique of instrumental reason unfortunately reduced "the Enlightenment" to Cartesian dualism and Baconian science. This narrow characterization persists in much critical plant studies and environmental humanities work, which rarely returns to primary materials or to recent history of science, the latter having undermined the dichotomy of mechanism and vitalism in particular.

speculation that they were lifeless because their breathing was imperceptible. In *Vegetable Staticks* (1727), Hales demonstrated that plants exchange gases and liquids with the atmosphere, marveling that seeds respire: "what a wonderful provision nature has made, that the Seeds of Plants should be well stored with very active principles," and that they be capable of "vigorously to draw... sublimed and exalted food from the air" (Hales 1727, 349). In pronouncing the abilities of plants to find nourishment from the "sublimed and exalted" realm of air, Hales was subtly reversing Plato's famous figure of the *arbor inversa*, the idealist vision of human beings as upside-down trees rooted in and nourished by the eidetic realm of Ideas (on Plato, see Marder 2011). Hales' mechanistic method, instead of reducing plants to hydraulic machines, drew them into the circle of breathing beings. By intertwining innovative experimental science with Plato's *arbor inversa*, Hales strengthened the affinities between human and plant—at that moment in history when too often we read of Enlightenment sciences severing them.

A few decades earlier, in *Micrographia* (1665), Robert Hooke also applied novel instruments in order to investigate vegetal life and its alien temporal and spatial scales. Hooke revealed for the first time "new Worlds and Terra-Incognitas to our view" by enlarging the human senses through the use of instruments (Hooke 1665, Preface n.p.). The effect he sought, for himself and the reader/viewer, was intense pleasure: "I do not only propose this kind of *Experimental Philosophy* as a matter of high *rapture* and *delight* of the mind, but even as a *material* and *sensible Pleasure*" (Hooke 1665, Preface, orig. emphasis). Where we have learned to expect an empiricism driven by disembodied, instrumental rationalism, in *Micrographia* we see instead, according to Alexander Wragge-Morley, "a form of empiricism that depended on the pleasures of sensory perception," as a conduit to a "pleasurable encounter with the wisdom of God" (Wragge-Morley 2020, 13).

In order to gain such an embodied and pleasurable knowledge of God's creation, Hooke plunged readers into a minute world by exploring those "curious mechanical contrivances" at the boundaries of life: seeds. Hooke revealed the remarkable forms and textures of the seeds of corn-violet, poppy, and purslain, and glimpsed the even tinier sporangia of fungi, seeing seeds as "little *automatons* or Engines" that are themselves tiny "Cabinets" of curiosities, wherein Nature "has laid up her Jewels and Master-pieces" (Hooke 1665, 154). Hooke reserved his most philosophical reflections in his discussion "Of the Seeds of *Tyme*," punning on the contemporary spelling of *tyme* meaning both thyme and time:

> The Grain affords a very pretty Object for the *Microscope*, namely, a Dish of Lemmons plac'd in a very little room; should a Lemmon or Nut be proportionably magnify'd to what this seed of Tyme is, it would make it appear as bigg as a Hay-reek and it would be no great wonder to see *Homers Iliads*, and *Homer* and all, cramm'd into such a Nut-shell. (Hooke 1665, 153)[2]

Hooke's composed still life with thyme seeds peered inside Nature's cabinet of curiosities to reveal "multitudes of curious Mechanick contrivances" with "ornaments

[2] In Shakespeare's *Macbeth*, Banquo says to the Weird Sisters, "If you can look into the seeds of time and say which grain will grow, and which will not, Speak then to me" (I.iii. 58–60).

… in the highest perfection": an artistic and scientific marvel akin to Homerian epic and containing divine wisdom: "who knows but the Creator may, in those characters, have written and engraven many of his most mysterious designs and counsels" (Hooke 1665, 154).

Hooke opens our minds to the incommensurable and interlinked scales of plant space and time, by performing the vegetative sublime in a series of calculations. In the receding scales of "contrivances and Mechanisms … crowded into an exceedingly" small space, he describes a minute "Vegetable body" (Fig. 5.1):

> A Plant growing on Rose leaves, that is abundantly smaller then [sic] Moss; insomuch, that neer 1000 of them would hardly make the bigness of one single Plant of Moss. And by comparing the bulk of Moss, with the bulk of the biggest kind of Vegetable we meet with in Story… we shall find that the bulk of the one will exceed the bulk of the other, no less than 2,985,984 Millions [i.e., nearly three trillion]. (Hooke 1665, 134)

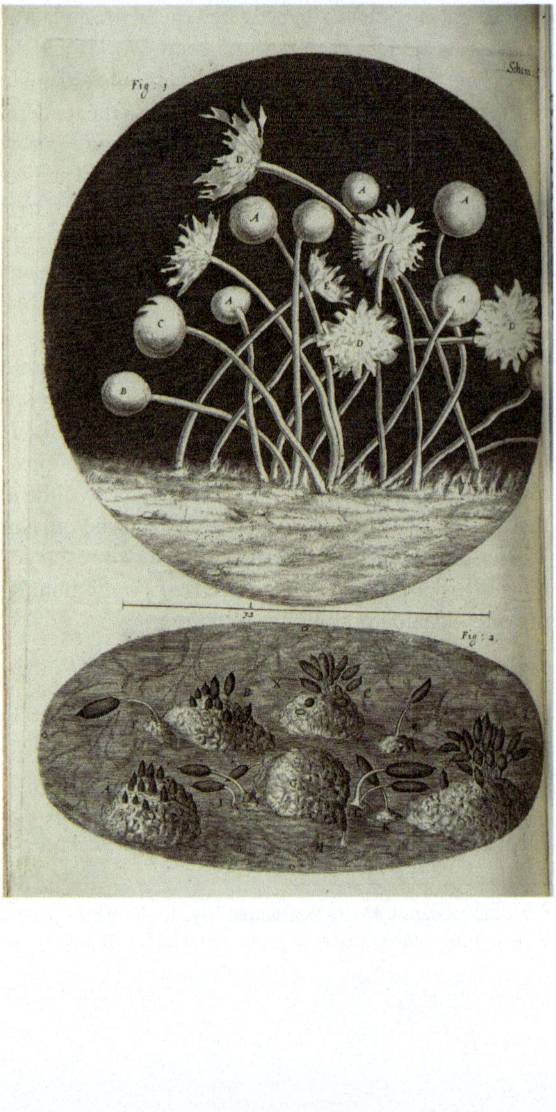

Fig. 5.1 Hooke, Schem. XII. *Micrographia* (1665) 124. National Library of Wales. Public domain. Microfungi growing on a leather book cover (above) and rose leaf (below), the first published images of microorganisms

Hooke asks us to consider that the microfungus growing on a rose leaf in his illustration is 1000 times smaller than a tiny moss sporophyte, which at 1/60th of an inch is itself nearly three trillion times smaller than the largest trees in the tropics (20 feet in diameter, he reckons), an astonishing calculation of the vegetal sublime.

The extreme scales of vegetal life he observed—a kingdom of organisms apparently spanning scale differences of trillion folds—opened for Hooke a window into the sacred scale of time. "So prodigiously various are the works of the Creator, and so All-sufficient is he to perform what to man would seem impossible" in creating living machines within machines at scales of both the largest trees and smallest microflora, that for God these are all "alike easie to him, even as one day, and a thousand years are to him as one and the same time" (Hooke 1665, 135). No wonder then that, as limit objects at the furthest reaches of human perception, the tiniest seeds were for Hooke potential examples of the seminal or corpuscular nature of matter itself, waiting only for the mysterious animating principle of heat to propel them into motion, like the microscopic clockwork they resembled. He contemplated compressing clockwork into ever more minute mechanisms: "I know not whether all the contrivances and *Mechanisms* requisite to a perfect Vegetable, may not be crowded into an exceedingly less room then [sic] this of Moss" (Hooke 1665, 134, orig. emphasis). The vertiginous shifts in spatial and temporal scales disclosed by the microscopic plant world revealed to Hooke that the seeds of time opened towards infinity, and millennia and days unfolded simultaneously, as they did for his god.

Hooke's language of seeds as machines within machines ad infinitum approaches the threshold of the opposing views of mechanism by vitalists like Leibnitz. In *Monadology* (1714), Leibnitz had argued that "the organic body of each living being is a kind of divine machine, or a natural automaton, which infinitely surpasses all artificial automata," its every part being a machine within a machine "ad infinitum" (Leibnitz 1898, 254). But as Jessica Riskin argues in *The Restless Clock*, Leibnitz' micromachines were "intrinsically active, self-moving" and contained their "own energies, forces, and powers of action" derived from an animating soul (Riskin 2016, 107). Hooke's and Boyle's exquisite micromachines, in contrast, were animated by the agency of God alone. Riskin reveals the shared language that accommodated the restless, unruly clockwork of Leibnitz's mechanical nature on the one hand, and the precise regularity of inanimate clockwork in the mechanism of Descartes, Hooke, or Boyle on the other hand (Riskin 2016, 104–105). The eclipse of the former by the latter in our histories of mechanistic thought has impoverished our understanding of natural philosophy, Riskin argues.

Natural philosophers like Hooke and Boyle saw in plant seeds a special occasion to interrogate the nature of all figurative seminal particles (the germs, seeds, eggs, or sperm of matter) and the dangerous possibilities that they are self-activating and self-organizing (and thus in no need of a deity, the road to atheism). As with the language of mechanism that Riskin explores, the language of seeds allowed for a breathtaking range of conflicting meaning, from Boyle's "curious engines" of plant machines, to Margaret Cavendish's and Leibnitz's ensouled particles no different from all of nature's minute animated parts. As Sheehan and Wahrman argue in their major reassessment of eighteenth-century materialism, "After 1650, when the sciences of life emerged as an expansive area of inquiry, the seed was rediscovered, an apparent panacea for the paradoxes of living nature" (Sheehan and Wahrman 2015, 152). This veritable "seed fetish" focused intense scientific, philosophical, aesthetic, and religious attention on the seeds of plants in particular, as housing nature's greatest secrets—her "Jewels and Master-pieces," in Hooke's words.

Enlightenment observers were particularly curious about the so-called sensitive plants or sensible seeds, which exhibited an animalistic motility, which they recognized as forms of volition that exceeded the vegetative. The seeds of Oat, Moscatum, and other Geraniums appeared to move and sow themselves, and thus became the focus of keen experimental interest. The awns (hair-like filaments) of some seeds enjoyed the ability to use changes in relative humidity (hygroscopic motion) as a means of dispersal and self-cultivation, by crawling along the earth and then screwing themselves into the soil. Thus, the seed of a wild oat (*Avena barbata*) was well known in early modern experiments as a curious contrivance, as in *Micrographia*:

> If you take one of these Grains, and wet the Beard [awn] in Water, you will presently see the small bended top to turn and move round, as if it were sensible; and by degrees, if it be continued wet enough, the joint or knee will streighten it self; and if it be suffer'd to dry again, it will by degrees move round another way, and at length bend again into its former posture. (Hooke 1665, 147)

What better way to understand the curious clockwork of these sensitive seeds, some thought, than by replicating them. Like Della Porta before him, Hooke built a hygrometer, an instrument measuring moisture, by harnessing an oat awn, which functioned like a tiny, animated spring that would wind and unwind itself "as if it were sensible"—"the very first footstep of Sensation, and Animate motion" said Hooke (Hooke 1665, 147–151). Hooke's illustration of and instructions for building the oat hygrometer are an early example of a hybrid vegetal instrument. By the early nineteenth century, one could purchase such vegetable hygrometers for home entertainment or field science. For example, in Maria Edgeworth's popular *Harry and Lucy* series of domestic science tales, Harry teaches his sister Lucy about hygrometers using different grass seeds. Lucy then recollects an extraordinary wooden artificial oat seed that could walk across the floor of a house instead of a field. Perhaps the first artificial seed, this gigantic "progressive hygrometer," as its inventor Richard Edgeworth called it, was given to Erasmus Darwin as a gift in 1790, who then made it famous in Book II of *The Botanic Garden: The Loves of the Plants* (Darwin 1790, 106–107).[3] Edgeworth's artificial plant automaton, a phytoid counterpart to the *androïde* already in circulation since the seventeenth century,[4] walked across his room self-propelled, using hygroscopic motion:

> It was a wooden automaton, with a long back, and four feet, with iron-pointed shoes, which clawed on little by little, so that it walked, or crept, or clawed, quite across the floor of its master's room, in a month's time, from the changes of damp and dry. (Edgeworth 1825, 1–58)

---

[3] See Darwin's and Edgeworth's correspondence in Edgeworth (1820, 130–131).

[4] On early androids (a term coined by Gabriel Naudé in 1625), see Riskin (2016, 113–150).

Darwin described the seed automaton as "creeping like a worm" across "the room which it inhabited," and asked with his characteristic zeal for application, "Might not this machine be applied as an Hygrometer to some meteorological purpose?" (Darwin 1790, 107).

Natural philosophers saw these ingenious vegetal machines, seeds, as instruments that could be improved, copied, and incorporated into ever larger projects. Hooke, the Edgeworths, and Darwin were fascinated by moving seeds, because these were rare instances in which plant locomotion became perceptible in human time. Most plants fall beneath the threshold of animation under the Aristotelean schema because of their apparent immobility—they are the kind of object that Bacon says "hides from the senses" because they are asynchronous with us: "It is plain that action of the sense is carried through in motion and that motion is carried through in time. So if the motion of any body is so slow or quick as to be out of phase with the time it takes for the sense to act, the object goes quite undetected" (Bacon 2004, 11–357). But "The Seeds of Tyme" revealed that the world and its plants are both pluritemporal and thus plurispatial, i.e., existing on radically different scales at once, not necessarily a singular Newtonian absolute. Only a minute sliver of the lives of seeds is perceptible to us, Hooke revealed through his instruments, the rest unfolding on scales a trillion-fold beyond. Microscopes can help us glimpse some of those secrets, but so can constructing gigantic artificial seeds that walk among us—two instrumental encounters with seeds that can help us perceive the time of plants in new ways.

5.2 Manipulating Plant Time

Accelerating plant time, both seasonally and diurnally, became an obsession of early plant sciences (and has remained so in the arts). The early eighteenth-century botanist Georg Andreas Agricola was a master manipulator of the time of plants: he developed a "Project for the universal *Multiplication* of *Trees*, *Shrubs* and *Flowers*" using innovative grafting, heating, composting, and germination techniques to achieve "the true art of rendring [sic] Nature obedient, and of forcing her to give us what we require of her" (Agricola 1721, 131). As Yvonne Gaspar writes, these innovations "manipulated the passage of time in ways that were remarkable and might previously have been deemed impossible" (Gaspar 2013, 107). Agricola's Baconian project was fundamentally devoted to mastering plant time: as Bacon had written in his *Sylva Sylvarum*, to "speed germination was 'a rich Experiment for profit'" (Gaspar 2013, 110), as was the desirability of conducting experiments "how by *Art* to make *Plants* more *Lasting* than the ordinary Period."[5] In Bacon's utopia *The New Atlantis*,[6] a futuristic

[5] Bacon's *Sylva Sylvarum* (1670) has an unusual structure, so I cite that here along with page numbers: Century V, Experiment 402 (90); Century VI, Experiment 587 (120).

[6] Bacon's *New Atlantis* is included in *Sylva Sylvarum* (Bacon 1670, 24) but it restarts with new pagination at the end of the volume.

institution boasts of the ability to "make (by Art) in… Orchards and Gardens, Trees and Flowers to come earlier or later than their seasons, and to come up and bear more speedily then by their natural course they do."[7] For Bacon, the project of extending plant time was profitable in multiple ways because, as Doina Rusu shows, he was committed to "us[ing] the conclusion from experimenting with plants to establish operations that could lead to the prolongation of human life" (Rusu 2020, 394).

In *The Artificial Gardiner: Being a Discovery of a New Invention for the Sudden Growth of all Sorts of Trees and Plants* (1717), an author purporting to be Agricola and beating him to press made even more extravagant claims in accelerating plant time: "In the space of One Hour, by the help of Fire and my Vegetative Mixture, I have made Twelve Slips of Limon Trees to produce Roots, Branches, and Leaves" (Anon. 1717, 18). For the sum of 25 gilders, he offered to reveal his secret to the first 160 people to send him the money. But that's not all—he also offered to reveal "How to Raise a Sallad in Two Hours Time" from seed, also for a price (Anon. 1717, 28). These fraudulent claims "to expedite the growth of Vegetables" exaggerated, and claimed credit for, Agricola's authentic researches in the "universal multiplication of plants," which relied on sophisticated grafting techniques in order to multiply one tree into many at an accelerated rate. In one segmented drawing of a pear tree branch transforming into an orchard, Agricola showed a time lapse series (bottom right to left) of how one small section of the branch above will root and grow into an entire tree, midway losing some leaves but "after some months" being restored to vitality and now a new individual tree (Agricola 1721, 184) (Fig. 5.2). In the background we see a miniature orchard of odd lemon trees, each grown from a single bud on a "mother leaf," an experiment he described taking over a year—much faster than a tree from seed, but much slower than the one day claimed in *The Artificial Gardiner* (Anon 1717, 18). Agricola's innovative experiments probed the "totipotent" and reiterative abilities of plants as collective, not individual, beings, two extraordinary aspects of plant life that most animals lack. Totipotent cells can reform as any cells in the organism, including new placental and embryonic cells, granting plants the ability to completely regenerate a new organism. Agricola described this as an alchemical "Transmutation of vegetables," "whereby they may be turn'd into the other" (Agricola 1721, 174). Reiteration also radically expands the abilities of plants to transcend individual, animalistic notions of being and dying. With a modular architecture that allows most plants "to split apart without dying" (the defining characteristic of a colonial organism), it turns out that "plants are primarily colonial, with unitary ones probably not more than 20% of all the species" (Hallé 2002, 116–117, 141–142). In *Phytologia* (1800), Erasmus Darwin summed up this extraordinary extra-individual nature of plants: "every bud of a tree is an individual vegetable being; … a tree therefore is a family or swarm of individual plants, like the polypus" (Darwin 1800, I. Section 5.2, p. 2).

[7] Bacon's savants also boasted of growing plants without the use of natural seeds, and of turning one plant into another.

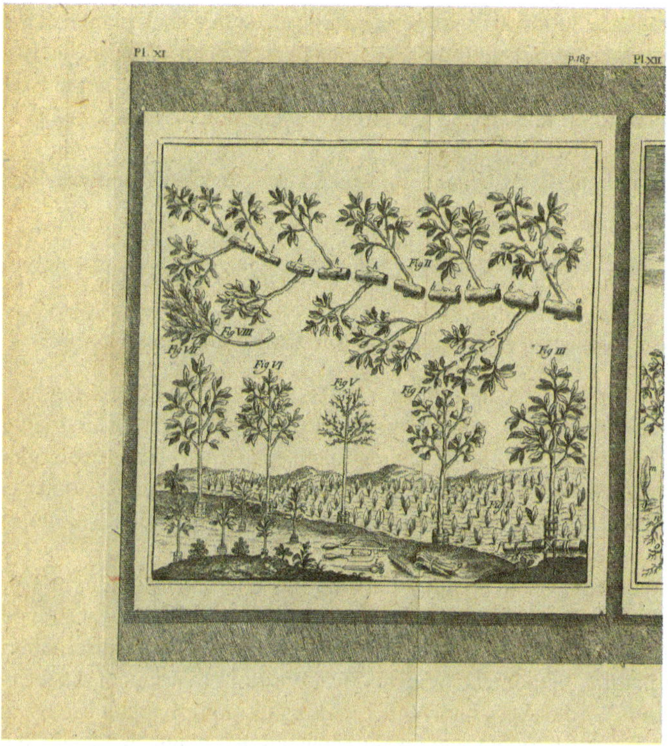

Fig. 5.2 Agricola, "How by Means of Fire and Mummy, Leaves, Twigs, Buds and Branches, May be turned unto Shrubs and Trees," *Philosophical Treatise of Husbandry* (1721) Pl. 11. With permission of Wellesley College Library, Special Collections

The seemingly infinite capacity for vegetal growth that Agricola, Bacon, and others sought to harness had economic as well as larger spiritual and cultural significance. Denouncing the claims made in his name for lettuce growing in two hours, and trees in one hour (87–88), Agricola marveled at the genuine power of vegetal nature to proliferate: from one seed, he calculated, a single plant can produce potentially 360,000 seeds, and if these all germinate, they would potentially produce 129,600,000,000 (129 + billion) seeds (Agricola 1721, 90).

This quantification of the vegetal sublime was often performed. In *Natural History of the Earth* (1726), John Woodward wondered why pre-diluvial plant fossils were dominated by ferns. Ferns had no value to humans and were "the meanest upon Earth […] nor Thistles, nor indeed any one kind of Weed whatsoever, has so great a Share of the Globe in its possession—both in the present and in the fossil past" (Woodward 1726, 53). Cultivated wheat, central to human civilization, is nowhere as productive as ferns or thistles; that disturbed Woodward, given humanity's centrality to god's plans, and given the Biblical importance of subduing the earth to bringing about this human dominion. Woodward calculated that each *Acanthium vulgare* (Scottish thistle) produces 300–400 seeds per flowerhead, so that each plant can make 20,000 seeds, for a total of 576 million seeds per plant if each of these seeds germinates (Woodward 1726, 53). "Thistles are, in their Nature, dispos'd to put in Execution that Curse" upon fallen humankind, inverting the fecundity of Eden (Woodward 1726, 58).

The consequences of such unceasing plant growth taxed Woodward's imagination and frightened him:

> This is an Increase so enormous as could hardly be imagined: and tis plain that … if not hindered by some Means, but carried regularly on, every Way, would in a very short Time, stock the whole Globe so as to scarcely leave Room for the Growth of any Thing else; and all these descendants of only one single original Seed. (Woodward 1726, 58)

A planet entirely of plants and for plants, with no usefulness for human needs, appeared to Woodward an ungodly nightmare, the inverse of the luxuriant and effortless growth he and many other botanists imagined in Eden. The idea of a global empire of ferns and thistles that so horrified Woodward was in fact understood as a potential reality and investigated, with increasing interest, in early plant studies probing the unsteady nature and limits of human dominion over plants.

Enlightenment botanical study pursued this alien plant vitality, pushed it to its limits and attempted accelerating it, whether from entirely mechanistic perspectives that saw plants as restless machines, or from concoctions of vitalist, materialist, neo-Platonist, natural theological, or alchemical positions that sought out the elusive seat of plant souls, as did Agricola. While working to multiply plant life by sundering it into many pieces, Agricola remained convinced that plants had souls, that this soul was material, and that he had located it, as indicated in his illustration of an almond seed identifying "that marvellous Point (Z), where is the Residence of the Vegetative Soul, as we may observe it with a Microscope" (Agricola 1721, 14). The vegetative soul was Aristotle's appetitive soul driving nourishment and growth, the lowest of his tripartite soul, the others being the sentient soul and the rational soul. And yet Agricola ventured to ask "the more curious than useful Question": "Whether the Vegetative Soul, as a Creature that acts and suffers in this World, may not hope for a Recompense in another, and a Place among the *Trees* of the Celestial Paradise?" (Agricola 1721, 75). By remaining open to questions of plant suffering, agency, and the sacred (indeed, the afterlife), this revolutionary Enlightenment botanist continues to speak to central concerns we have today in the vegetal turn.

5.3 Conclusion

Contemplating plant time opened the strangest possibilities when the human/plant boundary began to become more blurred. La Mettrie, in *L'Homme-Machine* (1747) and its sequel *L'Homme-Plante* (1748), scandalized Europe with the radical materialism of his vision of human life as a restless, self-animated, thinking, sentient machine, and then as a plant. Unlike caricatures of Cartesian mechanism and instrumental rationalism, La Mettrie's machine man was a self-organized clockwork suffused with lust, pleasure, reason, and imagination, all of it generated in an animate material body with no need of spiritualization. *L'Homme-Machine* got La Mettrie banned from his place of exile in Holland, and from his next exile he wrote the sequel, *L'Homme-Plante,* pushing his readers to make a bolder identification between human and plant life along a materialist continuum. That plants lack an immaterial soul and consciousness La Mettrie sees as virtually enviable ("no worries, no cares, no steps to take"), as is their ability to "make love with no difficulty" (La Mettrie 1994, 86). He wrote of wind pollination: "Sometimes they engage in coitus at a distance even when far apart," thanks to "the winds, those messengers of vegetable love" that "carry the sperm of males to the female plants" (La Mettrie 1994, 85–6). La Mettrie took up the legend of amorous palms, popularized in Pliny's *Natural History* and Renaissance verse, and transformed it into a figure of libertine erotic abandon, exemplary of his era's fascination with the outer limits of vegetal possibilities.

An outrageous British satirical response, *The Man-Plant* (1752), evoked the Paracelsian legend of the homunculus and mandrake, but in an updated and materialist La Mettrien vein. The satire described removing a human embryo from the womb vaginally and planting it in specially prepared soil to produce a baby while "ridding the Women of the Plagues and Fatigue of Gestation" (Miller 1752, 35). This method, alluding to recent discoveries in artificial incubation of chicken eggs, would be sexually liberating for women, who would no longer have to bear the reproductive burden, and more importantly, it would expand the British empire globally by allying it with the empire of plants. Each woman would be able to produce approximately 130 to 140 babies in this painless way, multiplying at the scale of plants: "we shall see infinite Broods of Subjects thus pullulating, thus serve to repeople and enrich, as well our Island as those vast Tracts in *North America*, which are so thinly habited" (Miller 1752, 34–35). But why stop there: "We may also then more reasonably grasp the Conquest of both the Indies, especially when we shall have completed the discovery of the Northwest Passage," and by "superior numbers" of "swarms of our own Subjects we shall become as populous as *China* itself" (Miller 1752, 36).

Limitless colonial expansion "of the True-Born Englishman" (Miller 1752, 34) appears here as a monstrous vegetal growth that turns the economic botany of its day into a nightmarish physiocracy and plantocracy. "Plantations of Men are as susceptible of Improvement as well as those of Trees," reasoned pseudonymous author of *The Man-Plant* (Miller 1752, 5)—in an absurd and incisive double entendre on "plantations of men" encompassing both the empire's global plantations worked by enslaved Africans, and the logical next step in such an imperial botany, i.e., reproducing more subjects and slaves through "improving" human fertility by appropriating the reproductive capacity of plants. Experiments, in thought and in deed, in accelerating the time of plants to suit diverse human desires, some of them truly horrifying, were thus widely pursued in germination, growth, and fructification during the Enlightenment.

Today, engineers and startups are hard at work similarly "improving" plants by building artificial leaves and trademarked MechanicalTrees [sic] to capture atmospheric carbon, the latter's selling point being that they are "1000 times faster" than natural trees at sequestering carbon.[8] These are the latest "improvements" in a long history of constructing artificial plants and seeds, as we saw with Edgeworth's mechanical seed—the most infamous examples being the 1990s sterile "terminator" seeds and other engineered organisms patented by Monsanto and other corporations. Terminator seeds have an "engineered seed death" time encoded into their DNA (often chemically triggered), to force farmers to purchase new seeds each year: these represent "the complete enclosure of the seed by the market," according to Thom van Dooren (van Dooren 2007, 79). More recently but with less public fanfare, "artificial seeds" or "synthetic seeds" (synseeds) allow for mass micropropagation of plants in agriculture, horticulture, and conservation, by using genetically identical "encapsulated plant tissues such as shoot buds, axillary buds, somatic embryos, shoot tips, [or] cell aggregate" to expand greatly the scale of its ability to reproduce (Qahtan et al. 2019, 2). Rather than wait for the plant's seed time, respecting its cycles of growth and rest, bioscientists mass-produce synseeds by accelerating seed time to a continuous process in controlled laboratory spaces. The most extreme form of such spaces are "phytotrons" (or ecotrons) which aim for complete computerized control of every aspect of the plant's environment (light intensity, photoperiod, temperature, humidity) in order to shape development (Munns 2017). Today's technologies like phytotrons, synseeds, or "vertical farms" leave no room to consider the possibility that, "[l]ike humans, plants can be violated by uninterrupted light that forces them to grow without nocturnal, as well as seasonal, breaks," as Marder maintains (Marder 2015b, 93). Sophie Gerber and Quentin Hiernaux trace this "phytotronic era" directly back to "Descartes' notorious view of organisms as machines," which while solely focused on animals, they see as disseminated throughout a uniform mechanistic tradition of Enlightenment sciences that denied plants life, agency, and ethical consideration (Gerber and Hiernaux 2022, 1).[9]

These phytotechnologies share in the centuries long effort to capture the time of plants. Agricola and the authors of *The Artificial Gardiner* and *The Man-Plant* contemplated similar technologies three centuries ago, in their efforts to accelerate the time of plants and seeds—but with notable differences. Agricola had also insisted upon the souls of plants as living beings, even as he intensified our ability to re/

[8] Silicon Kingdom Holdings has trademarked the "MechanicalTree" [sic] and plans to sell mechanical tree farms globally (https://mechanicaltrees.com/mechanicaltrees/) (accessed February 1, 2021). Different forms of artificial leaves are currently in development, all of them "faster" than plants; see Prajapati and Singh (2019).

[9] Relying largely on secondary sources, Gerber and Hiernaux group a heterogeneous number of savants together as advancing a Cartesian "plant as machine" thesis. In one case they misrepresent the source as a primary one when they are actually quoting from Delaporte's *Nature's Second Kingdom*: thus they say regarding Boerhaave that he wrote: "plants are only machines and are not alive. By plant we mean hydraulic body" (4); but the first statement was actually Delaporte's (84), not Boerhaave's, and indeed Boerhaave went on to also say that "un animal est un corps hygraulique" (Boerhaave 1748, 74). My point is that there was not one "idea of the plant as machine," and that we should be as nuanced in readings of the past as we are in readings of contemporary plant studies.

produce them through new technologies such as the shattering of one "mother" plant into an orchard of clones. As "a Creature that acts and suffers in this World," Agricola noted, a plant is potentially an ethical subject whose afterlife remains an open question to us. *The Man-Plant's* chilling vision of plantations of enslaved vegetal/human subjects fused the author's fictional science with his satirical political critique of the British empire, showing how the science of the day (i.e., the recent development of artificial incubation of chicken eggs) was imbricated within social and political life to nightmarish ends. And La Mettrie's plant machines and man-plants enjoyed voluptuous sex at a distance with no difficulty or cares—a mechanistic and materialist vision, but one awash in an alien "vegetable love." Thus, social, philosophical, and ethical questions were visibly imbricated with these earlier investigations into the lives of plants, at a time when "machines were more than machines" (Riskin 2016, 108), and seeds were curious engines, perhaps ensouled masterpieces of divine secrets as well.

Thus, we should be careful not to draw a causal line of inevitability between these conflicted Enlightenment inquiries and today's research on synthetic seeds or mechanical trees. Today's scientific literature on these phytotechnologies lacks a social or ethical component (beyond the now anodyne goal of "sustainable agriculture" or climate mitigation) and appears wholly concerned with how to commercialize the technology as quickly, profitably, and intensively as possible (see Lotte Asveld's essay in this volume for a wider range of phytotechnologies). This work thus does not follow in one unbroken line the economic, imaginative, and scientific passion of an Erasmus Darwin, who enthusiastically encouraged the industrial applications of Edgeworth's vegetal "progressive hygrometer," the monstrous wooden seed crawling across his floors, but did so within a politically radical vision of vegetal life as part of a vital world suffused with sentience. For Darwin, we must remember, *The Loves of the Plants* was the inseparable and equal counterpart to *The Economy of Vegetation,* the twin volumes of his masterpiece, *The Botanic Garden.*

Plant love, plant soul, plant suffering, were all matters of concern for these Enlightenment savants, sometimes at their most instrumentally manipulative moments. This humanistic and ethical dimension is absent from most plant science today, but has recently re-emerged in the revolutionary research into plant intelligence and behavior, a field facing often vigorous resistance from within biology. By working within this revitalized Enlightenment genealogy of the vegetal turn, however, we can avoid the burden of reinventing such "new" modern discourses as plant agency, sentience, and ethical standing—for it is really a vegetal return. Likewise, we should set aside the continued reliance on ahistorical notions of the Enlightenment as the occasion for the "ecological crisis of reason" plotting the death of nature and the lifelessness of plants—a picture that sits uneasily with the (often internally conflicted) body of work explored here. The Enlightenment vegetal turn could in some cases be, and was, driven by passion, moved by beauty, and riddled with sensations like desire, fear, and pleasure, often inspired by the curious properties of the time of plants.

Reference Lists

Agricola, Georg Andreas. 1721. *A Philosophical Treatise of Husbandry and Gardening*, revised by Richard Bradley. London: P. Vaillant

Anonymous. 1717. *The Artificial Gardiner: Being a Discovery of a New Invention for the Sudden Growth of all Sorts of Trees and Plants.* London: E. Curll.

Bacon, Francis. 1670. *Sylva Sylvarum; or, a Natural History in Ten Centuries* , 9th ed. London: W. Lee

Bacon, Francis. 2004. *The Oxford Francis Bacon*, ed. Rees. Oxford: Oxford University Press.

Boerhaave, Herman. 1748. *Elemens de Chymie*, ed. J. Nealume, vol. 1. Paris.

Boyle, Robert. 2000. Essay on Spontaneous Generation. In *The Works of Robert Boyle: Unpublished Writings*, eds. Michael Hunter and Edward Davis, vol. 13, pp. 273–288. London: Pickering and Chatto.

Boyle, Robert. 1954. The Sceptical Chymist, reprinted in Marie Boas, "An Early Version of Boyle's Sceptical Chymist". *Isis* 45: 153–168.

Darwin, Erasmus. 1790. *The Botanic Garden. Book II: The Loves of the Plants*, 2nd ed. London: J. Johnson.

Darwin, Erasmus. 1800. *Phytologia*: *Or the Philosophy of Agriculture and Husbandry*. London: P. Byrne.

Delaporte, François. 1982. *Nature's Second Kingdom: Explorations of Vegetality in the Eighteenth Century*. Cambridge: MIT Press.

Edgeworth, Richard Lovell. 1820. *Memoirs of Richard Lovell Edgeworth*, Esq., vol. 2. London: R. Hunter, Baldwin, Craddock & Joy.

Edgeworth, Maria. 1825. *Harry and Lucy Concluded*, vol. 1 of 4. London: R. Hunter, Baldwin, Cradock, and Joy.

Gaspar, Yvonne. 2013. *Between the Plant Sciences: Richard Bradley's Study of Horticulture, Botany and Landscape Design*. PhD dissertation, University of Notre Dame.

Gerber and Hiernaux. 2022. Plants as Machines: History, Philosophy, and Practical Consequences of an Idea. *Journal of Agricultural and Environmental Ethics* 35: 1–24.

Hales, Stephen. 1727. *Vegetable Staticks: Or, an Account of Some Statical Experiments on the Sap in Vegetables*. London: Printed for W. and J. Innys, and T. Woodward.

Hall, Matthew. 2011. *Plants as Persons: A Philosophical Botany.* SUNY Press.

Hallé, Francis. 2002. *In Praise of Plants*. trans. David Lee. London and Portland, OR: Timber Press.

Henry, John. 1986. Occult Qualities and the Experimental Philosophy: Active Principles in Pre-Newtonian Matter Theory. *History of Science* 24: 335–381.

Holberg, Ludvig. 1742. *A Journey to the World Under-Ground, by Nicholas Klimius*. London: Printed for T. Astley.

Hooke, Robert. 1665. *Micrographia, or Some Physiological Descriptions of Minute Bodies*. London: Royal Society

Keller, Evelyn Fox. 1985. *Reflections on Gender and Science*. Yale UP.

Koller, Dov. 2011. *The Restless Plant*, ed. Elizabeth Van Volkenburgh. Cambridge: Harvard University Press.

La Mettrie, Julien Offray de. 1994. *Man a Machine/Man a Plant*, trans. Richard Watson and Maya Rybalka. Indianapolis: Hackett.

Leibnitz, Gottfried. 1898. *The Monadology*, trans. Robert Latta. Oxford: Oxford University Press.

Linnaeus, Carl. 1811. *Lachesis Lapponica: Or, A Tour in Lapland*, trans. James Edward Smith, vol. 1. London: White & Cochrane.

Linnaeus, Carl. 2003. *Linnaeus' Philosophia Botanica*, trans. Stephen Freer. Oxford: Oxford University Press.

Marder, Michael. 2011. Plant-Soul: The Elusive Meanings of Vegetal Life. *Environmental Philosophy* 8: 83–99.

Marder, Michael. 2013. *Plant-thinking: A Philosophy of Vegetal Life*. 103. New York: Columbia University Press.

Marder, Michael. 2015a. The Place of Plants: Spatiality, Movement, Growth. *Performance Philosophy* 1: 185–194.
Marder, Michael. 2015b. The Sense of Seeds: Or Seminal Events. *Environmental Philosophy* 12: 87–97.
Meeker, Natania, and Antónia Szabari. 2019. *Radical Botany: Plants and Speculative Fiction*. New York: Fordham University Press.
Merchant, Carolyn. 1990. *The Death of Nature: Women, Ecology, and the Scientific Revolution*. NY: Harper.
Merchant, Carolyn. 2005. *Radical Ecology*, 2nd ed. London: Routledge.
"Miller, Vincent" [pseud.]. 1752. *The Man-Plant, or, Scheme for Increasing and Improving the British Breed*. London: M. Cooper.
Munns, David. 2017. *Engineering the Environment: Phytotrons and the Quest for Climate Control in the Cold War*. Philadelphia: University of Pittsburgh Press.
Plumwood, Val. 2001. *Environmental Culture: The Ecological Crisis of Reason*. London: Routledge.
Prajapati, Aditya, and Meenesh Singh. 2019. Assessment of Artificial Photosynthetic Systems for Integrated Carbon Capture and Conversion. *ACS Sustainable Chemistry & Engineering Journall* 7 (6): 5993–6003.
Pratt, Mary Louise. 2008. *Imperial Eyes: Travel writing and Transculturation*, 2nd ed. London: Routledge.
Qahtan, A.A., et al. 2019. An Introduction to Synthetic Seeds. In *Synthetic Seeds: Germplasm Regeneration, Preservation and Prospects*, eds. Faisal and Alatar, 1–20. London: Springer.
Riskin, Jessica Riskin. 2016. *The Restless Clock: A History of the Centuries-Long Debate Over What Makes Living Things Tick*. Chicago: University of Chicago Press.
Rusu, Doina-Cristina. 2020. Using Instruments in the Study of Animate Beings: Della Porta's and Bacon's Experiments with Plants. *Centaurus* 62: 393–405.
Schaffer, Simon. 1987. Godly Men and Mechanical Philosophers: Soul and Spirits in Restoration Natural Philosophy. *Science in Context* 1: 55–85.
Sheehan, Jonathan, and Dror Wahrman. 2015. *Invisible Hands: Self-Organization and the Eighteenth Century*. Chicago: Chicago University Press.
Trewavas, Anthony. 2014. *Plant Behaviour and Intelligence*. Oxford University Press.
van Dooren, Thom. 2007. Terminated Seed: Death, Proprietary Kinship, and the Production of (Bio)Wealth. *Science as Culture* 16: 71–93.
Wragge-Morley, Alexander. 2020. *Aesthetic Science: Representing Nature in the Royal Society of London, 1650–1720*. Chicago: University of Chicago Press.
Woodward, John. 1726. A *Natural History of the Earth*. London.

Chapter 6
The Power of Movement in Plants: Vegetal Dimensions in Darwin's Theory of Evolution

Claudia Rosciglione

Abstract This chapter chronicles Charles Darwin's research on plants, with particular attention to his 1880 *The Power of Movement in Plants*, co-authored with his son Francis. Darwin's interest in plants was consistent throughout his life, and his research brought new knowledge to botany, provided grounds to reassert the importance of the plant kingdom in connection with the animal kingdom, and gave evidence for many fundamental elements of his theory of evolution. This view allows to highlight Darwin's significant contribution to the debate and the research on the nature of plants, by proving that they possess sensibility, movement and a complex ability of interaction with the environment through their root tips. Darwin's work on plants, however, was largely overlooked or underestimated by many botanists of his time, who never recognized him as a scientist in their field and wrongly assumed that his theory of evolution would apply only to the animal kingdom. Darwin's research on plants was, instead, precisely proof that evolution was universal and constitutive of all living beings that co-evolve and coexist in the same ecosystems; and some of his most courageous theses on the vegetal-that plants possess sensibility, movement, and a complex ability to interact with their environments through their root tips—are now widely acknowledged among the bases of contemporary plant neurobiology.

6.1 Botany and Evolution in *The Origin of Species*

Charles Darwin's scientific interest in plants is present from the very beginning of his thinking about the origin of life and its evolution. Before embarking on his journey on the *Beagle*, Darwin took a three-years botany course taught by Professor Henslow.[1] During his time on the *Beagle*, in the notes from which would come *The Origin of Species*, Darwin writes not only about the animal world that he observes in the new

[1] See Kohn et al. (2005).

C. Rosciglione (✉)
University of Palermo, Palermo, Italy
e-mail: claudia.rosciglione@unipa.it

M. Di Paola (ed.), *The Vegetal Turn*, The International Library of Environmental, Agricultural and Food Ethics 38, https://doi.org/10.1007/978-3-031-68671-9_6

environments he explores, but also about the plant world, which he sees as an integral part of those same natural environments and whose role in the larger evolutionary process he recognizes clearly.[2]

In his notes Darwin wonders why flora differs so much from one island to another, and suggests it might have to do with the physical isolation of the atolls. He also speculates that if there were "land-birds" scattering seeds on the beach by pecking and carrying them around, the flora would be much more abundant than what he observed during his travels.[3] These scattered observations reveal Darwin's early awareness of the crucial role of flora in ecosystems and the relations of interdependence between the animal and plant worlds, seen as subject to the same laws of evolution. Later, in the Introduction to *The Origin of Species*, Darwin will argue for unity in diversity and the common descent of all species.

Darwin emphasizes that variety and variation are characteristics of plants, and that the maintenance of this diversity depends, like in the animal world, on natural selection, the struggle for existence, and the internal organization of each living thing. Plant organisms, just like animal organisms, struggle for existence, as exemplified by the competition for resources. Just as "two canine animals in a time of death, may be truly said to struggle with each other which shall get food and live" equally "a plant on the edge of a desert is said to struggle for life against the drought [...] A plant which annually produces a thousand seeds, of which on an average only one comes to maturity, may be more truly said to struggle with the plants of the same and other kinds which already clothe the ground" (*Ivi*, p. 62).

In *The Origin of Species*, Darwin will make it explicit that fundamental aspects of evolution apply to both the animal and the plant world, and he will also emphasize their co-evolutionary relationship.[4]

Plants' integral role in Darwinian theory becomes evident when considering Darwin's prolific output between 1860 and 1880. During this period, Darwin authored no fewer than seven books dedicated to plants. These texts served as a deepening and revisiting of his earlier observations recorded in *The Origin of Species*. He acknowledges the necessity of making it public, if only installments, recognizing

[2] Cf. Darwin (1839). We know that Darwin visited the Brazilian virgin forest, the Tierra del Fuego, and that he observed the plants of the Galapagos Islands, New Zealand, Australia, and various other places. But we know, above all, that he was fascinated by the variety of flora as it changed from one ecosystem to the other. The study of plants constitutes the privileged subject from which the first Darwinian ideas of unity in diversity arose as argued by Kohn: "Darwin did not become an evolutionist on Galápagos, but the basis for a profound shift in his understanding of species was established there—and it began with plants. [...] Darwin's Galápagos plant specimens, numbering well over 200, constitute the single most influential natural history collection of live organismus in the entire history of science. It is not just that Darwin's plants represent the foun-dational collection for the entire Galápagos flora. They also would turn out to be Darwin's best documented example of the evolution of species on islands" (Kohn 2008, 15–16).

[3] Cf. *Ivi*, p. 484 ff.

[4] This is what happens, for example, between flowers, insects, and the environment in which they are both found and evolutionarily explains the fertilization of a plant with flowers by pollination. About the role of coevolution for Darwin between flowers (plant world) and insects or birds (animal world) see Mancuso (2014, 57–58).

that ordaining it into a comprehensive work would require many years and that his health might not permit such an endeavor.

In fact, everything published by Darwin after *The Origin of Species*—encompassing insights, factual reports, experimental evidence, and new observations spanning both the plant and animal worlds—can be interpreted as a development, continuation, and even a completion of the "imperfect extract", as Darwin called that *The Origin of Species* in its Introduction. This perspective is crucial when approaching Darwin's compelling post-1860 observations on topics such as orchids and their fertilization by insects, the polymorphism of flowers, and the movements of climbing plants.[5] These contributions paved the way for Darwin's later work, *The Power of Movement in Plants*, co-authored with his son and professional botanist Francis, and published in 1880.

But the world of botany initially showed indifference, if not hostility, towards the Darwins' experiments and results on plants.[6] This resistance stemmed from the fact that their methodology was not recognized as scientifically sound in a botanical context primarily focused on plant taxonomy.[7] The Darwin's approached the study of plants in the same holistic manner as Charles did with all living beings: through an historical and evolutionist methodology. This allowed Charles and Francis to unveil previously unknown plant features, such as movement and sensibility and to show that they live by evolving according to the same laws that apply to all the other living beings, with which plants have constant interaction.[8]

[5] Cf. Darwin (1862, 1875a, 1875b).

[6] Many scholars of Darwin, as well as evolutionary post-Darwinist scientists, did not take Darwin's interest in botany in due account either and often ignored his findings. See Desmond and Moore (1992), Eldredge (2005), Pievani (2012). Darwin's interest in plants was often taken as amateurish and secondary. For the argument that Darwin's attention to the plant kingdom was instead due to a clear scientific interest in identifying the specific characteristics of plants in comparison to other living beings, rather than a mere pastime, see Kohn (2008), Smocovitis (2009), Cristofolini (2009), Mancuso (2005, 57–58), Mancuso and Viola (2013).

[7] One of Darwin's most relentless opponents was botanist Julius Sachs, who argued that the Darwins' claims in *The Power of Movement in Plants* were not valid in scientific terms. He stated that "the conclusion, as wonderful as it was sensational, that the growing point of the root, like the brain of an animal, dominates the various movements in the root" is reached "on the basis of experiments which were unskillfully made and improperly explained" (Sachs 1887, 689). Sachs, a highly esteemed botanist of the time, admitted neither to the methods nor to the findings of Darwin, and called him "a country-house experimenter". On the Sachs-Darwin dispute see De Chadarevian (1996). On the other hand, before that dispute erupted Darwin's work on orchids fertilization had been publicly commended by renowned botanist Joseph Hooker (cfr. Huxley 1918).

[8] On the validity of Darwinian botanical findings, see the volume edited by G. Cristofolini and A. Managlia (2009), which compares Darwinian findings with the current state of botanical research.

6.2 Movement and Sensitivity in Plants: The Role of the Radicle Tip

Whether plants move, and if so how, are pivotal questions in the history of scientific and biological thought. The question of the attribution of movement to plants is diriment in the history of scientific and particularly biological thought since, before Darwin, the prevailing hypothesis was that the plant world differed fundamentally from the animal kingdom, precisely due to plants' perceived lack of movement. Plants were considered less complex and inferior to animals mostly on these grounds. Aristotle, in *De partibus animalium*, based his explanation of plants' food intake on the premise that plants lack movement. Having no stomach, which is exclusive to animals endowed with locomotion, plants have roots that extract pre-processed nourishment from the soil.[9]

Darwin challenged the Aristotelian view already in the *Origin of Species* and, before *The Power of Movement in Plants*, also in a 1875 paper titled "The Movements of Climbing Plants." There he asserted that

> The conclusion is forced on our minds that the capacity of revolving, on which most climbers depend, is inherent, though undeveloped, in almost every plant of the vegetable kingdom. (Darwin 1875a, 205–206)[10]

Plant movement, according to Darwin, is not a permanent characteristic; instead, plants possess the faculty of movement, deploying it when advantageous for survival or co-evolution within an ecosystem. Darwin concludes by asserting that where plants do not move, it is not because they are by nature devoid of movement, but only because it will not be advantageous to them in that context.[11]

The Power of Movement in Plants delves into the evolutionary perspective on why plants need movement. The term "power" underscores Darwin's goal to illustrate what plants can achieve through movement that would be impossible otherwise. The text introduces the innovative notion that it is the roots that play the decisive role. Roots demonstrate the intricate relationship between a plant and its environment, showcasing sensitivity, information processing, and movement towards advantageous stimuli. Darwin, in the Introduction, introduces the terms "circumnutation and circumnutate" to describe plant movement. This is how Darwin describes such movement:

---

[9] Cf. Aristotle (2018 II, 3, 650a).

[10] See Darwin (1875a, 5). As a further demonstration that the plant world also falls fully under that world of life that evolves according to principles and rules that are the same for all species (struggle for life, adaptation, self-regulation, co-evolution, etc.), at the end of the *Preface* Darwin states that the many cases of climbing plants that he observes are "the beautiful adaptations as can be found in any part of the kingdom of nature", and then that "It is, also, an interesting fact that intermediate states between organs fitted for widely different functions, may be observed on the same individual plant of *Corydalis claviculate* and the common vine" concluding that "these cases illustrate in a striking manner the principle of the gradual evolution of the species" (Ibid.).

[11] Cf. *Ivi* (pp. 205–206).

If we observe a circumnutating stem, which happens at the time to be bent, we will say towards the North, it will be found gradually to bend more and more easterly, until it faces the east; and so onwards to the south, then to the west, and back again to the north. If the movement had been quite regular, the apex would have described a circle, or rather, as the stem is always growing upwards, a circular spiral. But it generally describes irregular elliptical or oval figures; for the apex, after pointing in anyone direction, commonly moves back to the opposite side, not, however, returning along the same line. Afterwards irregular other ellipse or ovals are successively described, with their longer axes directed to different points of the compass. Whilst describing such figures, the apex often travels in a zigzag line, or makes small subordinate loops or triangles. (*Ivi*, pp. 1–2)

In the next page the Darwin's state: "it will be shown that apparently every growing part of every plant is continually circumnutating, though often on a small scale" (Darwin and Darwin 1880, 3). Darwin and his son also illustrate that plant movements, particularly circumnutation, occur purposefully. The direction and amplitude of these movements "have only to be modified for the good of the plant in relation to internal or external stimuli" (*Ivi*, p. 4). Movement, in its various forms, occurs when beneficial for the plant's survival. It responds to external stimuli while simultaneously showcasing an internal capacity for reorganization. Just like in other living organisms, plant movements are not random; they predominantly lead to enhanced survival capacities for the individual and its species.

Darwin provides the example of heliotropic movements or positive phototropism, where plants position themselves to maximize exposure to sunlight, facilitating the efficient decomposition of carbon dioxide. This example illustrates that plant movements, even seemingly simple ones, contribute significantly to the survival and adaptation of the species. Many plants exhibit a purposeful movement toward sunlight as a strategic adaptation for better survival in environments where light predominantly comes from the sun. Simultaneously, there exists an opposite movement known as negative heliotropism. For instance, roots exhibit this movement by growing away from sunlight, a response shaped by the plant's prioritization of seeking nourishment from underground sources. The evolutionary perspective, according to Darwin, dictates that movements advantageous for survival are conserved and selected for over time, as responses of plants both above and below ground seem to be.

Darwin's exploration of plant movements reveals their active and purposeful engagement with the environment. And the recognition of circumnutation and the adaptability of movements in response to stimuli mark a crucial turning point in understanding the complexity and evolution of the plant kingdom.

No one can look at the plants growing on a bank or on the borders of a thick wood, and doubt that the young stems and leaves place themselves so that the leaves may be well illuminated. They are thus enabled to decompose carbonic acid. (*Ivi*, p. 449)

Darwin's keen observations also extend to other plant species, like *Drosera rotundifolia* or *Dionea*, that display no trace of heliotropism. Notably, these plants are all insectivorous, reinforcing the idea that their lack of heliotropic movement aligns with their unique survival strategies. Darwin asserts that the absence of heliotropism in these species is, from an evolutionary standpoint, not accidental:

> We can understand the reason why these insectivorous plants should not be heliotropic, as they do not live chiefly by decomposing carbonic acid; and it is much more important to them that their leaves should occupy the best position for capturing insects, than that they should be fully exposed to the light. (*Ivi*, p. 4)

Darwin's evolutionary perspective highlights that each movement serves a purpose, contributing to the overall fitness and survival of the individual plant and its species. It also elucidates that all plant movements, including geotropism and sensitivity to gravity, are essentially modifications of the main movement, circumnutation. These modifications occur based on what is evolutionarily advantageous for the plant or its specific parts.

Geotropism, observed in underground and aerial roots, dictates root movement towards the depth of the soil in pursuit of essential moisture and nutrients required for the plant's sustenance.[12] In his studies, Darwin reveals that the tip of the radicle alone is sensitive to geotropism and that, when excited, this tip transmits the influence or stimulus to the adjoining parts, causing them to bend. Darwin underscores the pivotal role of these tips in influencing the overall movement of radicles under the sway of geotropism. The importance of these studies lies in unraveling the intricate structure and functioning of plant beings. Plants, beyond their visible movements and sensitivity, exhibit internal organization where parts interact, transmit information, and influence each other based on their positions and roles within the plant and its ecosystem.

Darwin puts strong emphasis on the tips of the radicle, comparing their role to that of the brain in the animal kingdom—as transmitters of impulses that guide the movement and development of the entire organism.[13] And while highlighting the internal organizing capacity of plant beings, Darwin illustrates that radicle tips are not passive responders to external stimuli. Instead, they possess a multiplicity of sensitivities that are activated based on the situation and what is most advantageous for plant life in that situation.

In addition to geotropism, radicle tips exhibit high sensitivity to moisture, influencing the plant's movement towards moist conditions, a vital aspect of their survival and adaptation strategies:

> Radicles are also sensitive to air which contain more moisture on one side than the other, and they bend toward its source. It is, therefore, probable that they are in quite manner sensitive to dampness in the soil. It was ascertained in several cases that this sensitiveness resides in the tip, which transmits an influence causing the adjoining upper part to bend in opposition to geotropism towards the moist object. We may therefore infer that the roots will be deflected from their downward course towards any source of moisture in the soil. (*Ivi*, p. 551)

---

[12] Cf. Darwin and Darwin (1880, 512).

[13] Darwin gives an example from the animal kingdom in order to better understand what happens in the plant kingdom: "To see anything of the above kind in the animal kingdom, we should have to suppose than an animal whilst lying down determined to rise up in some particular direction; and that after its head had been cut off, an impulse continued to travel very slowly along the nerves to the proper muscles; so that after several hours, the headless animal rose up in the predetermined direction" (*Ivi*, p. 543).

In a Darwinian perspective, the behavior of the plant's root tips exemplifies adaptability and a dynamic pursuit of what is immediately advantageous for survival. Contrary to a rigid adherence to geotropism, the root tips exhibit a flexible nature, abandoning this movement to follow more advantageous conditions, such as moist soil, should the opportunity arise. When the root apex encounters obstacles in its underground movement, it showcases a remarkable ability to change direction, deviating from geotropic movement to navigate towards what offers the least resistance Darwin's experiments placing cardboards of varying thickness near root tips illustrate this sensitivity: the root tips consistently move towards the thinner cardboard. Darwin concludes that:

> We may feel almost sure, that when the tip encounters a stone or other obstacle in the ground, or even earth more compact on one side than on the other, the root will bend away as much as it can from the obstacle or the more resisting earth and will thus follow with unerring skill a line of least resistance. (*Ivi*, p. 550)

The multifaceted sensitivity of the root tip allows it to suspend geotropic movement when more useful movements lead towards nourishment. Conversely, it can revert to following the influence of gravity when necessary and functional for the plant's life.

Darwin emphasizes that multiple exciting causes can simultaneously act on the root tip. This underscores the plant's ability to prioritize and respond to various environmental cues, increasing its chances of survival and adaptive capabilities in diverse conditions. In essence, the tip of the radicle serves as a dynamic and strategic navigator for the plant, constantly adjusting its movements to optimize survival in its ever-changing environment. This is the sense in which Darwin states that it can happen that "two (or perhaps more) of the exciting causes often act simultaneously on the tip, and one conquers the other, no doubt in accordance with its importance for the life of the plant" (*Ivi*, p. 573).[14]

6.3 Evolution in the Plant and Animal Kingdoms: A Comparison

The root tips' ability to navigate the soil, avoiding obstacles and choosing paths of least resistance, prompts Darwin to draw a comparison with borrowing animals like moles. Both the radicle and the mole exhibit a purposeful movement, adapting their course based on environmental challenges. About the mole, Darwin says:

---

[14] On the revolutionary significance of Darwin's analysis of the root tips and their function for plants development and survival see Mancuso and Viola (2013), Baluŝka et al. (2009). Furthermore, Darwin's findings provided the basis for all the studies on the physiology of the roots, which lead to talk of a multifarious plant intelligence; see Alpi (2007), Baluŝka et al. (2010), Chamovitz (2012), Firn (2004), Hodges (2009), Lindsay (1876), Trewavas (2003).

> By continuously moving its head from side to side, or circumambulating, he will feel any stone or other obstacle as well as any difference in the hardness of the soil, and he will turn from that side; if the earth is damper on one than on the other side, it will turn thitherward as a better hunting-ground. Nevertheless, after each interruption, guided by the sense of gravity, he will be able to recover his downward course and to burrow to a greater depth. (*Ivi*, p. 200)

By repeatedly linking the functioning and development of plants, especially root tips, to aspects observed in the animal kingdom, Darwin highlights the interconnectedness of all living beings. This interconnectedness is manifested in shared principles of movement, adaptability, and responsiveness to environmental stimuli, transcending the traditional boundaries between plant and animal life. In Darwin's evolutionary framework, recognizing these similarities becomes integral to understanding the overarching principles governing life forms. The comparison between plant root tips and borrowing animals serves as a testament to the universality of certain adaptive strategies, emphasizing the common threads that weave through the diverse tapestry of the natural world.

Darwin himself was astounded by the revolutionary implications of his studies on plants, acknowledging the magnitude of the shift from viewing plants as merely living beings with static characteristics to recognizing their intricate and dynamic nature.[15] This paradigm shift challenged the notion of plants as passive entities and positions them as active participants in their environment, shaping and responding to their surroundings. This led Darwin to reject the idea of the plant kingdom as separate or inferior to the animal kingdom.[16] Instead, he emphasized the numerous similarities and shared dynamics between the two kingdoms, celebrating their commonalities within the diversity of life forms.

A key aspect that unites plant and animal beings, according to Darwin, is the presence of multiple sensitivities. The crucial issue lies in localizing these sensitivities, wherein specific parts of the organism can be individuated to play the role of transmitting excitability and movement to other parts, ultimately influencing the entire organism. The root extremity, described by Darwin as the most marvelous structure in plants, directs various parts, especially the roots themselves, and serves as the focal point for transmitting movements and sensations to the whole organism. Remarkably, all these processes occur in plants without the presence of a nervous system or a central directing organ analogous to the brain of animals. This absence of a centralized control system underscores the uniqueness of plant life and challenges conventional notions about the prerequisites for complex behaviors and responses. It also marks a significant milestone in the evolution of biological thought.

In the closing lines of *The Power of Movement in Plants*, Darwin writes:

> It is hardly an exaggeration to say that the tip of thew radicle thus endowed and having the power of directing the movements of the adjoining parts, acts like the brain of one of a lower animals; the brain being seated within the anterior end of the body, placed at the front of

---

[15] There are of course many theoretical, scientific and bibliographical references to other scientists and scholars who published on this subject before or at the same time as Darwin: for example Sachs (1879), Frank (1870), Ciesielski (1873), Richter (1879).

[16] Cf. Darwin, *Ivi*, p. 571.

> the body, receiving the impressions from the sense- organs of the senses, and directing the several movements. (*Ivi*, p. 573)

So, while Darwin goes so far as to say that there are internal dynamics in living organisms that are the same for both plant and animal beings, the reference to the behavior of animals and to the organ that is the brain must be understood in terms of a comparison that does not imply an attribution of animal characteristics to plant beings such as consciousness, awareness, decision-making ability, etc. That is why Darwin's comparison between the behavior of animals and the root tips of plants is proposed cautiously. He acknowledges the potential exaggeration in equating the root tips to the brain. But Darwin also insists that the comparison is hardly exaggerated if it reinforces the understanding of plants as possessing a sophisticated, self-organizing capability thanks to the root's extremity. It becomes exaggerated if it leads to attributing to plants characteristics that are unique to animals, such as consciousness, awareness, and decision-making abilities. Darwin's intent is not to anthropomorphize plants, but to highlight their remarkable complexity and responsiveness within the context of evolutionary processes.

What should not be forgotten is that the Darwinian hypotheses on the nature and development of plants are located within the same and only paradigm of reference, which is the evolutionary paradigm. In this view, the plant kingdom and the animal kingdom are subject to the same rules of natural selection, inheritance of characters, struggle for life, and co-evolution. Certainly, in Darwin's time the attention of his contemporaries, especially in academia, was directed toward the consequences that the theories of common descent and continuity in evolution had on the view of human beings and their role within the natural world and not so much on the discoveries made in the botanical realm, although revolutionary with respect to the dominant paradigm.

6.4 Conclusion

Darwin's contributions to plant science represent a starting point and an irreplaceable foundation for botanical research from his time to the present. The contemporary evolution of plant studies aligns seamlessly with Darwin's observations and theories on plants, their movements, and their wider ecological roles, showing the enduring relevance of his insights on our understanding of plant life and its place within the broader context of evolution. Today, we can say that what Darwin wrote in *The Power of Movement in Plants* and also in his earlier writings has found and continues to find confirmation in contemporary studies that have certainly given due prominence to the Darwinian perspective. Today's plant science places its own research in line with what emerged from Darwin's work, which is explicitly cited as the one who first demonstrated the complexity and variety of plant activity, which is not limited to reacting passively to external stimuli thanks mainly to the function of the root apices and the communication between them and the rest of the organism. We know well how today there are branches of botany that are very much in vogue and at the same time much discussed such as the one called "plant neurobiology" that interpret the presence in plants of certain functions, sensitivities and movements in the terms of a system very close to the nervous system in which we talk about intelligence, memory, decision-making and behavioral abilities.

Bibliography

Alpi, A., et al. 2007. Plant Neurobiology: No Brain, No Gain? *Trends in Plant Science* 12 (4): 135–136.

Aristotle. 2018. *De partibus animalium.* in *La vita*, ed. D. Lanza and M. Veggetti. Milano: Bompiani.

Baluŝka, F., S. Lev-Yadun, and S. Mancuso. 2009. The 'Root-Brain' Hypothesis of Charles and Francis Darwin: Revival After More Than 125 Years. *Plant Signaling and Behavior* 12 (4): 1121–1127.

Baluška, František, Simcha Lev-Yadun, and Stefano Mancuso. 2010. Swarm Intelligence in Plant Roots. *Trends in Ecology and Evolution* 12 (25): 682–683.

Chamovitz D. 2012. *What a Plant Knows. A Field Guide to the Senses.* Scientific American, Farrar: Strauss and Giroux.

Ciesielski T. 1873. *Arbeiten des Botanischen Instituts in Würtzburg.*

Cristofolini G. 2009. Darwin e l'evoluzione delle piante. Nota introduttiva. In *Il giardino di Darwin. L'evoluzione delle piante*, ed. G. Cristofolini, and A. Managlia. Torino: Umberto Allemandi & C..

Darwin, C. 1839. *A Naturalist's Voyage round the World.* London: John Murray.

Darwin, C. 1859. *The Origin of Species.* New York: D. Appleton and Company.

Darwin, C. 1862. *The various contrivances by which orchids are fertilized by insects.* London: John Murray.

Darwin, C. 1875a. *Insectivorous Plants.* London: John Murray.

Darwin, C. 1875b. *The Movements and Habits of Climbing Plants.* London: John Murray.

Darwin, Charles, and Francis Ed Darwin. 1880. *The power of movement in plants.* London: John Murray.

De Chadarevian. 1996. Laboratory Science versus Country-House Experiments. The Controversy between Julius Sachs and Charles Darwin. In *The British Journal for the History of Science* 29 (1): 17–41.

Desmond, A., and J. Moore. 1992. *Darwin.* London: Penguin.

Eldredge, N. 2005. *Darwin. Discovery the Tree of Life.* New York: Norton & Company.

Firn, R. 2004. Plant Intelligence: An Alternative Point of View. *Annals of Botany* 93 (4): 345–351.

Frank, A.B. 1870. *Die natürliche Wagerechte Richtung von Pflanzentheilen.* Leipzig: Verlag von Herm. Weissbach.

Hodges, A. 2009. Root Decisions. *Plant, Cell and Environment* 32 (6): 628–640.

Huxley L. 1918. *Life and Letters of Sir J. D. H.*, vol. 2. London: John Murrey.

Kohn, D. 2008. *Darwin's Garden: An Evolutionary Adventure.* New York: The New York Botanical Garden.

Kohn, D., G. Murrell, J. Parker, and M. Whitehorn. 2005. What Henslow taught Darwin. In. *Nature* 436: 643–645.

Lindsay W. L. 1876. Mind in plants. In *British Journal of Psychiatry* 21(96): 513–532.

Mancuso S. 2005. La neurobiologia vegetale: sinapsi, neurotrasmettitori e memoria nelle piante. In *Annali. Accademia nazionale di agricoltura.* 117–152.

Mancuso, S. 2007. Alcuni aspetti della neurobiologia vegetale. *Silvae* 2 (4): 231–259.

Mancuso, S. 2014. *Uomini che amano le piante. Storie di scienziati del mondo vegetale.* Firenze: Giunti.

Mancuso, S., and A. Viola A. 2013. *Verde brillante. Sensibilità e intelligenza del mondo vegetale.* Firenze: Giunti.

Pievani T. 2012. *Introduzione a Darwin.* Roma-Bari: Laterza.

Richter K. 1879. Arbeiten des pflanzenphysiologischen Institutes der k.k. Wiener Universität. XVI., Untersuchungen über den Einfluss der Beleuchtung auf das Eindringen der Keimwurzeln in den Boden : XV. Sitzung vom 19. Juni 1879 / von Dr. Carl Richter. In *Enthalten in Sitzungsberichte der mathematisch-naturwissenschaftlichen Classe der Kaiserlichen Akademie der Wissenschaften Wien: Aus der K.K. Hof- und Staatsdruckerei, in Commission bei Carl Gerold's Sohn* 1 (6): 16–33.

Sachs, J. 1879. *Arbeiten des Botanischen Instituts in Würtzburg.* Leipzig: Engelmann.

Sachs, J. 1887. *Lectures on the Physiology of Plants,* trans. H. M. Ward. Oxford: Oxford University Press.

Smocovitis V. 2009. Darwin's Botany in the Origin of Species. In *The Cambridge Companion to the "Origin of Species"*, ed. M. Ruse, and R. J. Richards. Cambridge: Cambridge University Press.

Trewavas A. 2003. Aspects of Plants Intelligence. *Annals of Botany* 92 (1): 1–20.

Chapter 7
Plant Mind and the Emergence of Plant Physiology: Dutrochet on Plant Sensitivity and Motility

Quentin Hiernaux

Abstract While some philosophers have addressed plants in their writings, they have rarely done so for their own sake. Instead, the plant question often serves as a counterpoint to thinking about the animal and the human, or as a stand-in for the general concept of 'nature'. In this paper, I argue that this relegation of plants has resulted in unquestioned assumptions that have been perpetuated in modern science and even in contemporary physiology. At the same time, I show that the emergence of plant physiology by a figure such as René Joachim Henri Dutrochet (1776–1847) raised philosophical questions and positions on how to interpret plant behaviour, in a marked contrast to some later reductionist laboratory science. This whole narrative, in turn, sheds light on the very current controversies surrounding the attribution of sensations or a mind to plants.

7.1 Introduction

The 'vegetal turn', or 'plant turn', is a movement that aims to refocus on the plant world; developing since the beginning of the twenty-first century, it has become increasingly important over the past 20 years. Two primary developments serve to explain this renewed interest in the plant world: new experiments and discoveries in plant physiology (Trewavas 2014), and growing ethical concern for the complexity of the plant world and its interactions with ecosystems and humans (Kallhoff et al. 2018; Hiernaux 2021a). I focus here on the former development, analysing it from a historical–conceptual perspective. Based on disruptive experiments on plant physiology that show phenomena of communication, memory, and learning in plants, some biologists and philosophers now consider that plants could have a mind similar to that of some animals (Trewavas 2002, 2003, 2005; Calvo 2016; Calvo and Keijzer 2011; Calvo and Baluška 2015; Calvo and Trewavas 2020; Maher 2017; Hiernaux 2019, 2021b, 2023). As such, plants could make intelligent choices and perhaps even

Q. Hiernaux (✉)
Fund for Scientific Research—FNRS, Université libre de Bruxelles, Brussels, Belgium
e-mail: Quentin.Hiernaux@ulb.ac.be

M. Di Paola (ed.), *The Vegetal Turn*, The International Library of Environmental, Agricultural and Food Ethics 38, https://doi.org/10.1007/978-3-031-68671-9_7

have a form of consciousness (Calvo et al. 2021; Taiz et al. 2019; Mallatt et al. 2021). How can we interpret the application of such controversial concepts to plants, and what are the scope, limits, and epistemological issues at stake? While I do not take direct part here in the *pro* or *contra* debate on plant mind, I propose to analyse the historical constitution of the debate's very framework.

Before their study in the fields of physiology and ethology in the nineteenth century, and then in the cognitive sciences during the twentieth century, the concepts of behaviour and mind lay in the province of philosophy. Throughout the ancient, medieval, and modern periods, the distinction between science and philosophy was ill-defined; historically, writings on animal and even plant mind have mainly been composed by philosophers in their thinking about the soul (Ingensiep 2001; Baldassarri and Blank 2021). The philosophical treatment of plants in a given epoch itself depends on contemporary conceptions of life; if it remains difficult today to even consider the question of plant mind (without presuming to answer it), this is due in part to a long philosophical tradition of excluding plants from thought (Hall 2011). As Marder (2013) has highlighted, the assumptions underpinning this philosophical exclusion are twofold: plants do not think, and philosophers do not think about plants. While some philosophers have addressed plants in their writings, they have rarely done so for their own sake. Instead, the plant question often serves as a counterpoint to thinking about the animal and the human, or as a stand-in for the general concept of 'nature'. In this paper, I argue that this relegation of plants has resulted in unquestioned assumptions that have been perpetuated in modern science and even in contemporary physiology. At the same time, I show that the emergence of plant physiology by a figure such as René Joachim Henri Dutrochet (1776–1847) raised philosophical questions and positions on how to interpret plant behaviour, in a marked contrast to some later reductionist laboratory science. This whole narrative, in turn, sheds light on the very current controversies surrounding the attribution of sensations or a mind to plants.

7.2 Historical Elements

This section reviews the status of plants in the West and their link with the mind in the context of the history of philosophy and science.

7.2.1 The Ancient Roots

At the origins of Western philosophy, the pre-Socratics, such as Empedocles, do not seem to have held a hierarchical view of living things with which later thinkers were familiar (Hall 2011; Zatta 2019). Moreover, the boundary between what we consider living or non-living today was not necessarily relevant for them, as they were instead concerned with the self-organisation of things—be they cosmic, vegetal, or animal—according to a common fundamental principle. Empedocles, for instance, assumed a kinship between all living beings.

Plato, who took his conception of the soul and its transmigration from the Pythagoreans, was the first to theorise more specifically about biological life in relation to the rest of matter. In the *Timaeus*, the soul is identified as the immortal principle that allows the self-motion of a living body. Consequently, the movements of living beings, including those of plants, are attributable to a psychic cause (Gavray 2019). Plato thus shares with Empedocles the idea of the continuity of all living things. Moreover, Plato recognises in plants only the faculty of desire, capable of experiencing pleasures and pains (*Timaeus* 77a–c). However, Hall (2011, 17–22) argues, Plato makes plants purely passive beings without any purpose of their own; as their creation would be owed to their instrumental use for humans and animals as food (*Timaeus* 90a), Plato here introduces a differentiation and domination of humans over the rest of the living world. For Plato, humans are the only living beings in which the intellective faculty of the soul is fully expressed. Nevertheless, plants, like all other living beings, possess a soul that is capable of intellection, provided that it is incarnated in the right body following metempsychosis.

For Aristotle, a natural body's spontaneous and inherent tendency to move is necessary for living, but not sufficient to characterise life. Indeed, a specific kind of movement is required to define life—the nutritive movement of the soul, which pushes an organised body to feed itself, and thus to grow and decay. Unlike Plato, however, Aristotle does not believe that the soul can exist outside a living body, of which the soul is the actualisation. Nor did he recognise any desire in plants, believing that only animals are endowed with the sensitive faculty of the soul, which allows them to experience sensations of pleasure and pain, and to seek or avoid those sensations by movement (*De Anima* II, 413b1–4). Consequently, plants do not have even the sense of touch, considered the most elementary of the five senses. Aristotle does follow Plato in considering the intellective faculty to be typical of humans, and even states that it is exclusive to them. He thus sets out more clearly than Plato a hierarchy of functions for living beings along with a rejection of any form of mind in plants. This instrumental conception of plant (and animal) life in such a hierarchical framework is found, for example, in the *Politics* (1256b).

As a naturalist, Aristotle is known for his treatises on zoology and his classification of animals, but his study of plants was limited. While the manuscript tradition has preserved a treatise on plants that was long attributed to him, it is now recognised as apocryphal and was most probably written by Nicolaus of Damascus. This treatise, *De plantis*, is confused because of its successive translations (Fenderspiel and Cronier 2018, 40), combining ideas from Aristotle and his student Theophrastus. Describing plants as immobile (815b22) and insentient (817b22–24), the treatise's author concludes that plants have no activities similar to those of animals and no psychic life (818a68).

In the works of Theophrastus, the successor of Aristotle and the founder of botany as a science, plants are better documented but they have a more ambiguous status. As an Aristotelian, Theophrastus bows to his master's metaphysical doctrine, declaring in the preamble to his *History of Plants* that plants, unlike animals, have no behaviour (HP 1.1.1). His words are often more nuanced, however, as he elsewhere observes the diversity of plant movements and even suggests a form of sensitivity in some of them (*De causis plantarum*, 2.19.3.6–2.19.4.5). Despite these observations, Theophrastus does not attempt to further integrate plant sensation into Aristotelianism—an occurrence that would have to wait until the Renaissance[1] and the West's rediscovery of Theophrastus' botany. Following the Latin translation of his treatises in the sixteenth century, botany as a science autonomous of medicine was redeveloped and a series of European authors contributed to the foundations of morphology, systematics, and modern plant physiology (Greene 1983; Morton 1981).

The early chapters of *Genesis* also set out a strong ontological distinction between the plant and animal worlds, positing an instrumentalisation of plants by humans (and other animals) as their sole reason for existence. The conjunction of Judeo-Christian influence and Aristotelianism would lead to the medieval and modern chain of beings (Lovejoy 1964), which is based on two main principles: plenitude and continuity. In the logic of the chain of beings, everything that *could* be created was created (by God) and between each element of creation there is a perfect continuous gradation of intermediate forms. The world can therefore be ordered in a hierarchy determined by an object's relative complexity. At the bottom of the order are minerals and plants, followed by animals, then humans, and finally supra-sensible beings such as celestial bodies and angels. Each category in this hierarchy has its own essential defining properties and grafted onto this ontological order is a moral order, in which the value of and respect offered to different beings depend on their degree of complexity. Conditioned by such a mental framework, Western scientists apprehend plants as simple beings incapable of possessing sensation or mind, much less matching the aptitudes of those beings deemed superior. Yet, as modern botanists have made discoveries, this perspective has caused increasing confusion.

7.2.2 Early Modern Science

Among the botanists of the Renaissance, the physician and philosopher Andrea Cesalpino (1524–1603) is exceptional in this regard. In his *De plantis libri XVI* (1583), particularly in the first book, Cesalpino not only attempts to describe plants, but to explain their functioning through a combination of Aristotelianism and observations drawn from Theophrastus and from his own experience. He thus explicitly questions the problem of the insensitivity of plants, which makes a coherent explanation of their nutrition very difficult (Cesalpino 1583, I, Chapter 3), as it is unclear how plants would select their food effectively if they had neither taste nor touch nor any other sensation. Similarly, Cesalpino notes that the coherent movements of

[1] Albert the Great is noteworthy in this regard, as his treatise *De vegetabilibus* was an exception in the medieval period concerned almost exclusively with the pharmacological use of plants (Panarelli 2020; Cerrito 2021).

climbing plants seem to require the sense of touch in order to detect and cling to their support (Cesalpino 1583, Chapter 11). His fidelity to Aristotelian doctrine, however, does not allow him to assert consistently that plants are indeed sensitive (Hiernaux and Tresnie 2023a, 2023b).

This quasi-dogmatic adherence to the Aristotelian thesis of plant insensitivity continued into the eighteenth century. Botanists such as La Mettrie, and especially Linnaeus, affirmed the insensitivity of plants by explaining their movements as the results of mechanical causes. Some went even further, claiming that plants were not really living beings at all; thought to neither breathe nor reproduce sexually, plants were sometimes relegated to the status of minerals (Delaporte 1982). All these modern authors endorsed the hierarchical conception of the chain of beings, considering plants to be passive, insensitive, and simpler (and therefore less respectable) than animals; as such, it was inconceivable to them to attribute sensitivity and a mind to plants. On a scientific level, this assumption compromised the study of plant behaviour, which was thought to be non-existent; philosophically, it made the notion of plants' souls quite problematic in the Christian tradition (Baldassarri and Blank 2021).

Views began to change gradually in the nineteenth century as laboratory experiments and stricter research protocols were developed. From this period onwards, questions concerning the sensitivity, agency, and even the mind of plants returned in force and required new explanations (see, for example, Fechner 1848; Boscowitz 1867). But traditional philosophical presuppositions remained influential. The early days of experimental physiology is thus pivotal for understanding twenty-first century controversies surrounding plant behaviour, sensitivity, and mind. In the following section, I provide commentary on a seminal treatise on plant physiology to illustrate how the discipline has made important advances in the understanding of plant motility and sensitivity, while also demonstrating how little its epistemological and metaphysical framework had changed prior to the twenty-first century.

7.3 Dutrochet and Nascent Plant Physiology

R. J. Henri Dutrochet was one of the founders of experimental plant physiology and was responsible for the discovery of osmosis and the cell theory, which was subsequently taken up and developed by Theodor Schwann. Dutrochet's experiments contributed to the joint study of plant and animal physiology, thereby laying the foundations of future biology. In 1824, he published his *Recherches anatomiques et physiologiques sur la structure intime des animaux et des végétaux et sur leur motilité* (*Anatomical and physiological research on the interior structures of animals and plants and on their motility*), an invaluable treatise for understanding the developments and presuppositions that physiologists faced in their study of sensitivity, movement, and plant mind. As I demonstrate below, an examination of this treatise shows that the conceptual framework of the modern debate on plant behaviour, including the controversies surrounding plant neurobiology, was already present from the early nineteenth century.

7.3.1 *Commentary on Dutrochet's Research*

According to Dutrochet, physiologists define 'sensitivity' as an organism's ability to detect and react to external causes. As a botanist, however, Dutrochet considers this definition too vague, instead linking sensitivity directly and biunivocally to an awareness of one's existence. In his view, sensitivity implies that there is a self that lives and feels, and vice versa; as for plants, he remarks that 'no one, I think, will be tempted to grant them a self, and consequently, sensations' (Dutrochet 1824, 2).[2] Therefore, an alternative account for plants' reactions and movements 'on the occasion of the influence of certain external causes' (Dutrochet 1824, 2) is required; following Xavier Bichat, Dutrochet thus attributes the reactions of plants to an 'analogue of the sensation in animals' (Dutrochet 1824, 2), establishing a distinction between an animal sensitivity that is the source of sensation (i.e. of self-consciousness, even if confused) and an organic sensitivity linked solely to the physical (nervous) irritability of organs or plants.[3] This dichotomy is not entirely new to Dutrochet, although it had previously been applied to the understanding of only animal functioning. For example, the naturalist Lamarck writes,

> Plants are living bodies that are not irritable, incapable of instantly and iteratively contracting any of their parts on themselves, and lacking the ability to act, as well as the ability to move. (Lamarck 1815, 124)

Dutrochet, for his part, notes the reactionary movements of plants, and so classifies them as irritable in the same way that animal organs are, but not sensitive like animals themselves, which are capable of integrating the multitude of their sensations into a whole that provides them with a form of agency (what he calls 'consciousness' of themselves[4]).

Dutrochet thus demonstrates how physiologists' philosophical presuppositions hinder them from giving a coherent account of their observation of plants. One of these presuppositions, of Platonic origin, associates the mind with sensation (at least in its relation to desire) or even 'self-consciousness'—an option that Dutrochet rejects by telling us that no one would be tempted to admit it. The Aristotelian alternative denies plants any form of sensitivity at all, in direct opposition to nineteenth century scientific observations. Dutrochet's theory thus allows him to account for the common intuition (as conveyed by the chain of beings) that plants have no mind while still acknowledging the observable fact that they do react and move.

---

[2] All translations of Dutrochet's quotations are mine.

[3] For the sake of clarity, I write simply 'animal sensation' and (plant) 'irritability' in what follows, rather than 'sensation' and 'organic sensitivity'.

[4] It is unclear what Dutrochet meant by this expression, but it does in any case involve a form of mind.

Through this compromising approach, Dutrochet builds a dualistic epistemological framework in which the sensation of animal mind (the ancients' intellective and sensitive soul) is opposed to the irritability of (reactive) matter and plant activity (the ancients' nutritive soul). Physiology itself does not necessitate this dualism, of course, but the weight of Western metaphysics does. One of the most obvious manifestations of this dualism is the duplication of terms used for animal phenomena and their plant 'analogues', which helps to maintain the idea that the two types of phenomena are distinct. For instance, Dutrochet argues for physiology's abandonment of the term 'sensitivity' to avoid misunderstanding between 'true' sensation and irritability, proposing in its place the term 'nervimotion', which does not necessarily imply a sensation 'consciously' felt by the organism. The term should therefore be understood as a middle ground between irritability and true sensation. Both faculties imply 'nervimotion' in living beings, understood as the ability to react with movement. In plants, however, this movement is only internal, whereas in animals it is expressed as 'locomotility'—the ability to move parts of the body (Dutrochet 1824, 4–6). In discussing nervimotility and locomotility, Dutrochet coins scientific terms corresponding to Aristotle's historical distinction between local movement, which is proper to animals, and other types of movement that correspond to a change of state (such as growth or ageing) in living beings.

Of course, this lexical duplication does not prove the existence or absence of mind in plants, but as a rhetorical strategy it serves to incline thought towards a particular position, much as the use of a single term to designate both animal and plant phenomena suggests a unified original phenomenon. Similar concerns over lexical distinctions can be seen in current debates surrounding plant neurobiology, in which proponents favour the use of a liberal vocabulary that suggests a commonality between the phenomena of animal and plant life; certain cellular activities in plants, for example, are described in terms akin to the functioning of neurons or the nervous system (Baluska et al. 2006; Barlow 2008). Plant neurobiologists defend the attribution of a 'memory' or 'learning faculty' to plants (Trewavas 2014; Gagliano et al. 2014, 2016; Mancuso 2015), while their opponents instead refer to the phenomena described as 'a consequence of developmental progression' (Firn 2004) or 'adaptive response' (Taiz et al. 2019), considering the former terms to be metaphorical at best (Struik et al. 2008). The former criticise the latter for their reductionism; the latter criticise the former for their zoomorphism (or even anthropomorphism). Those opposed to the attribution of increased agency to plants sometimes refuse even the application of the terms 'behaviour' or 'sensation' to plants (Burgat 2020 exemplifies this position; for other examples, see Cvrčková et al. 2016). In this way, the dichotomy between animal sensation and organ or plant irritability promoted by Dutrochet in 1824 is debated anew by twenty-first-century plant physiologists. For example, a current detractor of plant behaviour can dismiss plant sensation and mind (learning ability for example) in comparing whole plants to animals organs (like a hand):

> A root undergoing gravitropism is actually a different root at the start and end of the process. If a root is exposed to an extreme gravitational stimulus twice, 12 hours apart […], the second stimulus will initiate a new response in cells that have not necessarily experienced the first

> stimulation, and hence they cannot usefully have learned. This is in contrast to an animal where learned behaviour resides at the organism level—the hand has no memory of being burned, but the brain controlling the hand might possess the appropriate knowledge after an adequate previous learning experience. (Firn 2004, 347)

For physiologists such as Firn, since a plant has no brain, it would lack the capacity for information centralisation that allows integrated behaviour, and would thus be comparable to a collection of independent irritable organs. Others, such as Trewavas (2004, 2014), challenge this argument by noting the other ways in which plants integrate information and make unified reactions.

The presumed hierarchy of living beings also influenced the nascent field of plant physiology; from the start, plants were considered simpler than animals. Remarkably, however, plants' simplicity no longer contributed to their disdain, but instead became an asset in the analytical approach to physiological observation and experimentation. Thus, Dutrochet, writing that the faculties of movement in plants are 'less energetic, much less developed' (Dutrochet 1824, 6) than in animals, welcomes this rudimentary nature: whereas the many internal and external causes of their movements may be an insurmountable difficulty to a study of the laws of motility in animals, the study of movement 'is much simplified in plants, and it is probably to them alone that we shall owe the solution of the principal problems of the science of life' (Dutrochet 1824, 6). For the first time, animals were not the absolute model of life, and zoology was no longer necessarily the paradigm of biology. This new attitude changed the biological view of life by emphasising the continuity between animal and plant life forms, rather than an essential division. Of course, differences were not denied, as the duplicity of vocabulary reminds us; by establishing a parallelism with animal physiology, however, plant physiology suggested a continuity between all forms of life, including between plants and animals, which comparative anatomy alone did not allow. It may be no exaggeration therefore to understand transformist biology and the emergence of Darwinian evolutionism in the nineteenth century as a product, in part, of the innovative intellectual context of nascent plant physiology.

In the field of comparative anatomy, there were isolated attempts to demonstrate this continuity, such as that of Robinet, who sought to find the common morphological structures of animals (including humans) and plants, but these efforts were more speculative (not to say far-fetched) than scientific, as his contemporaries were quick to note (Lovejoy 1964, 281). Rather than a community of structures, plant physiologists attempted to discover functional analogues between animal and plant anatomy. Indeed, it was apparent that plants have neither brains nor nerves as such, although they react and move: 'These phenomena [of irritability] tend, therefore, to prove that there is in plants, if not a nervous system, at least some elements of it' (Dutrochet 1824, 15). By studying the microscopic anatomy of plant tissues and organs and taking the sensitive plant (*Mimosa pudica*) as a model,[5] Dutrochet theorises that plants had a diffuse nervous system, dispersed in the tissues in 'globular corpuscles' or 'nervous corpuscles'—small cells that he observed arranged on the

[5] Dutrochet (1824, 81) mentions the work of his predecessors, Duhamel du Monceau and Dufay, *Note sur la sensitive* (Mémoire de l'académie royale des sciences 1736), in this regard.

walls of the large cells, but not joined together in masses as in animals (Dutrochet 1824, 14). From experiments on the reactionary movements of the *Mimosa pudica*, Dutrochet concludes that the sap (rather than the bark or pith) transmits nervimotion in plants: 'All those portions of the central system which contain tubes suitable for transmitting the sap, are equally suitable for transmitting the nervimotion' (Dutrochet 1824, 75). He thus establishes an analogy between the nervous system and the vascular system. As René Desfontaines had already demonstrated experimentally by stimulating the roots of *Mimosa pudica* with sulphuric acid, the transmission of the stimulus can take place through the whole plant from any of its parts. We can therefore note a community of functions, observed experimentally, to which physiologists sometimes propose anatomical analogues. While these analogues are not necessarily morphologically close, they may be considered similar for other reasons, such as their possession of the same chemical properties, for instance:

> As they [the globular corpuscles] compose all the organs of animals, this proves that it is not from their form that we must draw inductions to determine their functions; but, in animals, the chemical nature of these globular corpuscles is not everywhere the same. Thus, those which compose the muscles are soluble in acids, while those which compose the nervous system are insoluble in these same acids, but only soluble in alkalis. Now, such is also the chemical nature of the globular corpuscles which are observed in plants, as we have just explained. This may therefore authorise us to think that these globular corpuscles are nervous organs, or rather that they are the scattered elements of a diffuse nervous system, or one which is not united in masses, as it is in animals. This consideration, based on the analogy of the chemical nature of globular corpuscles, is further strengthened by the observation of the intimate structure of the nervous system of certain animals. (Dutrochet 1824, 14)

Neither the existence and function of auxin (and plant hormones) nor cytology were known to Dutrochet. Similarly, the statocysts and statoliths that enable plants to detect gravity, and the photoreceptors that enable them to detect light, were not discovered until well after his time. Some contemporary scientists had therefore hypothesised that light or gravity acted directly and mechanically on plants as they do on physical bodies. By experimenting with the positive gravitropism of stems and negative gravitropism of roots, however, Dutrochet is able to reject this type of reductionist explanation.

A particularly topical experiment conducted by Dutrochet concerns the orientation at a distance of the mistletoe seed's 'root' (Dutrochet 1824, 105–107).[6] Dutrochet studied the tropism of the germinating mistletoe; as a parasitic plant, its tropism is not perpendicular to the centre of the earth, but rather to the body on which it germinates. The origin of the movement's influence is therefore neither constant nor predetermined since it depends on the seed's place of germination. Furthermore, if the mistletoe seed is suspended on a wire and thus not in contact with a body when it germinates, it spontaneously directs its 'radicle' towards the nearest body, provided that it is no more than about one millimetre away.

Dutrochet's observation recalls recent experiments on the nutation movements of climbing plants, which, much like the mistletoe's tropism, change in the presence of

[6] In mistletoe (*Viscum album*), the 'radicle' described here is actually the hypocotyl, which will become a sucker and not a real root.

a support with which the plants are not in contact but somehow detect at a distance (Guerra et al. 2019; Raja et al. 2020; Simonetti et al. 2021). As the supports are inorganic, chemically neutral, and are lit by zenithal lighting to remove shadows, it remains unclear how these plants detect them. Some parasitic plants, such as dodder, are known to orient themselves towards their host at a distance after the latter emits volatile organic compounds (Kelly 1992). While mistletoe is also a parasitic plant, however, the germinating mistletoe seed in Dutrochet's experiment does not orient itself towards a living host, but towards 'a small wooden board' (Dutrochet 1824, 105). This phenomenon of mistletoe tropism at a distance is understood today as a combination of negative gravitropism and negative phototropism (the germinating mistletoe detects the near body thanks to differences in light) (Williams 1963; Sallé and Aber 1986).[7]

According to Dutrochet, the movements of plants in his experiments differ from the movements of the inorganic world because the former are not caused directly by stimuli acting on their 'weightable matter' (that is, the materiality of their body). These plants' movements are instead mediated by a 'nervimotor' sensitivity, which influences the movement of the organs in a second stage:

> There is therefore no doubt that this movement is *spontaneous*, that is, that it is due to an inner and vital cause brought into play by the influence of an external agent. This spontaneity of the direction of the radicle of the mistletoe under the influence of the attraction proves in an incontestable way that this attraction acted only on the nervomotility of the plant, and not at all on its weightable matter. The same is undoubtedly true of *terrestrial* plants. The unknown cause of attraction is only the occasional cause of the downward movement of the roots and the ascent of the stems; it is not the immediate cause; it acts, in this circumstance, as a nervimotor agent. We shall see below new proofs of the generality of this important fact in physiology, namely, that the visible movements of plants are all *spontaneous* movements, executed on the occasion of the influence of an external agent, and not movements *impressed* by this agent. (Dutrochet 1824, 107; italics in the original)

For Dutrochet, the reactions and movements of living things always depend on an inner, vital cause, whereas the physical cause is only the occasion for a response from the plant. He further generalises this observation on the root's movement to that of the leaf, and to all movements in which plant sensitivity intervenes (126ff.). Unlike the interaction between a physical body and the cause of its movement, there is no direct and immediate causal relationship between a stimulus and a plant's reaction; the external stimulus does not directly 'impress' the plant's reaction. Instead, a plant's produces a delayed response, mediated by its irritability on the occasion of a physical stimulus. This process corresponds to the very definition of a behaviour (Dretske 1988, 26–27), and some may consider this process sufficient evidence of a proto-mind or minimal cognition (Van Duijn et al. 2006; Thompson 2007). Whereas a stone falls owing to the direct influence of gravity, gravitropism involves an intermediary, the plant's agency, which interprets the external world through the cellular receptors that detect gravity, just as heliotropism involves the interpretation of light through photoreceptors. Dutrochet thus defends a non-reductionist explanatory conception of biology that refuses to reduce the phenomena of life—even plant life—to strictly

[7] I thank Pr. Pierre Meerts for these references.

physical explanations. This does not imply a belief in plant consciousness nor does it make Dutrochet a vitalist, but simply a biologist who recognises a form of agency in plants' behaviour.

To borrow terminology from the philosopher of ethology Vinciane Desprets (2014, 58),[8] we may say that Dutrochet sees plants' behaviour as a *response* to their environment (a 'spontaneity') where the strict mechanicist or behaviourist would simply see a physical *reaction*. A reaction, strictly understood, does not allow for error, whereas a response, by requiring minimal flexibility for information processing, does. Thus some genetically-modified plants have their gravitropic ability completely inhibited, causing a different response to gravitational factors than that in 'normal' plants. In contrast, stones will always react in the same way to gravitational constraints; they do not have theoretical alternatives to falling, or the capacity to respond to their situation. Of course, the 'weightable matter' of plants will experience a fall like any other body, but their sensitivity also makes them capable of interpreting this force and of responding to it in a specific way, through the growth of roots. This form of plant agency is not identified with a form of mind by Dutrochet, however, as he maintains an (artificially) dualistic epistemological framework by postulating that mind ('self-consciousness') is closely dependent on animal sensation.

7.4 Conclusion

One's interpretation of plants' activities thus depends on one's philosophical conception of life; in this respect, contemporary biology remains largely dependent on the history of Western philosophy. For Dutrochet, the study of true sensitivity, understood as the faculty by which we acquire knowledge, concerns the mind in its most elementary manifestation of self-consciousness. It is therefore the province of "psychology"—what we would today call cognitive science[9]:

> It is therefore contrary to sound reason, to good philosophy, to place in a science of observation, such as physiology, that of all the phenomena of nature which is most necessarily removed from our investigations; the study of sensitivity and sensation belongs exclusively to psychology. (Dutrochet 1824, 5)

The cognitive sciences thus study animal faculties, but conventionally (if not arbitrarily) disregard plant activities, which are 'relegated' to physiology. Although this dualistic disposition is a fundamental metaphysical postulate that conditions scientific methodology, it is not itself a scientific result; as a postulate, attempts to demonstrate this disposition tend to beg the question, as can be seen by the argument that Dutrochet himself puts forward. As he explains, if irritability were equivalent to sensation, one would have to admit that the organs and the different irritable parts of

---

[8] Desprets was herself inspired by Eileen Crist's (1999) contrast between the ethological approaches of Darwin and Lorenz.

[9] In 1824, psychology was not an experimental or even empirical science; Aristotle's study of the faculties of the soul, for example, is described as 'psychology'.

a living being (such as its muscles) would each have a consciousness. Considering this conclusion 'inadmissible', Dutrochet thus denies the validity of an equivalence between irritability and sensation, and therefore between plants and animals. Yet this reasoning rests solely on the assumption of a close association between sensation and consciousness—an association that is not necessarily self-evident, and which is itself a position that should have to be justified.[10] Even if one should take for granted the dichotomy between animal sensation and irritability, this distinction is ultimately theoretical without a defined limit between the two and, above all, without set criteria by which one faculty is to be distinguished from the other. Biologists have fuelled debate in their search for such empirical criteria since the nineteenth century, and plant neurobiologists continue this questioning in the twenty-first century. It would seem, however, that the empirical study of plants and animals, although indispensable for the advancement of biology, cannot provide a decisive criterion for a metaphysical postulate on the nature of the mind and its association with certain vital phenomena. Recognition of this fact requires that we accept either position as an epistemological choice or an ethical commitment (concerning the acts we consider methodologically or morally acceptable), but no longer as a scientific pseudo-demonstration. At such a fundamental level, philosophy cannot be excluded from the scientific field, for it is at the field's origin.

References

Baldassarri, F., and A. Blank, eds. 2021. *Vegetative Powers: The Roots of Life in Ancient, Medieval, and Early Modern Natural Philosophy*. Cham: Springer.

Baluška, F., S. Mancuso, and D. Volkmann, eds. 2006. *Communication in Plants: Neuronal Aspects of Plant Life*. Berlin: Springer-Verlag.

Barlow, P.W. 2008. Reflections on 'Plant Neurobiology.' *Bio Systems* 92 (2): 132–147.

Boscowitz, A. 1867. *L'âme de la plante*. Paris: Ducrocq.

Burgat, F. 2020. *Qu'est-ce qu'une plante*. Paris: Seuil.

Calvo, P. 2016. The Philosophy of Plant Neurobiology: A Manifesto. *Synthese* 193 (5): 1323–1343.

Calvo, P., and F. Baluška. 2015. Conditions for Minimal Intelligence Across Eukaryota: A Cognitive Science Perspective. *Frontiers in Psychology* 6: 1329.

Calvo, P., F. Baluška, and A. Trewavas. 2021. Integrated Information as a Possible Basis for Plant Consciousness. *Biochemical and Biophysical Research Communications* 564: 158–165.

Calvo, P., and F. Keijzer. 2011. Plants: Adaptive Behavior, Root-brains, and Minimal Cognition. *Adaptive Behavior* 19 (3): 155–171.

Calvo, P., and A. Trewavas. 2020. Physiology and the (Neuro)biology of Plant Behavior: A Farewell to Arms. *Trends in Plant Science* 25 (3): 214–216. https://doi.org/10.1016/j.tplants.2019.12.016.

Cerrito, A. 2021. Disclosing the Hidden Life of Plants: Theories of the Vegetative Soul in Albert the Great's De vegetabilibus et plantis. In *Vegetative Powers*, eds. Fabrizio Baldassarri and Andreas Blank, 105–122. Cham: Springer.

[10] Several alternate postulates are possible: the idea that the mind is totally independent of sensation; the idea that sensation is expressed according to a gradation and that only certain degrees of sensation are associated with the mind; the idea that the mind emerges from sensation only when certain conditions are met; the idea that the mind is an illusion resulting from an integral physical determinism; etc.

Crist, E. 1999. *Images of Animals*. Philadelphia: Temple University Press.
Cvrčková, F., V. Žárský, and A. Markoš. 2016. Plant Studies May Lead Us to Rethink the Concept of Behavior. *Frontiers in Psychology* 7: 622.
Delaporte, F. 1982. *Nature's Second Kingdom: Explorations of Vegetality in the Eighteenth Century*, Engl. transl. Arthur Goldhammer. Cambridge (MA) and London: MIT Press.
Despret, V. 2014. *Que diraient les animaux, si... on leur posait les bonnes questions?*. Paris: La Découverte.
Dretske, F. 1988. *Explaining Behavior*. Cambridge, MA: MIT Press.
Duhamel du Monceau, H., and C. Dufay. 1736. *Note sur la sensitive*. Paris: Mémoire de l'académie royale des sciences.
Dutrochet, R. 1824. *Recherches anatomiques et physiologiques sur la structure intime des animaux et des végétaux et sur leur motilité*. Paris: Baillière.
Fechner, Gustav Theodor. 1848. *Über das Seelenleben der Pflanzen*. Leipzig: Leopold Voß.
Fenderspiel, M., and M. Cronier. 2018. *Pseudo-Aristote. Du Monde ; Positions et dénominations des vents; Des plantes*. Paris: Les Belles Lettres.
Firn, R. 2004. Plant Intelligence: An Alternative Point of View. *Annals of Botany* 93: 345–351.
Gagliano, M., M. Renton, M. Depczynski, and S. Mancuso. 2014. Experience Teaches Plants to Learn Faster and Forget Slower in Environments Where it Matters. *Oecologia* 175 (1): 63–72.
Gagliano, M., V.V. Vyazovskiy, A.A. Borbély, M. Grimonprez, and M. Depczynski. 2016. Learning by Association in Plants. *Scientific Reports* 6: 38427.
Gavray, M. 2019. En quoi l'intelligible est-il (un) vivant? *Les Cahiers Philosophiques De Strasbourg* 45: 71–89.
Greene, E.L. 1983. *Landmarks of Botanical History II*. Stanford: Stanford University Press.
Guerra, S., A. Peressotti, F. Peressotti, et al. 2019. Flexible Control of Movement in Plants. *Scientific Reports* 9: 16570.
Hall, M. 2011. *Plants as Persons: A Philosophical Botany*. Albany, NY: State University of New York Press.
Hiernaux, Q. 2019. History and Epistemology of Plant Behaviour: A Pluralistic View? *Synthese* 198 (4): 3625–3650.
Hiernaux, Q. 2023. *From Plant Behavior to Plant Intelligence?*, 1st ed. 2020. *Du comportement végétal à l'intelligence des plantes?* Versailles: Quae.
Hiernaux, Q. 2021a. The Ethics of Plant Flourishing and Agricultural Ethics: Theoretical Distinctions and Concrete Recommendations in Light of the Environmental Crisis. *Philosophies* 6 (4): 91.
Hiernaux, Q. 2021b. Differentiating Behavior, Cognition, and Consciousness in Plants. *Journal of Consciousness Studies* 28 (1–2): 126–135.
Hiernaux, Q., and C. Tresnie. 2023a. Aristotelian Metaphysics of the Vegetative Soul and Early Modern Plant Physiology: Comparison between Plant Functions in Aristotle, Pseudo-Aristotle, and Cesalpino. In *Andrea Cesalpino: An Aristotelian Natural Philosopher in the Renaissance*, eds. Fabrizio Baldassarri and Craig Martin. London: Bloomsbury.
Hiernaux, Q., and C. Tresnie. 2023b. *Andrea Cesalpino's De Plantis Libri XVI (1583) and the Transformation of Medical Botany in the 16th century, Edition, Translation, and Commentary on Book I*. Boston-Berlin: De Gruyter.
Ingensiep, H.W. 2001. *Geschichte der Pflanzenseele*. Stuttgart: Alfred Kröner Verlag.
Kallhoff, A., M. Schörgenhumer, and M. Di Paola, eds. 2018. *Plant Ethics, Concepts and Applications*. London and New York: Routledge.
Kelly, C.K. 1992. Resource choice in *Cuscuta europea*. *Proceedings of the National Academy of Sciences* 89: 12194–12197.
de Lamarck, J.-B.C. 1815. *Histoire naturelle des animaux sans vertèbres*. Paris: Verdière.
Lovejoy, A. 1964 [1936]. *The Great Chain of Being: A Study of the History of an Idea*. Cambridge, Massachusetts: Harvard University Press.
Maher, C. 2017. *Plant Minds: A Philosophical Defense*. New York: Routledge.

Mallatt, J., M.R. Blatt, A. Draguhn, et al. 2021. Debunking a Myth: Plant Consciousness. *Protoplasma* 258: 459–476.
Mancuso, S., and A. Viola. 2015. *Brilliant Green: The Surprising History and Science of Plant Intelligence*, trans. J. Benham. Washington, DC: Island Press.
Marder, M. 2013. *Plant-thinking a Philosophy of Vegetal Life*. New York: Columbia University Press.
Morton, A.G. 1981. *History of Botanical Science: An Account of the Development of Botany from Ancient Times to the Present Day*. London: Academic Press Inc.
Panarelli, M., 2020. Albert the Great's *De vegetabilibus* and its Unique Position Among the Medieval Commentaries on *De plantis*. In *Peri Phyton. Greek Botanical Treatises in the West and the East*, eds. Maria Fernanda Ferrini and Guido Giglioni, 137–162. Macerata: edizioni Università di Macerata.
Raja, V., P.L. Silva, R. Holghoomi, et al. 2020. The Dynamics of Plant Nutation. *Scientific Reports* 10: 19465.
Sallé, G., and M. Aber. 1986. Les phanérogames parasites: biologie et stratégies de lutte. *Bulletin de la Société Botanique de France. Lettres Botaniques* 133 (3): 235–263.
Simonetti, V., M. Bulgheroni, S. Guerra, et al. 2021. Can Plants Move Like Animals? A Three-Dimensional Stereovision Analysis of Movement in Plants. *Animals* 11: 1854.
Struik, P.C., X. Yin, and H. Meinke. 2008. Plant Neurobiology and Green Plant Intelligence: Science, Metaphors and Nonsense. *Journal of the Science of Food and Agriculture* 88: 363–370.
Taiz, L., D. Alkon, A. Draguhn, A. Murphy, M. Blatt, C. Hawes, G. Thiel, and D.G. Robinson. 2019. Plants Neither Possess nor Require Consciousness. *Trends in Plant Science* 24 (8): 677–687.
Thompson, E. 2007. *Mind in Life*. Belknap: Cambridge, MA.
Trewavas, A. 2002. Mindless Mastery. *Nature* 415 (6874): 841.
Trewavas, A. 2003. Aspects of Plant Intelligence. *Annals of Botany* 92: 1–20.
Trewavas, A. 2004. Aspects of Plant Intelligence: An Answer to Firn. *Annals of Botany* 93: 353–357.
Trewavas, A. 2005. Green Plants as Intelligent Organisms. *Trends in Plant Science* 10 (9): 413–419.
Trewavas, A. 2014. *Plant Behaviour and Intelligence*. Oxford: Oxford University Press.
Van Duijn, M., F. Keijzer, and D. Franken. 2006. Principles of Minimal Cognition: Casting Cognition as Sensorimotor Coordination. *International Society for Adaptive Behavior* 14 (2): 157–170.
Williams, C.N. 1963. Development of *Tapinanthus bangwensis* (Engler and Krause) Danser and Contact with the Host. *Annals of Botany* 27 (108): 641–646.
Zatta, C. 2019. *Interconnectedness: The Living World of the Early Greek Philosophers. International Pre-Platonic Studies*, vol. 7. Sankt Augustin: Academia Verlag.

Chapter 8
Different Kinds of Minds: From Bacteria to Plants and Animals

Andrea Nani and Gabriele Volpara

Abstract Exciting times characterize the scientific panorama, in which multidisciplinary approaches (neuroscientific, mathematical, philosophical, psychological, etc.) are used to tackle difficult topics, such as the nature and origin of consciousness, as well as to study the living world under new perspectives. In particular, the field of plant biology has seen interesting developments in the last few years. It is not unusual now to find terms like *mental*, *cognitive,* or *intelligent* used to refer to the internal and external behaviors of plants, and to debate as to what degree plants can be considered sentient beings. Plants seem to show sophisticated mechanisms of interaction at different biological levels, which have been compared to similar mechanisms showed by animals. This chapter provides a review of these mechanisms and discuss whether they can endow the vegetal world of the capacity to perform cognitive operations in order to create adequate responses to environment and social dynamics within plant populations. Particular attention is given to the systems of communication within a singular vegetal organism, or between plants of the same or of different vegetal and animal species. Despite the prodigious adaptations they can develop, plants do not seem to have the representational minds that animals have; they nonetheless show complex behavioral patterns and form synergistic networks with other organisms (such as bacteria, fungi and animals), which may support the hypothesis that they might constitute *distributed minds* as well as the need to redefine the concept of *cognition.*

> [...]
> We wouldn't run out of topics
> for so much connects us.
> The same star keeps us in reach.

A. Nani (✉)
Department of Psychology, University of Turin, Turin, Italy
e-mail: andrea.nani@unito.it

G. Volpara
ICoN Cognitive Neuroscience Center, Institute for Advanced Studies IUSS, Pavia, Italy

M. Di Paola (ed.), *The Vegetal Turn*, The International Library of Environmental, Agricultural and Food Ethics 38, https://doi.org/10.1007/978-3-031-68671-9_8

We cast shadows according to the same laws.
Both of us at least try to know something,
each in our own way,
and even in what we don't know
there lies a resemblance.
[…]
Wisława Szymborska, *The Silence of Plants*, translated by Joanna Trzeciak (2002)

8.1 Introduction

In the current scientific panorama, an increasing number of studies have been exploring the cognitive and behavioral characteristics of different living organisms, from bacteria to plants and animals, especially with regard to their capacities to perform complex cognitive processes, including those supporting consciousness and self-awareness (Volpara et al. 2022; Nani et al. 2021; Segundo-Ortin and Calvo 2021; Trewavas 2014). The neural mechanisms of cognition have been widely investigated both from the point of view of a single organism (Branco and Regrave 2020; Kempermann 2019; Zhou et al. 2019) and from the point of view of groups of the same species within the same environment (Dwortz et al. 2022; Ferreira-Fernandes and Peça 2022; Olsson et al. 2020).

With regard to the plant kingdom, studies about plant behavior and signaling (sometimes improperly called "Plant Neurobiology"), which have their root in the first experiments performed in the first half of the 18th century by De Mairan (1729), are currently stirring a lively debate among biologists, philosophers, psychologists, and neuroscientists (Brenner et al. 2006; Segundo-Ortin and Calvo 2021). Indeed, there are very different theoretical positions regarding how plants communicate, behave and appear to possess a form of sentience and/or consciousness and a level of sensitivity and adaptiveness (Nani et al. 2021). We have in fact discovered that subtle communication can occur between plants of the same species (Kumar et al. 2020) as well as between plants of different species (Bonato et al. 2021). Plants can cooperate but also fiercely vie for natural resources, as it is clearly observed in the "parasitic" behaviors shown by *Cuscuta pentagona* (Alakonya et al. 2012). Based on these complex relationships, some theoretical proposals claim that plants possess "intentions" and the ability to "make decisions" (Trewavas 2014), and are even endowed of some sort of conscious sentience (Calvo 2017; Calvo et al. 2017).

In this chapter, we discuss the complex web of relations between living organisms not only within the plant kingdom but also across kingdoms. Our analysis will especially focus on cognition. We will explore different cognitive domains, which may characterize different kinds of minds.

8.2 The Multisensory Processing of Plants

Plants are autotrophic organisms, which is to say that they can feed on inorganic substances by transforming them into assimilable organic compounds, a unique capacity that animals do not have. The extremely complex process of photosynthesis makes plants extremely apt to colonize almost every spot of the Earth surface. They do not rely on animal organisms to survive; while, in contrast, animals could not survive without plants (Baluska and Mancuso 2020). Another crucial difference between animals and plants lies in their structural organization. Animals have a central structure (the nervous system) dedicated to sense the world and act accordingly. Plants have mostly a distributed organization, in which none of its parts is strictly essential to the functioning of the whole. For example, the synthesis of hormones does not occur in specific glands (as it is the case of animals), but in unspecialized tissues such as apices and roots, which are also important for communication with neighboring plants (Baluska et al. 2010). Moreover, the quantity and types of hormones are significantly less than in animals and, in some cases, different concentrations of the same hormone can have different effects on the same organ (Li et al. 2017).

In plants, the transfer of information occurs through electrical, chemical, and hydraulic signals operating in combination with each other. The system that uses electrical transmission, which has been compared to animal nerves, is useful for communicating quickly over both short and long distances. Multiple types of electrical activations can be used: action potentials, potential variations, or system potentials (Zimmermann and Mithöfer 2013).

The biochemical system, on the other hand, is essential for the construction of nutrients (Blankenship 2021). The walls of plant cells have physical and chemical properties that allow communication through various molecules. In particular, cell walls maintain a proper continuum between plant structure and plant function (plasma membrane, cytoskeleton, and mechano-sensitivity) by transducing mechanical signals (Rodakowska et al. 2009).

As for the hydraulic system, studies carried out with positron emission tomography have observed real-time modulations of components necessary for plant life (Hubeau and Steppe 2015) through two vascular systems, one that allows the ascending flow of water and mineral salts from roots to leaves ("xylem") and another that allows the descending flow of nutrients from leaves to all the other parts of the plant ("phloem"). The first absorbs substances from the soil and transports them to the centers of the photosynthesis; the second distributes the products of the photosynthesis.

Plants show a great variety of responses to environmental stimuli. They sense surroundings and direct their growth toward light sources. This phenomenon, called phototropism, involves the production and transmission of auxin, a class of plant hormones or plant-growth regulators (Liscum et al. 2014). This sensitivity to light, especially on the leaves, makes plants orient toward light as well as heat sources (Buda et al. 2003; Gagliano et al. 2016). Light then triggers the regulation of enzymatic activity and the plant's internal clocks, which establish the circadian rhythm (photoperiodism) and the longer metabolic periods based on the seasonal cycle (Queiroz and Morel 1974). Plants are less static than one might think; they develop, seeking out light sources and growing in those directions with their leaves, and with their roots they grow in the areas of the soil that have the greatest amounts of salts and the nutrients they need (Ferreira et al. 2017; Muthert et al. 2020).

It is well known that plants emit odors that animals can perceive in order to influence their behavior. Furthermore, chemical substances can be detected by plants and used by them to start internal reactions (Mancuso and Viola 2015). Communicative exchanges between plants and animals occur through biogenic volatile organic compounds (BVOCs) (Bonato et al. 2021); plants therefore might be thought of as possessing a sort of *chemical sense of smell.*

It is much less intuitive the attribution of touch and taste to plants. Broadly speaking, if we conceive these senses not in strictly animal terms, plants might present some sort of touching and tasting behaviors (Mancuso and Viola 2015). For example, climbing and parasitic plants probe the environment and adhere to the surfaces that can better sustain their growth. Moreover, plants can explore the soil or the surfaces of objects and walls to arrange their leaves in such a way that they can receive as much light as possible. *Hedera elix* can move along the walls of houses by reducing the maximum relative temperatures and humidity (Sternberg et al. 2011). Roots and leaves absorb and produce nutrients, while carnivorous plants attract, trap and digest their prey for the nutrients they contain (Freund et al. 2022; Greenwood et al. 2011; Volkov et al. 2008). The ascidia, in particular, are very sensitive to the touches of their prey, which rapidly activate the trap. In a sense, carnivorous plants can be considered as having a sort of taste for certain organisms; they mostly consume insects, but some larger plants are capable of digesting reptiles and small mammals. Smaller carnivorous plants are specialized in eating single-celled organisms (such as bacteria and protozoa) or aquatic preys such as crustaceans, mosquito larvae and small fish (Mancuso & Viola 2015). Since touch is the fundamental sensorium for the construction of agency and a sense of self in animals, the tactile behavior of plants might similarly be the basis for a very rudimentary form of self and agency.

Although plants do not possess an auditory system, they can grow, such as the roots of *Arabidopsis thaliana*, toward a sound source (200 Hz), showing preference for the direction of ion fluxes (Rodrigo-Moreno et al. 2017). This behavior is referred to as "phonotropism", but the recent scientific field of phytoacoustics reveals that plants are capable of much more (Frongia et al. 2020). For example, with the help of small, highly sensitive sound receivers, it has been found that plants can emit sound from the xylem (Jung et al. 2018) and even feeble ultrasound waves when stressed (Khait et al. 2018). Plants can perceive caterpillar's chewing and prepare appropriate chemical defenses (Appel & Cocroft 2014); they can also "hear" the sounds emitted by the pollinators' wings and increase within minutes the sugar concentration in their nectar (Veits et al. 2019).

The list of tropisms of plants is long, including thigmotropism (oriented growth in response to mechanical contact), gravitropism or geotropism (oriented growth in response to gravity), hydrotropism (oriented growth in response to water), halotropism (oriented growth in response to soil salinity), thermotropism (oriented growth in response to soil temperature), chemotropism (oriented growth or movement in response to a chemical stimulus), electrotropism (oriented growth in response to electric currents), magnetotropism (oriented growth in response to magnetic fields) and oxytropism (oriented growth in response to oxygen concentrations) (Ferreira et al. 2017; Muthert et al. 2020). This variety of behaviors demonstrates that plants are capable of very complex internal processes, which can analyze the environment in order to appropriately respond to it.

This evidence suggests that plants may have a form of cognition based on the processing of information regarding both their internal mechanisms of chemical regulation and the implementation of responses to the environment and to other organisms (Blankenship 2021; Zimmermann and Mithöfer 2013; Rodakowska et al. 2009). Furthermore, although the cognitive processes of plants tend to be distributed rather than embedded in specific processing systems, as is the case with animals, they are perfectly able to respond to the environment in a structured way with long-term strategies (Novoplansky 2016).

8.3 Communication Among Plants: Roots, Branches and "Dialects"

In the rhizosphere (Gupta and Sharma 2021), multiple relationships between plants are organized through communication occurring via roots and chemical signals. Analogously to the "world wide web", plants are supposed to take part in a sort of "wood wide web" (Sen 2000). Within this wide network, numerous fungi, bacteria, and plants live in mutual relationship. In the symbiotic association of mycorrhiza, fungi form sophisticated "meta-community hubs" (Toju et al. 2018), which constitute important nodes in the wood wide web by allowing faster transfers of information between roots of different plants, thus influencing rooting depth, rates of photosynthesis, germination, and gene regulation of defense chemistry (Simard 2018). Within the web, plants can also transfer water among intact conspecifics that are in need of nutrients, as it has been observed in trees of *Agathis australis* (Bader and Leuzinger 2019). These findings have significant implications for the hypothesis that plants, along with other organisms, form superorganisms that depend on complex ways of communication (Bader and Leuzinger 2019).

Even within these trans-kingdoms' assemblages, the roots of individual plants typically develop without touching each another, so that individual plants appear to have a peri-personal space, which may rely on a sort of self-recognition (Mahall and Callaway 1996). Plants undoubtedly recognize the presence of familiar or unfamiliar plants nearby: root growth is in fact increased when they are next to unfamiliar plants. This phenomenon has been observed in *Glycine max*, which produces fewer roots and more pods when it lives in the same pot with other plants (Gersani et al. 2001). On the other hand, seeds of different parent plants of *Cakile edentula* will develop many roots, trying to take possession of as much territory as possible in order to not run out of nutrients; whereas the seeds derived from the same mother plant will develop fewer roots and privilege aerial growth (Dudley and File 2007; Gersani et al. 2001). These physiological mechanisms are supposed to prevent "inbreeding" in plants (Haring et al. 1990).

These phenomena seem to support not only the existence of a primary form of biological plant cognition (Keijzer 2017), but also some capacity for identity discrimination (self-other/familiar-unfamiliar). Sensitivity to other plants also affects the airspace, particularly when it comes to branches and fronds, whose competition and distribution for light appear to be regulated by the emission of BVOCs (Bonato et al. 2021; Kegge and Pierik 2010). Although many trees intertwine their branches, others do not, as if exhibiting some sort of "frond shyness" (Hallé 2005): this is especially evident in the *Pinaceae*, which share needle-like leaves, and in the *Fagaceae* such as *Fagus*, *Quercus*, and *Castanea*.

Plants can also warn each other of dangers, for example by emitting chemical signals when harmed by herbivores (Moreira et al. 2016). Indeed, there is a variety of chemical components that are used to signal danger, depending on the species and where the plants grow; these may be seen as constituting different "dialects" (Bonato et al. 2021). So different plant populations, closer or further away from the one emitting warning signals, will respond differently. After being alerted by conspecifics, plants of *Artemisia tridentata* which are close to the signal respond better to insect damage than distant ones (Heil and Karban 2010; Karban et al. 2006).

8.4 Communication Between Plants and Animals: Food, Reproduction, and "Alliances"

The relationship between plants and animals is especially associated with the need for plant reproduction on the one hand, and the animal need for nourishment on the other. For this purpose, a variety of BVOCs signals is used in plant-animal communication. These substances can contain nitrogen, as well as aromatic compounds, and are specific to the animal and plant that communicate (Bonato et al. 2021; Kumar et al. 2020). Terpenoids, a generic name for the oxygenated derivatives of terpenes (alcohols, aldehydes, carotenoids, ketones), are emitted by plants in response to both internal (genetic and biochemical) and external (ecological) factors. They also appear to have dose-dependent effects and vary both qualitatively and quantitatively at the intraspecific level (Langenheim 1994). This characteristic has been widely observed in the relationship between orchid species and the pollinator euglossine bee: more than 50 different volatile compounds emitted by orchids have been found to play a role in this relationship (Dodson 1969).

Apart from the wind, which is one of the main vectors for seed and pollen transport, some plants require entomophilous pollination provided by insects such as bees, wasps, and ants, or a zoogamous pollination, provided by animals such as reptiles, rodents, birds, fishes and bats (Anderson et al. 2011; Ratto et al. 2018; Simon et al. 2011). The relationship is so strict that the extinction or a survival threat to the pollinator would dramatically endanger the plant, as it is the case for bats in certain parts of Oceania, where the bats are the only existing vector for *Pteropus natalis* (Todd et al. 2022).

Communication between plants and their animal pollinators occurs by means of colors, flower shapes, and odors. The variety and combination of these elements exhibited by plants is astonishing; they can reflect light in different wavelengths and emit an incredible range of fragrances and odors, from pleasant aromas to disgusting ones. A telltale example is offered by *Amorphophallus titanium* and its episodic, large, gorgeous inflorescence that attracts flies with an odor similar to that of rotting meat (Shirasu et al. 2010).

With regard to shape, some plants get as far as to mimic the aspect of stones, such as *Blossfeldia liliputana* and *Lithops aucampiae* (Mancuso 2018), while others can resemble other plants, such as *Boquila trifoliolata* (Gianoli & Carrasco-Urra 2014), or (weirdly) even human lips, such as *Psychotria elata.* Orchids can also imitate the shape and behavior of animals: for example, *Drakaea glyptodon* and *Ophrys apifera* mimic the shape and pheromones of female wasp *Zaspilothynnus trilobatus* and of the hymenopteron *Eucera longicornis*, respectively (Bohman et al. 2016).

Opposite versions of the same techniques are used in self-defense. Plants like *Acacia drepanolobium* release substances that make their leaves unpalatable or even poisonous for animals (Robbins et al. 1987). Of note, this process can affect a few leaves, thus demonstrating high specificity in the self-monitoring system of plants (Walters 2011). Some plants, such as the *Phaseolus lunatus*, are even able to attract with certain substances the predators or parasites of the animal that feed on them (Dicke et al. 1999). The Delprim maize, in particular, can attract entomopathogenic nematodes (*Heterorhabditis megidis*) that eat the larvae of the beetle *Diabrotica virgifera virgifera*, which usually attack its roots (Rasmann et al. 2005).

8.5 Discussion

Plants have been populating the Earth for millions of years, and without them no animal life would be possible (Baluska and Mancuso 2020). As we have seen, plants are capable of an incredible variety of biological processes, evolved in intimate relationship with other organisms, both Eukariota (Animalia, Fungi, Piantae, Chromista, Protista) and Prokaryota (Bacteria, Archea) (Cavalier-Smith 1998).

We have already seen how strictly intertwined are the relations between animals and fungi and plants. Similarly, plants form with bacteria complex synergies. Bacteria constitute the most successful and adaptive life form on Earth (Dennett 2017). Bacteria create intimate synergies with both plants and animals, so intimate that the partners cannot exist without the other. They also influence the behavior and the responses to the environmental stimuli of plants and animals. Bacteria interact with plants by means of molecular compounds which are important for plant growth, health and stress resilience, as they can modulate gene expression and metabolism (Compant et al. 2019; Ortíz-Castro et al. 2009). In turn, bacterial interactions with animals are even subtler, as microorganisms can alter the host's nervous system and modify its behavior by using molecular mechanisms (Masuzzo et al. 2020; Rohrscheib and Brownlie 2013).

Some bacteria live in strong symbiosis with certain organisms, so much so that we can consider the two organisms as parts of an integrated system. For example, *Bradyrhizobium japonicum* receives nitrogen compounds from the *Glycine max* (soybean) and, in turn, provides nutritive substances to its mutualist (Egamberdieva et al. 2018). Bacteria can even change the environment by altering its chemical balance, thus producing their own affordances (Dennett 2017). To do so, they are able of some sort of collective computation, which has been defined as "quorum sensing"; in other words, they can regulate their gene expressions in response to cell density and chemical gradients (Miller and Bassler 2001). This ability is a sophisticated collective behavior; in fact, the dynamics of quorum sensing follow parameters that are significant for pattern formation (Majumdar et al. 2022). From the evolutionary point of view, it has been hypothesized that quorum sensing may have favor the development of multicellular organisms (Miller and Bassler 2001). It could therefore be claimed that these simple life forms operate like computer algorithms, as they analyze certain inputs in order to produce certain outputs. These simple operations may be considered as proto-cognition.

Furthermore, plants have numerous channels of perception, as they are highly sensitive to oxygen levels, chemicals, gravity, light, water, salinity, and temperature (Mancuso and Viola 2015; Mancuso 2018; Muthert et al. 2020). The efficiency of this functional organization is maximized by the synergy between the networks of roots and of fungi and the communities of bacteria (Toju et al. 2018; Sen 2000). A system that allows for a better use of resources (Bader and Leuzinger 2019), for creating suitable responses based on previous experience (Simard 2018), for responding to dangers by emitting warning signals and producing protective chemical substances (Robbins et al. 1987), for influencing animals' behavior for reproductive purposes (Ratto et al. 2018), for interacting with the environment in articulate ways with long-term strategies (Novoplansky 2016) can be considered intelligent (Trewavas 2014) and endowed with particular forms of cognition (Keijzer 2017). The intimate collaboration and synergies with other organisms (bacteria, fungi, insects, and animals) allows for cognitive-distributed operations (e.g., decision making, learning, signaling) that the single organisms would not be able to realize in isolation (Whiten et al. 2021), such as it happens with insect colonies, which can be viewed as single superorganisms (Sasaki and Pratt 2018).

However, these distributed functional operations are not sufficient to create the mental representations that we are used to consider the ingredients of animal minds. There are in fact no representations in plant minds (Maher 2017). Still, the functional operations described above compel at least a redefinition of the concept of "cognition". In fact, synergies between microorganisms and plants and animals, which are necessary for their reciprocal survival, can be considered to be on a cognitive continuum, ranging from the operations carried out by the simplest living creatures, such as single-celled eukaryotes, to complex living structures, like plants and ourselves. To better understand these relations, we can either employ new categories, that have not yet been developed, or we can expand the extant conceptual tools (Nani et al. 2021). If the hypothesis of the cognitive continuum is true, we think the second option is more reasonable and in line with other proposals suggesting that all living organisms may have a form of reasoning (Lyon 2006; Lyon and Keijzer 2007).

In this sense, cognition could be conceived broadly as any elaboration of information implemented by a functionally organized substrate (be it natural or artificial). Furthermore, within the category of *cognition* it is fruitful to distinguish between unconscious and conscious cognitive elaborations, with the former characterized by a much wider set of possible operations than the latter. As a consequence, a mind can be formed by many different cognitive processes; plausibly, there are minds made of solely unconscious processing and other minds made of both unconscious and conscious elaborations.

Thus, with the help of this perspective turn regarding the concept of *cognition* and with the adoption of the conceptual framework that we have proposed in Nani et al. (2021), we could characterize all functions and behaviors of living organisms along three dimensions: adaptiveness, sensitivity, and sentience. Every living creature, in fact, can be described dynamically along these three coordinates.

The dimension of adaptiveness refers to the level of flexibility that organisms have in adapting to the environment and its changes by varying metabolic, physiological, and behavioral processes. The dimension of sensitivity refers to the capacity that organisms manifest in responding adequately to environmental stimuli through different sensorial structures. These two dimensions characterize different aspect of living beings, as the former describes how well an organism can adapt to environmental changes, while the latter describes the range of sensible interactions between an organism and its ecological niche. It is therefore apparent that the two dimensions are independent, as an organism can have a high level of adaptiveness but a low level of sensitivity (a perfect adaptation to a narrow environment, like a thermophile), or vice versa (like a koala or a panda, which, despite their good level of sensitivity, are dependent on a strict diet: eucalyptus leaves or bamboo branches, respectively). Finally, the dimension of sentience describes the level of perception that organisms feel and requires an interplay of complex feedforward and feedback signals.

Sensitivity and sentience differ because the former refers to the number of sensory channels and their more or less developed receptiveness, while the latter refers to the capacity of self-monitoring perception, which is fundamental in order to avoid what can perturb the integrity of the organism. This capacity is heavily based on feedback processes, which have been developed and refined along the cognitive continuum that we have discussed before. The development goes from the unconscious elaboration of information to consciousness and self-consciousness. The dimension of sentience, therefore, includes unconscious processes (low levels of sentience) and conscious ones (high levels of sentience). Associating sentience with a cognitive continuum that moves from unconscious to conscious processing may be considered as odd, as the property of being sentient is usually regarded as a synonym of being conscious. But if we are inclined to expand the meaning of *cognition*, we should consider a wise move to expand the meaning of *sentience* as well, so much so to encompass within it not only conscious perceptions but also unconscious ones and their corresponding cognitive products. Conceiving sentience this way is the best option to account for the complex and large variety of information processing and behaviors in living organisms, as well as to appreciate adequately subtle distinctions in biological processes.

Plants have undoubtedly highly complex physiological structures, which allow them to reach good levels of adaptiveness, sensitivity, and sentience (Calvo and Friston 2017). Compared to animals, however, the range of their behaviors and cognitive operations is quite limited. In fact, even the most astounding plant behaviors, as those discussed above, cannot bear comparison with the impressive variety of actions and conducts exhibited by animals, nor can plant communications be compared with the dynamical and complex systems of communications developed by animals. This is so because plants and animals have different biological needs, which require different survival strategies based on different cognitive processes. And these different sets of cognitive elaborations characterize different kinds of minds. Animal minds are profoundly oriented to a wide range of movements, while plant minds are principally formed to operate with a minimum of movement. As we have seen, another radical difference is that animals must get nutrients from other organisms, whereas plants can produce their own nourishment. All these aspects are reflected in the functional connectivity of their internal systems. The embodied and centered cognition of animals, therefore, is deeply different from the distributed cognition of plants. They simply have different kinds of minds, which can be adequately appreciated within the framework of adaptive, sensitive, and sentient coordinates.

8.6 Conclusion

In this chapter, we have explored the hypothesis of different kinds of minds according to the nature of their cognitive processing, with a special focus on plants. We have proposed that all organisms can be fruitfully described according to three coordinates: adaptiveness, sensitivity, and sentience. Plants are complex living beings that exhibit highly adaptive abilities, sensitivity to the environment, and sentience (Nani et al. 2021). The complexity of the relationship between plants and the environment as well as between plants and other organisms (bacteria, fungi, and animals), enhance the power of their functional organization. These complex relationships create the affordances for the emergence of distributed minds, which are capable of different forms of cognition as well as dynamical behavioral patterns. In fact, plant synergies make possible to organize and elaborate information in a highly distributed manner, so as to allocate the cognitive effort among different organisms. Furthermore, most cognitive elaborations are shared with (if not demanded to) other organisms with which plants establish complex synergies. Unlike animals, the highly distributed functional organization of plants, in which none of their parts is strictly essential to the functioning of the whole, make plant unable to construct internal representations of the external world (Maher 2017). On the one hand, this cognitive distributed organization is a biological advantage for resident organisms, but, on the other, prevents them from developing a conscious sentience characterized by psychological and phenomenological variety. Plants appear therefore to be sentient but unconscious parts of trans-kingdoms distributed minds.

References

Alakonya, Amos, Ravi Kumar, Daniel Koenig, Seisuke Kimura, Brad Townsley, Steven Runo, Helena M. Garces, Julie Kang, Andrea Yanez, Rakefet David-Schwartz, Jesse Machuka, and Neelima Sinha. 2012. Interspecific RNA Interference of SHOOT MERISTEMLESS-Like Disrupts Cuscuta pentagona Plant Parasitism. *The Plant Cell* 24 (7): 3153–3166. https://doi.org/10.1105/tpc.112.099994.

Anderson, Jill T., Tim Nuttle, Joe S. Saldaña Rojas, Thomas H. Pendergast, and Alexander S. Flecker. 2011. Extremely Long-distance Seed Dispersal by an Overfished Amazonian Frugivore. *Proceedings of the Royal Society B: Biological Sciences* 278 (1723): 3329–3335.https://doi.org/10.1098/rspb.2011.0155.

Appel, Heidi M., and Reginald B. Cocroft. 2014. Plants Respond to Leaf Vibrations Caused by Insect Herbivore Chewing. *Oecologia* 175 (4): 1257–1266. https://doi.org/10.1007/s00442-014-2995-6.

Bader, Martin, K.-F., and Sebastian Leuzinger. 2019. Hydraulic Coupling of a Leafless Kauri Tree Remnant to Conspecific Hosts. *iScience* 19: 1238–1247. https://doi.org/10.1016/j.isci.2019.05.009.

Baluška, Franstišek, and Stefano Mancuso. 2020. Plants, Climate and Humans: Plant Intelligence Changes Everything. *EMBO reports* 21 (3): e50109. https://doi.org/10.15252/embr.202050109.

Baluška, Franstišek, Stefano Mancuso, Dieter Volkmann, and Peter W. Barlow. 2010. Root Apex Transition Zone: A Signalling–response Nexus in the Root. *Trends in Plant Science* 15 (7): 402–408. https://doi.org/10.1016/j.tplants.2010.04.007.

Blankenship, Robert E. 2021. *Molecular Mechanisms of Photosynthesis*. Hoboken: John Wiley and Sons.

Bohman, Björn., Gavin R. Flematti, Russell A. Barrow, Eran Pichersky, and Rod Peakall. 2016. Pollination by Sexual Deception—It Takes Chemistry to Work. *Current Opinion in Plant Biology* 32: 37–46. https://doi.org/10.1016/j.pbi.2016.06.004.

Bonato, Bianca, Francesca Perressotti, Silvia Guerra, Qiuran Wang, and Umberto Castiello. 2021. Cracking the Code: A Comparative Approach to Plant Communication. *Communicative and Integration Bilogy* 14 (1): 176–185. https://doi.org/10.1080/19420889.2021.1956719.

Branco, Tiago, and Peter Redgrave. 2020. The Neural Basis of Escape Behavior in Vertebrates. *Annual Review of Neuroscience* 43 (1): 417–439. https://doi.org/10.1146/annurev-neuro-100219-122527.

Brenner, Eric D., Rainer Stahlberg, Stefano Mancuso, Jorge Vivanco, František Baluška, and Elizabeth Van Volkenburgh. 2006. Plant Neurobiology: An Integrated View of Plant Signaling. *Trends in Plant Science* 11 (8): 413–419. https://doi.org/10.1016/j.tplants.2006.06.009.

Buda, Anieszka, Tadeusz Zawadzki, Maciej Krupa, Maria Stolarz, and Witold Okulski. 2003. Daily and Infradian Rhythms of Circumnutation Intensity in Helianthus Annuus. *Physiologia Plantarum* 119: 582–589. https://doi.org/10.1046/j.1399-3054.2003.00198.x.

Calvo, Paco. 2017. What is it Like to be a Plant? *Journal of Consciousness Studies* 24 (9–10): 205–227.

Calvo, Paco, and Karl Friston. 2017. Predicting Green: Really Radical (Plant) Predictive Processing. *Journal of the Royal Society Interface* 14 (131): 20170096. https://doi.org/10.1098/rsif.2017.0096.

Calvo, Paco, Vaidurya P. Sahi, and Anthony Trewavas. 2017. Are Plants Sentient? *Plant, Cell & Environment* 40 (11): 2858–2869. https://doi.org/10.1111/pce.13065.

Cavalier-Smith, Thomas. 1998. A Revised Six-kingdom System of Life. *Biological Reviews* 73 (3): 203–266. https://doi.org/10.1017/S0006323198005167.

Compant, Stéphane., Abdul Samad, Hanna Faist, and Angela Sessitsch. 2019. A Review on the Plant Microbiome: Ecology, Functions, and Emerging Trends in Microbial Application. *Journal of Advanced Research* 19: 29–37. https://doi.org/10.1016/j.jare.2019.03.004.

De Mairan, Jean J. O. 1729. Observation Botanique, Histoire de l'Academie royale avec les mémoires de mathématique et de physique tirés des registres de cette. *Académie* 31: 35–36.

Dennett, Daniel C. 2017. *From Bacteria to Bach and Back: The Evolution of Minds*. New York: WW Norton and Company.

Dicke, Marcel, Rieta Gols, Daniel Ludeking, and Maarten A. Posthumus. 1999. Jasmonic Acid and Herbivory Differentially Induce Carnivore-attracting Plant Volatiles in Lima Bean Plants. *Journal of Chemical Ecology* 25 (8): 1907–1922. https://doi.org/10.1023/A:1020942102181.

Dodson, Calaway H., Robert L. Dressler, Harold G. Hills, Ralph M. Adams, and Norris H. Williams. 1969. Biologically Active Compounds in Orchid Fragrances: Function of Natural Plant Products in Orchid Flower Odors and the Attraction of Specific Pollinators are Described. *Science 164* (3885): 1243–1249. https://doi.org/10.1126/science.164.3885.1243.

Dudley, Susan A., and Amanda L. File. 2007. Kin Recognition in an Annual Plant. *Biology Letters* 3 (4): 435–438. https://doi.org/10.1098/rsbl.2007.0232.

Dwortz, Madeleine F., James P. Curley, Kay M. Tye, and Nancy Padilla-Coreano. 2022. Neural Systems that Facilitate the Representation of Social Rank. *Philosophical Transactions of the Royal Society B* 377 (20200444): 1–15. https://doi.org/10.1098/rstb.2020.0444.

Egamberdieva, Dilfuza, Dilfuza Jabborova, Stephan J. Wirth, Pravej Alam, Mohammed N. Alyemeni, and Parvaiz Ahmad. 2018. Interactive Effects of Nutrients and Bradyrhizobium Japonicum on the Growth and Root Architecture of Soybean (Glycine max L.). *Frontiers in Microbiology* 9 (1000): 1–11. https://doi.org/10.3389/fmicb.2018.01000.

Ferreira, Cristiane S., Maria T. F. Piedade, and Augusto C. Franco. 2017. Submergence, Seed Germination, and Seedling Development of the Amazonian Floodplain Tree Pseudobombax Munguba: Evidence for Root Oxytropism. *Trees* 31 (2): 705–716. https://doi.org/10.1007/s00468-016-1501-7.

Ferreira-Fernandes, Emanuel, and João. Peça. 2022. The Neural Circuit Architecture of Social Hierarchy in Rodents and Primates. *Frontiers in Cellular Neuroscience* 16 (874310): 1–17. https://doi.org/10.3389/fncel.2022.874310.

Freund, Matthias, Dorothea Graus, Andreas Fleischmann, Kadeem J. Gilbert, Qianshi Lin, Tanya Renner, Christian Stigloher, Victor A. Albert, Rainer Hedrich, and Kenji Fukushima. 2022. The Digestive Systems of Carnivorous Plants. *Plant Physiology* 190 (1): 44–59. https://doi.org/10.1093/plphys/kiac232.

Frongia, F., L. Forti, and L. Arru. 2020, December 1. Sound Perception and its Effects in Plants and Algae. *Plant Signaling & Behavior* 15 (12): 1828674. https://doi.org/10.1080/15592324.2020.1828674.

Gagliano, Monica, Vladyslav V. Vyazovskiy, Alexander A. Borbély, Mavra Grimonprez, and Martial Depczynski. 2016. Learning by Association in Plants. *Scientific Reports* 6 (1): 1–9. https://doi.org/10.1038/srep38427.

Gersani, Mordechai, Joel S. Brown, Erin E. O'Brien, Godfrey M. Maina, and Zvika Abramsky. 2001. Tragedy of the Commons as a Result of Root Competition. *Journal of Ecology* 89 (4): 660–669. https://doi.org/10.1046/j.0022-0477.2001.00609.x.

Gianoli, Ernesto, and Fernando Carrasco-Urra. 2014. Leaf Mimicry in a Climbing Plant Protects Against Herbivory. *Current Biology* 24 (9): 984–987. https://doi.org/10.1016/j.cub.2014.03.010.

Greenwood, Melinda, Charles Clarke, Ch'ien C. Lee, Ansou Gunsalam, and Rohan H. Clarke. 2011. A Unique Resource Mutualism Between the Giant Bornean Pitcher Plant, Nepenthes Rajah, and Members of a Small Mammal Community. *PLoS One* 6 (6): e21114, 1–6. https://doi.org/10.1371/journal.pone.0021114.

Gupta, Vadakattu V. S. R., and Anil K. Sharma. 2021. *Rhizosphere Biology: Interactions Between Microbes and Plants*. Springer Singapore: Springer.

Hallé, F. 2005. *Plaidoyer pour l'arbre*. Arles: Actes Sud.

Haring, Volker, Julie E. Gray, Bruce A. McClure, Marilyn A. Anderson, and Adrienne E. Clarke. 1990. Self-incompatibility: A Self-recognition System in Plants. *Science* 250 (4983): 937–941. https://doi.org/10.1126/science.2237440

Heil, Martin, and Richard Karban. 2010. Explaining Evolution of Plant Communication by Airborne Signals. *Trends in Ecology and Evolution* 25 (3): 137–144. https://doi.org/10.1016/j.tree.2009.09.010.

Hubeau, Michiel, and Kathy Steppe. 2015. Plant-PET Scans: In Vivo Mapping of Xylem and Phloem Functioning. *Trends in Plant Science* 20 (10): 676–685. https://doi.org/10.1016/j.tplants.2015.07.008.

Jung, J., S. Kim, J. Kim, M. Jeong, and M. Ryu. 2018. Beyond Chemical Triggers: Evidence for Sound-evoked Physiological Reactions in Plants. *Frontiers in Plant Science* 9: 25. https://doi.org/10.3389/fpls.2018.00025.

Karban, Richiard, Kaori Shiojiri, Mikaela Huntzinger, and Andrew C. McCall. 2006. Damage-induced Resistance in Sagebrush: Volatiles are Key to Intra-and Interplant Communication. *Ecology* 87 (4): 922–930. https://doi.org/10.1890/0012-9658(2006)87[922:DRISVA]2.0.CO;2.

Kegge, Wouter, and Ronald Pierik. 2010. Biogenic Volatile Organic Compounds and Plant Competition. *Trends in Plant Science* 15 (3): 126–132. https://doi.org/10.1016/j.tplants.2009.11.007.

Keijzer, Fred A. 2017. Evolutionary Convergence and Biologically Embodied Cognition. *Interface Focus* 7 (3): 20160123, 1–9. https://doi.org/10.1098/rsfs.2016.0123.

Kempermann, Gerd. 2019. Environmental Enrichment, New Neurons and the Neurobiology of Individuality. *Nature Reviews Neuroscience* 20 (4): 235–245. https://doi.org/10.1038/s41583-019-0120-x.

Khait, Ithzak, R. Sharon, R. Perelman, A. Boonman, Yossi Yovel, and Lilach Hadany. 2018. The Sounds of Plants–plants Emit Remotely-detectable Ultrasounds that Can Reveal Plant Stress. *Suppl Mat*. https://doi.org/10.1101/507590.

Kumar, Amit, Maurizio Memo, and Andrea Mastinu. 2020. Plant Behaviour: An Evolutionary Response to the Environment? *Pant Biology* 22 (6): 961–970. https://doi.org/10.1111/plb.13149.

Langenheim, Jean H. 1994. Higher Plant Terpenoids: A Phytocentric Overview of Their Ecological Roles. *Journal of Chemical Ecology* 20 (6): 1223–1280. https://doi.org/10.1007/BF02059809.

Li, Jiayang, Chuanyou Li, and Steven M. Smith. 2017. *Hormone Metabolism and Signaling in Plants*. London: Academic press.

Liscum, Emmanuel, Scott K. Askinosie, Daniel L. Leuchtman, Johanna Morrow, Kyle T. Willenburg, and Diana R. Coats. 2014. Phototropism: Growing Towards an Understanding of Plant Movement. *The Plant Cell* 26 (1): 38–55. https://doi.org/10.1105/tpc.113.119727.

Lyon, Pamela. 2006. The Biogenic Approach to Cognition. *Cognitive Processing* 7 (1): 11–29. https://doi.org/10.1007/s10339-005-0016-8.

Lyon, Pamela, Keijzer, Fred. 2007. The Human Stain: Why Cognitivism Can't Tell Us What Cognition Is and What It Does. In *The Mind, the World and the Body*, ed. B. Wallace, 132–165. Exeter: Imprint Academic.

Mancuso, Stefano. 2018. *The Revolutionary Genius of Plants: A New Understanding of Plant Intelligence and Behavior*. New York: Simon and Schuster.

Mancuso, Stefano, and Alessandra Viola. 2015. *Brilliant Green: The Surprising History and Science of Plant Intelligence*. Washington DC: Island Press.

Mahall, Bruce E., and Ragan M. Callaway. 1996. Effects of Regional Origin and Genotype on Intraspecific Root Communication in the Desert Shrub Ambrosia dumosa (Asteraceae). *American Journal of Botany* 83 (1): 93–98. https://doi.org/10.1002/j.1537-2197.1996.tb13879.x.

Maher, Chauncey. 2017. *Plant Minds: A Philosophical Defense*. New York: Routledge.

Majumdar, Sarangam, Sisir Roy, and Rodolfo R. Llinás. 2022. Study of the Bacterial "Conversations" and Pattern Formation in the Quorum Sensing System using Numerical Simulation. *Journal of Human, Earth, and Future* 3: 30–45. https://doi.org/10.28991/HEF-SP2022-01-03.

Masuzzo, Ambra, Martina Montanari, Léopold. Kurz, and Julien Royet. 2020. How Bacteria Impact Host Nervous System and Behaviors: Lessons from Flies and Worms. *Trends in Neurosciences* 43 (12): 998–1010. https://doi.org/10.1016/j.tins.2020.09.007.

Miller, Melissa B., and Bonnie L. Bassler. 2001. Quorum Sensing in Bacteria. *Annual Review of Microbiology* 55 (1): 165–199. https://doi.org/10.1146/annurev.micro.55.1.165.

Moreira, Xoaquín, Luis Abdala-Roberts, Sergio Rasmann, Bastien Castagneyrol, and Kailen A. Mooney. 2016. Plant Diversity Effects on Insect Herbivores and Their Natural Enemies: Current Thinking, Recent Findings, and Future Directions. *Current Opinion in Insect Science* 14: 1–7. https://doi.org/10.1016/j.cois.2015.10.003.

Muthert, Lucius W. F., Luigi G. Izzo, Martijn Van Zanten, and Giovanna Aronne. 2020. Root Tropisms: Investigations on Earth and in Space to Unravel Plant Growth Direction. *Frontiers in Plant Science* 10 (1807): 22. https://doi.org/10.3389/fpls.2019.01807.

Nani, Andrea, Gabriele Volpara, and Andrea Faggio. 2021. Sentience With or Without Consciousness. *Journal of Consciousness Studies* 28 (1–2): 60–79. https://www.ingentaconnect.com/content/imp/jcs/2021/00000028/f0020001/art00004.

Novoplansky, Ariel. 2016. Future perception in plants. In *Anticipation Across Disciplines*, ed. N. Mihai, 57–70. New York: Springer.

Olsson, A., Ewelina Knapska Andreas, and Björn. Lindström. 2020. The Neural and Computational Systems of Social Learning. *Nature Reviews Neuroscience* 21 (4): 197–212. https://doi.org/10.1038/s41583-020-0276-4.

Ortíz-Castro, Randy, Hexon Angel Contreras-Cornejo, Lourdes Macías-Rodríguez, and José López-Bucio. 2009. The Role of Microbial Signals in Plant Growth and Development. *Plant Signaling & Behavior* 4 (8): 701–712. https://doi.org/10.4161/psb.4.8.9047.

Queiroz, Orlando, and Claudine Morel. 1974. Photoperiodism and Enzyme Activity: Towards a Model for the Control of Circadian Metabolic Rhythms in the Crassulacean Acid Metabolism. *Plant Physiology* 53 (4): 596–602. https://doi.org/10.1104/pp.53.4.596.

Rasmann, Sergio, Tobias G. Köllner, Jörg. Degenhardt, Ivan Hiltpold, Stefan Toepfer, Ulrich Kuhlmann, Jonathan Gershenzion, and Ted C. J. Turlings. 2005. Recruitment of Entomopathogenic Nematodes by Insect-damaged Maize Roots. *Nature* 434 (7034): 732–737. https://doi.org/10.1038/nature03451.

Ratto, Fabrizia, Benno I. Simmons, Rebecca Spake, Veronica Zamora-Gutierrez, Michael A. MacDonald, Jennifer C. Merriman, Constance J. Tremlett, Guy M. Poppy, Kelvin S.-H.. Peh, and Lynn V. Dicks. 2018. Global Importance of Vertebrate Pollinators for Plant Reproductive Success: A Meta-analysis. *Frontiers in Ecology and the Environment* 16 (2): 82–90. https://doi.org/10.1002/fee.1763.

Robbins, Charles T., Thomas A. Hanley, Ann E. Hagerman, Olav Hjeljord, Dan L. Baker, Charles C. Schwartz, and William W. Mautz. 1987. Role of Tannins in Defending Plants Against Ruminants: Reduction in Protein Availability. *Ecology* 68 (1): 98–107. https://doi.org/10.2307/1938809.

Rodakowska, Ewelina, Marta Derba-Maceluch, Anna Kasprowicz, Pawel Zawadzki, Agnieszka Szuba, Daniel Kierzkowski, and Przemyslaw Wojtaszek. 2009. Signaling and cell walls. In *Signaling in Plants*, ed. František Baluska and Stefano Mancuso, 173–193. Berlin, Heidelberg: Springer.

Rodrigo-Moreno, Ana, Nadia Bazihizina, Elisa Azzarello, Elisa Masi, Daniel Tran, François Bouteau, František Baluska, and Stefano Mancuso. 2017. Root Phonotropism: Early Signalling Events Following Sound Perception in Arabidopsis Roots. *Plant Science* 264: 9–15. https://doi.org/10.1016/j.plantsci.2017.08.001.

Rohrscheib, Chelsie E., and Jeremy C. Brownlie. 2013. Microorganisms that Manipulate Complex Animal Behaviours by Affecting the Host's Nervous System. *Springer Science Reviews* 1: 133–140. https://doi.org/10.1007/s40362-013-0013-8.

Sasaki, Takao, and Stephen C. Pratt. 2018. The Psychology of Superorganisms: Collective Decision Making by Insect Societies. *Annual review of entomology* 63: 259–275. https://doi.org/10.1146/annurev-ento-020117-043249.

Segundo-Ortin, Miguel, and Paco Calvo. 2021. Consciousness and Cognition in Plants. *Wiley Interdisciplinary Reviews: Cognitive Science* 13 (2): e1578, 1–23. https://doi.org/10.1002/wcs.1578.

Sen, Robin. 2000. Budgeting for the Wood-wide Web. *New Phytologist* 145 (2): 161–163. https://doi.org/10.1046/j.1469-8137.2000.00585.x.

Shirasu, Mika, Kouki Fujioka, Satoshi Kakishima, Shunji Nagai, Yasuko Tomizawa, Hirokazu Tsukaya, Jin Murata, Yoshinobu Manobe, and Kazushige Touhara. 2010. Chemical Identity of a Rotting Animal-like Odor Emitted from the Inflorescence of the Titan Arum (Amorphophallus Titanum). *Bioscience, Biotechnology, and Biochemistry* 74 (12): 2550–2554. https://doi.org/10.1271/bbb.100692.

Simard, S.W. 2018. Mycorrhizal Networks Facilitate Tree Communication, Learning, and Memory. In *Memory and learning in plants*, ed. F. Baluska, M. Gagliano, and G. Witzany, 191–213. Cham: Springer. https://doi.org/10.1007/978-3-319-75596-0_10.

Simon, Ralph, Marc W. Holderied, Corinna U. Koch, and Otto von Helversen. 2011. Floral Acoustics: Conspicuous Echoes of a Dish-shaped Leaf Attract Bat Pollinators. *Science* 333 (6042): 631–633. https://doi.org/10.1126/science.1204210.

Sternberg, Troy, Heather Viles, and Alan Cathersides. 2011. Evaluating the Role of Ivy (Hedera Helix) in Moderating Wall Surface Microclimates and Contributing to the Bioprotection of Historic Buildings. *Building and Environment* 46 (2): 293–297. https://doi.org/10.1016/j.buildenv.2010.07.017.

Szymborska, Wislawa. 2002. *Miracle Fair: Selected Poems of Wislawa Szymborska*, trans. Joanna Trzeciak. New York: WW Norton & Company.

Todd, Christopher M., David A. Westcott, John M. Martin, Karrie Rose, Adam McKeown, Jane Hall, and Justin A. Welbergen. 2022. Body-size Dependent Foraging Strategies in the Christmas Island Flying-fox: Implications for Seed and Pollen Dispersal Within a Threatened Island Ecosystem. *Movement Ecology* 10 (1): 1–15. https://doi.org/10.1186/s40462-022-00315-8.

Toju, Hirokazu, Akifumi S. Tanabe, and Hirotoshi Sato. 2018. Network Hubs in Root-associated Fungal Metacommunities. *Microbiome* 6 (1): 1–16. https://doi.org/10.1186/s40168-018-0497-1.

Trewavas, Anthony. 2014. *Plant Behaviour and Intelligence*. Oxford: OUP Oxford.

Marine, Veits, Itzhak Khait, Uri Obolski, Eyal Zinger, Arjan Boonman, Aya Goldshtein, Kfir Saban, Rya Seltzer, Udi Ben-Dor, Paz Estlein, et al. 2019. Flowers Respond to Pollinator Sound Within Minutes by Increasing Nectar Sugar Concentration. *Ecology Letters* 22 (9): 1483–1492. https://doi.org/10.1111/ele.13331.

Volkov, Alexander G., Tejumade Adesina, Vladislav S. Markin, and Emil Jovanov. 2008. Kinetics and Mechanism of Dionaea Muscipula Trap Closing. *Plant physiology* 146 (2): 694–702. https://doi.org/10.1104/pp.107.108241.

Volpara, Gabriele, Andrea Nani, and Franco Cauda. 2022. The Reflected Face as a Mask of the Self: An Appraisal of the Psychological and Neuroscientific Research About Self-face Recognition. *Topoi* 41 (4): 715–730. https://doi.org/10.1007/s11245-022-09815-z.

Walters, Dale. 2011. *Plant Defense: Warding Off Attack by Pathogens, Herbivores and Parasitic Plants*. Manhattan: John Wiley and Sons.

Whiten, Andrew, Dora Biro, Nicolas Bredeche, Ellen C. Garland, and Simon Kirby. 2021. The Emergence of Collective Knowledge and Cumulative Culture in Animals, Humans and Machines. *Philosophical Transaction of Royal Society B* 377: 20200306. https://doi.org/10.1098/rstb.2020.0306.

Zimmermann, Matthias R., and Axel Mithöfer. 2013. Electrical Long-distance Signaling in Plants. In *Long-distance Systemic Signaling and Communication in Plants*, ed. J. Shah, R. Chaturvedi, and F. Baluska, 291–308. Berlin, Heidelberg: Springer.

Zhou, Zheng, Xuemei Liu, Shanping Chen, Zhijang Zhang, Yuanming Liu, Q. Quentin Montardy, Yongqiang Tang, et al. 2019. A VTA GABAergic Neural Circuit Mediates Visually Evoked Innate Defensive Responses. *Neuron* 103 (3): 473–488. https://doi.org/10.1016/j.neuron.2019.05.027.

Chapter 9
Seeing Through the Eyes of Plants

Aaron M. Ellison and Eric Zeigler

Abstract To understand, empathize with, and "become-with" other species, we need to see the world through their eyes. We might imagine ourselves looking at the world through the eyes of a leopard, a gazelle, or a frog, but what would it mean to look at the world through the eyes of a plant? Imagining this is challenging because people often fail even to see plants; we first need to acknowledge the existence of plants and our dependence on them before we can begin to understand how they see. In this chapter, we illustrate how we use a range of photographic techniques combined with creative visualizations of scientific data to explore two aspects of plant "vision." Plants perceive ("see") much more of the radiant spectrum than people do, and they use this radiation to eat (photosynthesis) and breathe (photorespiration). Plants also experience time differently from us. Their lifespans vary from minutes to millennia. In becoming-with plants, we need to step outside the decades-to-century lifespan that frames many of our interactions with the world. We suggest that seeing in plant time can unify Michael Marder's three interpretations of plant temporality: cyclic temporality of iteration, repetition, and reproduction; seasonal changes; and the apparent infinite temporality of plant growth engendered by iteration and repetition. Our analysis and juxtaposition of plant vision and time points toward an hermeneutics of plant vision that could be used to see into the being of a plant's leaves.

9.1 Introduction

We are only beginning to imagine how, and what it might mean, to fully "become-with" (Haraway 2008) any of the myriad species with which we share the Earth (e.g., Armstrong et al. 2019, 2021). The path to becoming-with—that is, entering into a true symbiotic relationship with another species (as opposed to domesticating

A. M. Ellison (✉)
Harvard University Herbaria, Harvard University, Cambridge, MA, USA
e-mail: amellison17@gmail.com

E. Zeigler
Department of Art, University of Toledo, Toledo, OH, USA

M. Di Paola (ed.), *The Vegetal Turn*, The International Library of Environmental, Agricultural and Food Ethics 38, https://doi.org/10.1007/978-3-031-68671-9_9

or collaborating with it)—was forged by Deleuze and Guattari (1987) and Haraway (2008). Haraway (2008) joined the concepts of companionship (Latin *cum panis* ["with bread"]) + respect (*respicere* ["to look again"]) to suggest how to move beyond the subservient domestication of animals by humans and toward a richer set of inter-species relationships and symbiosis (Tsing 2012; see also Margulis 1998). Although Tsing extended the idea of becoming-with outside of the animal kingdom, Haraway and most others who have thought deeply about becoming-with consider only animals, especially mammals (Rees 2017). Far fewer have asked what it would take to become-with a plant (Deleuze and Guattari 1987; Houle 2011; Marder 2013; 2014; Armstrong et al. 2019).

In conceptualizing becoming-with, Haraway (2008) noted that the root of *respicere* is *specere* ("to look" or "to behold") but chose to emphasize that *specere* also is the root of "species." Haraway's joining of companionship + species in becoming-with clearly recalled the interspecific partnership implied by Derrida's assertion that "[o]ne never eats entirely on one's own" (Derrida 1991, 115). However, to become-with another species, we need to do more than eat (or eat with) them (*cf*. Pollan's description of the Mayan "corn people" and North Americans as "corn chips with legs;" Pollan 2011, 19 and 23, respectively). Rather, on the road to becoming-with another species, we need to more completely "be" them (e.g., Nagel [1974] provocatively asked, "What is it like to be a bat?" and *cf.* Marder [2014, 139], who asked "What can we know about plants?"). We need to empathize with them more fully, "grow a mile" in their roots and rhizomes,[1] and see the world through their eyes while using their cognitive systems (or analogs thereof) to translate different wavelengths of light into usable, visual information (Chamovitz 2017; Zeigler & Ellison 2022). While respecting the approach of Haraway *et seq*., we suggest here that refocusing attention on the visual meaning of *specere* can provide an alternative turn toward becoming-with plants.

We might imagine ourselves looking at the world through the eyes of a leopard, a gazelle, a frog, or even Nagel's sonar-seeing bat, but what would it mean to look at the world through the eyes of a plant? Imagining this is challenging because people often fail to even see plants or recognize their instrumental or aesthetic value.[2] Conversely, scholars of plant studies who have worked to imagine the characteristics ("essences" in Marder 2014) through which we could enter into a "heterogeneous alliance … of plants or plant-lives" (Houle 2011, 97) do not consider that plants see. For example, in her deep exegesis on becoming-with plants, Houle (2011) drew attention to the importance of plant communication. For Houle (2011), communication (in this case, via volatile chemical compounds; Baldwin and Schultz 1983) is one of six ways that plants differ from animals, none of which involved how either interact with light. Similar sets of plant characteristics are included in scientific studies of plant

[1] The origin of the aphorism "walk a mile in his moccasins," as a pathway to empathy and compassion is usually credited to Lathrap's (1895) poem "Judge Softly."

[2] This all-too-common trait, usually referred to as "plant blindness" (Allen 2003), is more appropriately called "plant awareness disparity" (Parsley 2020).

behavior (e.g., McNickle et al. 2009; Ellison 2020) or discussions of becoming-with plants (e.g., Marder 2013; Stark 2015; but see Ruggles 2017). The absence of vision is striking, as Deleuze (1994, 211) had described an organism (a category that presumably included his rhizomatous plant) as "nothing if not the solution to a problem, as are each of its differentiated organs, such as the eye which solves a light 'problem'."

In this chapter, we examine how plants "see" and the temporality of plant "vision." In considering how plants see, we do not re-hash the processes by which leaves, with their photosynthetic systems, have solved the problem of using light energy to convert carbon dioxide and water to the food they need to eat (this solution has been well articulated by botanists and plant physiologists; Chamovitz 2017). Rather, we provide an initial sketch of an hermeneutics of plant vision[3] that we can use to put ourselves as fully as possible into the being of a plant's leaves.[4] To do this, we use a range of photographic techniques combined with creative visualizations of scientific data to explore how plants see (perceive) light in wavelengths different from humans (Zeigler and Ellison 2022). Like Marder (2013; 2014), we recognize that this approach is only an approximation, and that we are still at a stage where we cannot put ourselves entirely in the visual system of a plant. Nonetheless, our approximations can be used by us and others working to expand aesthetic perception and principles into the vegetal realm (e.g., Brady and Prior 2020; Ellison and Zeigler 2021).

We then use our elaborated sense of plant vision to examine its temporality. For Marder (2013, 95), "the meaning of vegetal being *is time*" (emphasis added). We suggest that seeing in plant time can unify Marder's three interpretations of plant temporality: cyclic temporality of iteration, repetition, and reproduction; seasonal changes; and the apparent infinite temporality of plant growth engendered by iteration and repetition. Although Marder (2013) located much of this temporality in the seed, its germination, growth-through-reiteration into a mature plant, and subsequent reproduction of new seeds (2013, 96), he also noted (citing Heidegger 1962, 287–288) that "the fruit does not actively 'come into ripeness of its own accord'" but rather "is indebted for its maturation to the rays of the sun" (1962, 100) and that in their temporal progression, plants "crave water and sunlight" (1962, 130).

Plant vision—how a plant sees—is central to plants' cravings and the maturation of their fruits, but "sunlight" barely scratches the surface of what a plant sees and craves,[5] or any of the other multiple alliances plants need to flourish. In sum, through

[3] As scientists and photographers focused on understanding and interpreting "nature," we are using hermeneutics here in an expanded Heideggerian sense. We take as a given that there is an objective nature, for which science can provide a reasonable explanation and photography a reasonable documentation; and that our scientific understanding is internally coherent and compatible with the ontological implications of our day-to-day actions (see also Blattner 1994, 199*ff*).

[4] Marder (2013, 10) similarly emphasized the importance of leaves but then followed Deleuze's (1994, xvii) in wanting to approximate the ideal of putting ourselves as wholly as possible in a plant's rhizomes.

[5] Our thinking about what a plant sees and craves differs from what Marder (2013, 130) identified as a cross-kingdom analogy with a person's suffering from thirst and the deprivation of social contact.

a deeper understanding of plant vision, we aim to make perceptible the vegetal temporality that Marder (2013, 103) considered imperceptible.

9.2 Consider a Bristlecone

There are hundreds of thousands of plant species, most with seemingly uncountable individuals, and all with some common characteristics and many unique essences. Our exemplar is the Great Basin Bristlecone Pine (*Pinus longaeva*), one of the longest-lived individual organisms on Earth (Fig. 9.1).

Trees in any of the three recognized species of bristlecone pines can live for at least 5000 years (Schulman 1954; Bailey 1970). In this Methuselan time span, they see—and through their rings, their wood remembers—many events that are invisible to humans and others that are rapidly disappearing from human memory. At the same time, bristlecone pines can reveal these events and teach these memories to people. For example, bristlecone pines remain alive and thrive on mountainsides and mountaintops that are eroding beneath them. The trees maintain their spatial rootedness while time inexorably slips by because the growth rate of the trees' roots matches (or just exceeds) the rate of erosion. Indeed, in a nascent human/vegetal partnership, human geologists have learned the rates of mountaintop erosion from the trees themselves (LaMarche 1968) and bristlecones have engendered new appreciation for the aesthetics of nature (Ferguson 1968). The latter is especially important because, as Stark (2015, 183) pointed out, "plant[s] often operate as a placeholder for nature itself."

9.3 Do Plants See?

Plants require light, but do they see it? Marder (2014, 182; see also Chamovitz 2017) noted that "[w]hat largely determines vegetal activity above ground level is the plants' tending toward and photosynthetic processing of sunlight." But unlike Derrida's interspecific omnivorous partnerships, plants in Marder's view are physically (ontically) open to light *as* light while being ontologically indifferent to it, storing the memory of light and interpreting it as needed in new spatiotemporal

Rather, we locate this vision and craving within the plant *as* plant; teleology (i.e., intention toward a particular end) is not implied.

Fig. 9.1 An ancient bristlecone pine (*Pinus longaeva*) growing on a steep hillside at 3100 m above sea level (a.s.l.) in the Schulman Grove, Inyo National Forest, California, USA. Digital image © Eric Zeigler & Aaron M. Ellison. Colour image available at https://www.ericzeigler.com/vegetal-turn

configurations and processes.[6] We suggest there is much more to the how and what plants see than simply light as light.

Plants sense far more of the spectrum than do people (Fig. 9.2). Our three retinal cones perceive reflected radiation from about 380–700 nm, in what is anthropocentrically referred to as the "visible spectrum." In contrast, plants have many photosensitive pigments that are grouped into six categories: UVR8, cryptochromes, phototropins, photosynthetic pigments (chlorophyll-a, -b, and the carotenoids), and phytochromes. These pigments absorb radiation from about 315–860 nm, which includes ultraviolet and infrared wavelengths that are invisible to the human eye. Note that few photosensitive pigments absorb green and yellow wavelengths ($\approx$500–600 nm; Fig. 9.2), which is why we see leaves as "green." That is, the green we see is invisible to plants because these wavelengths are reflected from a leaf's surface.[7]

Photographs of specific spectra reflected by plants illustrate what they actually see (Fig. 9.3). Within the human-visible spectrum, plants absorb most of the radiation at the opposite ends of the ranges sensed by our cones. Our green and red cones are most sensitive to the green and yellow wavelengths (540–560 nm)—those *least* absorbed by a plant's photosensitive pigments. As a result, when we look at a tree, we see the reflected light of "green" leaves on a contrasting background of darker bark (Fig. 9.3, centre). However, plants have three sets of photosensitive pigments that efficiently absorb ultraviolet radiation (Fig. 9.2). Thus, our ultraviolet images of plants have dark, nonreflective leaves that, because we only see reflected light, are barely visible to us (Fig. 9.3, left). In contrast, infrared wavelengths beyond 850 nm are not absorbed by phytochromes (Fig. 9.2). To us, infrared images of plants have bright, glistening leaves that appear starkly white (Fig. 9.3, right).

What does a bristlecone pine see in a forest? Like hawthorn leaves (Fig. 9.3), bristlecone needles absorb virtually all the ultraviolet radiation; they appear dark and dull to us (Fig. 9.4, top). The sky, though, is bright white, suggesting that these trees do not "see" the sky. In contrast, needles in full sunlight glow bright white in the

[6] Marder (2014, 183) was describing and elaborating on Heidegger's assessment of the "plant-character of the plant." It is not clear if Marder subscribes completely to Heidegger's assessment of what it is like to "be" a plant, but it is the only discussion of light (and photosynthesis) in *The Philosopher's Plant*. However, Marder's assertions about the ontological indifference (Hegel; see also Simonti 2015) of plants are not always consistent. On the one hand, a plant could be a good Hegelian and be ontologically indifferent to light while storing and later interpreting its memory. On the other hand, Marder (2014, 179) noted the etymological and conceptual links between the Greek *phuton* (plant) and *phusis* (nature): "So conceived, nature is the [Heidegger's] common 'ground of all beings,' which is also the case of the plant, whose principle of vitality potentiates the existence of all other living creatures." This is more along the lines of Descartes's ontological argument for the existence of God than Hegel's (and Heidegger's) ontological indifference (Simonti 2015). Finally, Marder (2014, 182–183) identified a plant's ontological indifference in Heidegger's "basic determination of Dasein to be open for being claimed by the presence [being] of something." But for Heidegger, Dasein was the uniquely human experience of being. Are Marder's plants, therefore, human (or *vice-versa*)?

[7] The role of green (and yellow) wavelengths in plant growth and metabolism is understudied. Early studies suggested that green light inhibited plant growth (e.g., Klein et al. 1965), but more recent ones have shown that green regulates many plant-physiological processes (e.g., Johkan et al. 2012; Golovatskaya and Karnachuk 2015). See Grant (1987) for additional discussion of leaf reflectance.

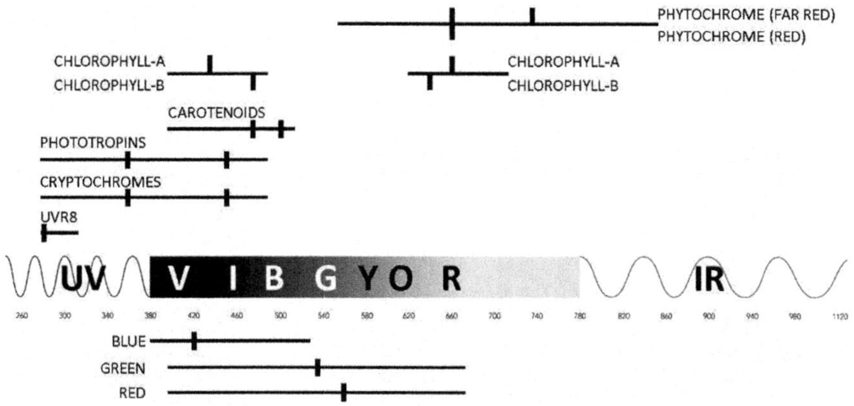

Fig. 9.2 Wavelengths (in nanometers) of light (spectrum) perceived by plants (above the spectrum) and humans (below the spectrum). The colours of the spectrum are indicated by one or two letters (**U**ltra**V**iolet, **V**iolet, **I**ndigo, **B**lue, **G**reen, **Y**ellow, **O**range, **R**ed, **I**nfra**R**ed). For each receptor (photosensitive pigments for plants, cones for humans), the horizontal line spans the range of wavelengths that can be perceived, and the thick vertical line indicates the wavelength for which each receptor has maximal sensitivity. The names of the photosensitive pigments or colours assigned to retinal cones are indicated adjacent to each horizontal line. Note that chlorophyll-a and chlorophyll-b have photosensitivity at both low and higher wavelengths but not intermediate ones. A colour version of this figure is available at https://www.ericzeigler.com/vegetal-turn. Data used to create the image are in Kami et al. (2010). Digital image © Eric Zeigler & Aaron M. Ellison

infrared (Fig. 9.4, bottom). Needles in shade and shadows appear greyer to us because most of the infrared radiation has already been reflected by the needles in the sun. The snow on the ground is brightest in the human-visible spectrum (Fig. 9.4, centre), but the plants also see the snow. Some ultraviolet, blue, and red light reflecting off the snow is absorbed by shaded needles, rendering the snow less bright in the ultraviolet and infrared and the shaded needles greyer in the infrared.

These images suggest that plants are more than open to light *as* light and do more than store its memory. Plants see different wavelengths of light, absorbing and retaining some while reflecting others unused and unstored. The light that is retained is used as energy by the plant to combine carbon dioxide and water into the sugars and starches it needs to grow—reiterate leaves and stems, roots and rhizomes, and flowers, fruits, and seeds—and to respire (breathe). What isn't respired (a heat-producing, infrared-emitting process, too) is transformed before being stored as memories; memories not only transmitted to future generations through seeds but also laid down as growth rings that preserve a record of the plant's lived experiences. One wonders: how do trees read these records? What mode of communication might trees use to describe the snow and the sky—the same chemical language they use to talk about the insects that are eating them, or an entirely different one? (Fig. 9.4).

Fig. 9.3 A Frosted Hawthorn tree (*Crataegus pruinosa*) imaged in different spectra: ultraviolet (340–380 nm; left); human-visible (380–750 nm; centre); infrared (850–1100 nm right). The ultraviolet and infrared images were made using a Nikon D3000 camera with a sensor that we modified to capture wavelengths from ≈325–1100 nm and either a Kolari Vision ultraviolet bandpass transmission lens filter or an 850-nm infrared lens filter. The human-visible spectrum image was made with an unmodified Nikon D800e camera; the colour image was desaturated (converted to grayscale) in Adobe Photoshop (the original colour image is online at https://www.ericzeigler.com/vegetal-turn). Because humans cannot see reflected light in ultraviolet or infrared wavelengths, we mathematically transformed the ultraviolet and infrared images into grayscale images that we can see. Digital images © Eric Zeigler & Aaron M. Ellison

9.4 Vegetal Time

For people, time is generally perceived as linear and progressive. However, when we see the world, time is erased. Our brain takes a set of images that are at least subtly different from one another and reconstructs them into a fixed, "smoothed" image that seems stable/unchanging.[8] Similarly, our reconstructions of plant time, preserved as memories in tree rings, are first and foremost linear and progressive, albeit interruptible by disturbances such as ice storms, insect outbreaks, or volcanic eruptions (Fig. 9.5; see also Ellison et al. 2014).

When we try to think rhizomatously in plant time, though, we first think of rhythms and cycles.[9] Marder (2013) linked a plant's seasonal changes (e.g., winter dormancy,

[8] The "noise" in the constant barrage of visual stimuli that become retinal images is a result of motion, changes in light and perspective, occlusions and discontinuities, and the passage of time (Manassi & Whitney 2022). Yet we do not see the world as a jittery movie. Rather, we perceive our world as stable in time, both in the instant and over much longer spans; Simons and Rensink (2005) have called this phenomenon "change blindness." Manassi & Whitney (2022) experimentally located the source of change blindness at the level of perceptual experience, not memory. That is, the human brain's view of the world is not ontologically indifferent but rather accords more with Kant's ontology of totalization (sensu Simonti 2015).

[9] What scientists call "phenology:" periodic phenomena such as flowering and fruiting that are related to daily, seasonal, or annual changes in climate.

Fig. 9.4 A stand of bristlecones pines growing at 3,100 m a.s.l. in the Schulman Grove, Inyo National Forest, California, USA. The three images are of the same scene but imaged in different spectra: ultraviolet (290–400 nm; top); human-visible (380–750 nm; centre); and infrared (850–1100 nm; bottom). The ultraviolet image was captured on Ilford HF5 + sheet film using a Linhof 4 × 5 camera outfitted with a Kodak 18A Wratten ultraviolet-pass filter. The desaturated human-visible (colour version available at https://www.ericzeigler.com/vegetal-turn) and infrared images were taken as described in the caption to Fig. 9.3. Digital images © Eric Zeigler & Aaron M. Ellison

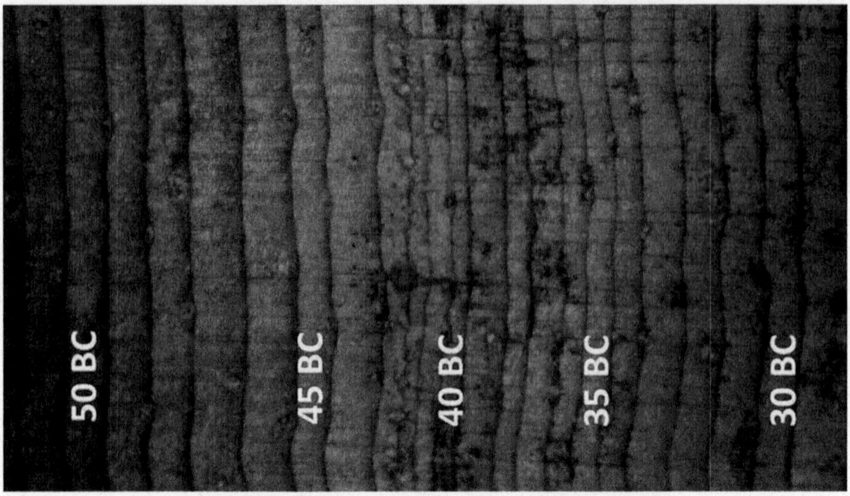

Fig. 9.5 A cross-section through a tree core taken from a bristlecone pine growing in the Snake Mountains, Great Basin National Park, Nevada, USA. The light areas between the darker lines are annual growth rings—how much the tree grew in diameter each year. Narrower growth rings indicate slower growth than wider ones, and the width is used by dendrochronologists (literally: "tree timers") to infer and reconstruct past climates and growing conditions. The narrow growth rings just after 44 BCE were associated with a large volcanic eruption. Photo: Desert Research Institute (https://www.flickr.com/photos/driscience/19314254419/; CC-BY-NC-ND 2.0)

spring flowering, summer leaf-out; Fig. 9.6) with its capacity for iteration and repetition into an infinite temporality of growth. If we overlay the seasonal cycles along and through time's arrow, the cycles extend into a helix, expressed in the traces of a tree's vessels and stored in the memory of its ever-expanding growth rings. Indeed, this helical structure is visible in the way that many trees grow (Fig. 9.7).

Rings are produced in roots, too. While the root remains underground, these rings, like those in the tree's trunk, are radial and (mostly) symmetrical. But in long-lived trees like bristlecones that also live on steep, eroding slopes, the ground literally disappears beneath their feet. Only if the roots can continue to grow downward along the slope can they continue to absorb water and nutrients. Because only the part of the root that is in contact with the soil continues to grow, its flattened rings appear semicircular (LaMarche 1968). These new tree-memories synchronize geological time with plant time. They are laid down over millennia and record in wood what we otherwise have learned to read only in stone. Like stone, trees and tree-memories embody the entire history of life on Earth (Nassar & Barbour 2019). This embodiment makes perceptible the heterogenous, non-contemporaneous (with respect to people) times of plants that Marder (2013, 103) considers imperceptible to conscious human observers.

Fig. 9.6 During a few weeks in the spring, the Ohio Buckeye (*Aesculus glabra*) emerges from winter dormancy (left), flowers (centre), and unfurls its leaves (right). These three images of the same tree were clipped and desaturated from a longer series of colour images assembled in an animated gif (both available online at https://www.ericzeigler.com/vegetal-turn). Digital images and associated gif © Eric Zeigler & Aaron M. Ellison

9.5 Toward a Hermeneutics of Plant Vision

Our consideration of how plants see with their leaves adds a new and important curve to the vegetal turn and suggests additional avenues for becoming-with plants. Most critical discussions of becoming-with plants have focused on roots and rhizomes. This emphasis has retraced a well-trodden path from Plato (for whom roots were closest to the realm of ideas), Aristotle (roots were the only part of the plant that could eat), and Plotinus (the One was firmly settled in the root and the root was the refuge for the Soul of All at the end of the world), through Maimonides and Leibniz (roots and runners [rhizomes] give rise to new plants), and culminating with Deleuze's rhizome-thought (i.e., a plant's epistemology), which he placed in opposition to arborescent thought (Deleuze 1994, xvii).[10]

The Deluzean rhizome is universally interconnected whereas the shoots (and leaves) of his trees are branched and weakly connected. His rhizomes are asexual and antigeneaological but his shoots with its fruits and seeds are the object of reproduction whose expression is a family tree with its memory of genealogy. Most importantly, his rhizomes are acentered, nonhierarchical, and non-signifying, whereas his trees are hierarchical: "[w]hat is at question in the rhizome is a relation to sexuality—but also to the animal, the vegetal, … things natural and artificial—that is totally different from the arborescent relation: all manner of "becomings" (Deleuze & Guattari 1987, 23–24).

We would argue that plant vision—and their shoots and leaves—are rhizomatic in the Deleuzean sense. However, *contra* Deleuze, we assert that shoots and leaves are not hierarchical. Rather, they are tightly interconnected with one another *and* with

[10] This path is traversed in the successive chapters of *The Philosopher's Plant* (Marder 2014).

Fig. 9.7 Helical growth visible in living (centre) and dead (right) ancient bristlecone pines (*Pinus longaeva*) growing at 3,200 m a.s.l., below Wheeler Peak, Great Basin National Park, Nevada, USA. Digital image © Eric Zeigler & Aaron M. Ellison. (Colour image available online at https://www.ericzeigler.com/vegetal-turn)

roots and literal rhizomes. Plants *as* plants are more than rhizomes and roots, fruits and seeds. Without leaves to see and to produce starch[11] through photosynthesis there would be no rhizomes or roots, no *cum panis*, no becoming-with. Deleuze & Guattari misinterpreted the structure of the parts of trees that are aboveground. They are not hierarchical but, like roots and rhizomes, infinitely fractal (Magalhães & Seifert 2015); the outmoded hierarchical concept illustrated by Aristotle's and Linnaeus's *scala naturae* no more applies to arborescence than it does to the infinitely reticulate web of life (Pérez et al. 2010). We need to embrace the fractal totality and temporality of plants, see as they see, feed as they feed, and grow as they grow (Haraway 2016; Ellison 2019).

9.6 Conclusion

We are only beginning to imagine what it might mean to fully "become-with" a plant (but see Ellison 2019). Symbioses between plants and animals, such as the pollination by male wasps of orchids whose appearance mimics female wasps[12] and the protective mutualism between ants and *Acacia*[13] plants heralded by Haraway (2016), could be models for botanical becomings-with (see also Ellison & Gotelli 2021). Both the orchid-wasp pollination system and the protective mutualism between ants and *Acacia* trees are true symbioses. They involve "reaching into the internal tissues of each participant, shaping genomes and developmental patterning of the structures and functions of both companion species…becoming-with each other in response-ability" (Haraway 2016, 124–125). Pertinent to our discussion herein, aimed at expanding the Deleuzean rhizome, neither symbiosis involves literal roots or rhizomes, and both are the product of co-evolution.[14] Such back-and-forth co-evolutionary dynamics (Ehrlich & Raven 1964) are at the heart of symbioses and

[11] The product of photosynthesis—starch—is the essence of *panis* and the root of becoming-with. Starch is also the key element shared by plants with the mycorrhizal networks of fungi-*cum*-roots that, in turn, share nitrogen, phosphorus, and other nutrients with the leaves and shoots.

[12] Deleuze and Guattari (1987, 11) consider this pollination symbiosis to be an example of becoming-with: "Wasp and orchid, as heterogeneous elements, form a rhizome. … At the same time, something else entirely is going on: … a veritable becoming, a becoming-wasp of the orchid and a becoming-orchid of the wasp."

[13] The swollen, spiny stipules ("thorns") at the base of each leaf provide shelter for ants, which in turn fiercely defend the plant from other herbivores (Janzen 1966; Chomicki et al. 2015). The stipules themselves are integral parts of the leaf itself and are modified from the "teeth" that are found on the edges of the leaves of many plant species (Sinnott and Bailey 1914). Finally, at the tips of their leaves, acacias produce "Beltian bodies"—small packets of food rich in fats—that nourish the ants.

[14] Although Deleuze and Guattari (1987, 7) are expansive in their definition of a rhizome, which need not be a literal underground stem (a rhizome is a system in which "any point can be connected to anything other, and must be"), they assert that rhizomes such as the orchid-wasp system cannot be a product of Darwinian evolution by descent: "Becoming is not an evolution, at least not an evolution by descent and filiation" (1987, 263). They do allow for evolution to give rise to becomings, but only in the "domain of symbioses that bring into play beings of totally different scales and kingdoms"

true becomings-with. These kinds of interactions are what humans need to undertake to move beyond the domestication of crops and harvesting of plants as resources—the one-way streets that characterize our relationships with plants and their world.

The ontological indifference of plants and their apparent powerlessness contains the seeds of their freedom (Marder 2013). By curving and climbing upward into the light, a plant forms and transforms the world (Lemm 2015). Marder (2013, 149) wrote that "plants will not be free unless the political and economic conditions responsible both for their oppression and for our appreciation of them change as well." Seeing and accepting temporal changes—social, political, economic, climatic—while remaining firmly rooted yet free in the places we live is the central challenge for 21st-century humanity. Becoming-with plants can help us meet this challenge. Seen through the eyes *and* roots of bristlecone pines, we have suggested how becoming-with plants can move us productively through the next rotation of the Vegetal Turn. Through this turn, plants will do more than simply operate as a metonymic placeholder for nature. Plants—and humans—must become companions in, with, and for nature.

Acknowledgements Photography and post-processing of images of bristlecone pines was facilitated by the Axon Lab Creative Residency in the Department of Art at the University of Toledo. Erik Johanson provided provocative comments on an early draft of the chapter; an anonymous reviewer and Marcello Di Paola provided additional comments on the submitted version of the chapter that led to substantial improvements in the final version. The manuscript was written while AME was a Natural Sciences Fellow at the Swedish Collegium for Advanced Study.

References

Allen, William. 2003. Plant Blindness. *BioScience* 53 (10): 926.

Armstrong, Josh, Alexandra Lakind, Chessa Adsit-Morris, eds. 2019. *becoming*—Botanical: A Post-modern Liber Herbalis. Glasgow: Objet-a Creative Studio.

Armstrong, Josh, Alexandra Lakind, Chessa Adsit-Morris, Rebekka Sæter, eds. 2021. *becoming*—Feral. Glasgow: Objet-a Creative Studio.

Bailey, Dana K. 1970. Phytogeography and Taxonomy of *Pinus* Section *Balfourianae*. *Annals of the Missouri Botanical Garden* 57 (2): 210–249.

Baldwin, Ian T., and Jack C. Schultz. 1983. Rapid Changes in Tree Leaf Chemistry Induced by Damage: Evidence for Communication Between Plants. *Science* 221 (4607): 277–279.

Blattner, William D. 1994. Is Heidegger a Kantian Idealist? *Inquiry* 37 (2): 185–201.

Brady, Emily, and Jonathan Prior. 2020. Environmental Aesthetics: A Synthetic Review. *People and Nature* 2 (2): 254–266.

(1987, 263) such as wasps and orchids, ants and *Acacia*. Yet, ample evidence exists that Darwinian evolution-by-descent has given rise to orchids that effect their own reproduction by fooling wasps into mating with them ("pseudocopulation"; Gaskett et al. 2008) and to ants that protect acacias in return for food and shelter. It is also not necessary that all such symbioses are as mutually beneficial as the ant-*Acacia* system. Although Deleuze and Guattari (1987, 324) assert that both the orchid and the wasp are liberated pieces of the other's reproductive system (the orchid "becomes the object of an orgasm in the wasp"), experimental data instead suggest that pseudocopulatory orchids are exploiting and even harming their wasp pollinators through their coevolutionary arms race (e.g., Wong and Schiestl 2002).

Chamovitz, Daniel. 2017. *What a Plant Knows: A Field Guide to the Senses* (updated and expanded edition). New York: Farrar, Straus and Giroux.

Chomicki, Guillaume, Philip S. Ward, and Susanne S. Renner. 2015. Macroevolutionary Assembly of Ant/Plant Symbioses: *Pseudomyrmex* Ants and Their Ant-housing Plants in the Neotropics. *Proceedings of the Royal Society B* 282 (1819): 20152200.

Deleuze, Gilles. 1994. *Difference and Repetition* (translated from the 1968 edition of *Différence et Repetition* by Paul Patton). New York: Columbia University Press.

Deleuze, Gilles, and Félix Guattari. 1987. *A Thousand Plateaus: Capitalism and Schizophrenia* (translated from the 1980 edition of *Mille Plateaux*, volume 2 of *Capitalisme et Schizophrénie* by Brian Massumi). London: Continuum.

Derrida, Jacques (with Jean-Luc Nancy). 1991. "Eating Well", or the Calculation of the Subject: An Interview with Jacques Derrida. In *Who Comes After the Subject?*, eds. Eduardo Cadava, Peter Connor, and Jean-Luc Nancy, 96–118. New York: Routledge.

Ehrlich, Paul R., and Peter H. Raven. 1964. Butterflies and Plants: A Study in Coevolution. *Evolution* 18 (4): 586–608.

Ellison, Aaron M. 2019. *Tsuga canadensis* (Eastern Hemlock). In *becoming*-Botanical: A Post-modern Liber Herbalis, eds. Josh Armstrong and Alexandra Lakind, 112–114. Glasgow: Objet-a Creative Studio.

Ellison, Aaron M. 2020. Foraging Modes of Carnivorous Plants. *Israel Journal of Ecology and Evolution* 66 (1–2): 101–112.

Ellison, Aaron M., and Nicholas J. Gotelli. 2021. Ants (Hymenoptera: Formicidae) and Humans: From Inspiration and Metaphor to 21st-century Symbiont. *Myrmecological News* 31: 225–240.

Ellison, Aaron M., and Eric Zeigler. 2021. Towards a Non-anthropocentric Aesthetic of Environmental Balance and Change. Paper presented at the 14th International Institute of Applied Aesthetics Biennial Conference, Lahti, Finland, 3–5 June 2021.

Ellison, Aaron M., Michael Lavine, Peter B. Kerson, Audrey A. Barker Plotkin, and David A. Orwig. 2014. Building a Foundation: Land-use History and Dendrochronology Reveal Temporal Dynamics of a *Tsuga canadensis* (Pinaceae) Forest. *Rhodora* 116 (968): 377–427.

Ferguson, Charles W. 1968. Bristlecone Pine: Science and Esthetics. *Science* 159 (3817): 839–846.

Gaskett, Anne C., Claire G. Winnick, and Mariella E. Herberstein. 2008. Orchid Sexual Deceit Provokes Ejaculation. *The American Naturalist* 171 (6): E206–E212.

Golovatskaya, Irina F., and R.A. Karnachuk. 2015. Role of Green Light in Physiological Activity of Plants. *Russian Journal of Plant Physiology* 62 (6): 727–740.

Grant, Lois. 1987. Diffuse and Specular Characteristics of Leaf Reflectance. *Remote Sensing of Environment* 22 (2): 209–322.

Haraway, Donna J. 2008. *When Species Meet*. Minneapolis: The University of Minnesota Press.

Haraway, Donna J. 2016. Sowing Worlds: A Seed Bag for Terraforming with Earth Others. In *Staying with the Trouble: Making Kin in the Chthulucene*, ed. Donna H. Haraway, 117–125. Durham: Duke University Press.

Heidegger, Martin. 1962. *Being and Time*, trans. John Macquarrie and Edward Robinson. New York: Harper & Row.

Houle, Karen L. F. 2011. Animal, Vegetable, Mineral: Ethics as Extension or Becoming? *Journal for Critical Animal Studies* 9 (1–2): 89–115.

Janzen, Daniel H. 1966. Coevolution of Mutualism Between Ants and Acacias in Central America. *Evolution* 20 (3): 249–275.

Johkan, M., K. Shoji, F. Goto, S. Hahida, and T. Yoshihara. 2012. Effect of Green Light Wavelength and Intensity on Photomorphogenesis and Photosynthesis in *Lactuca sativa*. *Environmental and Experimental Botany* 75 (1): 128–133.

Kami, Chitose, Séverine. Lorrain, Patricia Honitschek, and Christian Fankhauser. 2010. Light-regulated Plant Growth and Development. *Current Topics in Developmental Biology* 91: 29–66.

Klein, Richard M., Pamela C. Edsall, and Arthur C. Gentile. 1965. Effects of Near Ultraviolet and Green Radiations on Plant Growth. *Plant Physiology* 40 (5): 903–906.

LaMarche, Valmore C., Jr. 1968. Rates of Slope Degradation as Determined from Botanical Evidence, White Mountains, California. US Geological Survey Professional Paper, 352-I, 341–377.

Lathrap, Mary T. 1895. Walk a Mile in His Moccasins. AAANativeArts, 1 May 2014. https://www.aaanativearts.com/walk-mile-in-his-moccasins (Accessed 28 January 2023).

Lemm, Vanessa. 2015. What We Can Learn from Plants About the Creation of Values. *Nietzsche-Studien* 44 (1): 78–87.

Magalhães, Tarquinio, and Thomas Seifert. 2015. Below- and Aboveground Architecture of *Androstachys johnsonii* Prain: Topological Analysis of the Root and Shoot Systems. *Plant and Soil* 394 (1–2): 257–269.

Manassi, Mauro, and David Whitney. 2022. Illusion of Visual Stability Through Active Perceptual Serial Dependence. *Science Advances* 8: eabk2480.

Marder, Michael. 2013. *Plant-thinking: A Philosophy of Vegetal Life*. New York: Columbia University Press.

Marder, Michael. 2014. *The Philosopher's Plant: An Intellectual Herbarium*. New York: Columbia University Press.

Margulis, Lynn. 1998. *Symbiotic Planet*. New York: Basic Books.

McNickle, Gordon G., Colleen Cassady St. Clair, and James F. Cahill, Jr. 2009. Focusing the Metaphor: Plant Root Foraging Behaviour. *Trends in Ecology & Evolution* 24 (8): 419–426.

Nagel, Thomas. 1974. What is it Like to be a Bat? *The Philosophical Review* 83 (4): 435–450.

Nassar, Dalia, and Margaret M. Barbour. 2019. Rooted. Aeon, 16 October 2019. https://aeon.co/essays/what-can-an-embodied-history-of-trees-teach-us-about-life (Accessed 7 June 2022).

Parsley, Kathryn M. 2020. Plant Awareness Disparity: A Case for Renaming Plant Blindness. *Plants, People, Planet* 2 (6): 598–601.

Pérez, Julio E., Carmen Alfonsi, and Carlos Muñoz. 2010. Towards a New Evolutionary Theory. *Interciencia* 35 (11): 862–868.

Pollan, Michael. 2011. *The Omnivore's Dilemma: The Search for the Perfect Meal in a Fast-Food World*. London: Bloomsbury.

Rees, Amanda. 2017. Animal Agents? Historiography, Theory and the History of Science in the Anthropocene. *BJHS: Themes* 2: 1–10.

Ruggles, Laura. 2017. The Minds of Plants. Aeon, 12 December 2017. https://aeon.co/essays/beyond-the-animal-brain-plants-have-cognitive-capacities-too (Accessed 28 May 2022).

Schulman, Edmund. 1954. Longevity Under Adversity in Conifers. *Science* 119 (3091): 396–399.

Simons, Daniel J., and Ronald A. Rensink. 2005. Change Blindness: Past, Present, and Future. *Trends in Cognitive Sciences* 9 (1): 16–20.

Simonti, Jure. 2015. The Ontological Indifference. A Realist Reading of Kant and Hegel. *Filozofija i Društvo* 26 (2): 369–391.

Sinnott, Edmund W., and Irving W. Bailey. 1914. Investigations on the Phylogeny of the Angiosperms. 3. Nodal Anatomy and the Morphology of Stipules. *American Journal of Botany* 1 (9): 441–453.

Stark, Hannah. 2015. Deleuze and Critical Plant Studies. In *Deleuze and the Non/Human*, ed. Jon Roffe and Hannah Stark, 180–196. New York: Palgrave Macmillan.

Tsing, Anna. 2012. Unruly Edges: Mushrooms as Companion Species: For Donna Haraway. *Environmental Humanities* 1 (1): 141–154.

Wong, Bob B. M., and Florian P. Schiestl. 2002. How an Orchid Harms its Pollinator. *Proceedings of the Royal Society of London B* 269 (1500): 1529–1532.

Zeigler, Eric, and Aaron M. Ellison. 2022. Learning to See Differently. In *Seeing Across Disciplines: Visual Literacy and Education. The Book of Selected Readings 2022*, eds. Jung Lee, Stephanie Beene, Xiaoning Chen, Wanju Huang, Lee Okan, & Filipa Rodrigues, 146–169. Stockton: International Visual Literacy Association.

Chapter 10
Methods for Plant Ethics

Jenna Aarnio, Mireille Farzan, Laura Puumala, Mikko M. Puumala, Oskari Sivula, Tiia Topinoja, and Juho Vaiste

Abstract Plant ethics has its own set of questions, concepts, and philosophical assumptions, of which many are novel and call for original ways of thinking. This chapter focuses on how some common methods in philosophical theorizing, such as analogies and thought experiments, fare in the context of plant ethics. Based on plant ethics literature, we identify and examine two methodological strategies for plant ethics. The first strategy focuses on finding possible morally relevant similarities between plants and animals, including humans. Based on these similarities, familiar arguments from animal and traditional ethics can be extended to plant ethics. This can be called the extensionist strategy. Our focus will be on analogy argumentation and analysis. The second strategy focuses on highlighting an ontological division between plants and animals. According to this strategy, morally relevant properties are different for plants than they are for animals. This will be called the otherness strategy. Here our focus will be on conceptual analysis, and we will discuss a constructive and deconstructive approach.

10.1 Introduction

Plant ethics has its own set of questions, concepts, and philosophical assumptions, of which many are novel and call for original ways of thinking. How to determine our moral obligations concerning plants? When is it wrong to kill a plant? Are there moral reasons to value crop plants over wild plants or vice-versa? There is no clear methodological approach to these questions. This chapter focuses on how

J. Aarnio · M. Farzan · L. Puumala · M. M. Puumala (✉) · O. Sivula · T. Topinoja
Department of Philosophy, Contemporary History and Political Science, University of Turku, Turku, Finland

J. Aarnio
e-mail: jenna.k.aarnio@utu.fi

J. Vaiste
Department of Management and Entrepreneurship, University of Turku, Turku, Finland

M. Di Paola (ed.), *The Vegetal Turn*, The International Library of Environmental, Agricultural and Food Ethics 38, https://doi.org/10.1007/978-3-031-68671-9_10

some common methods in philosophical theorizing, such as analogies and thought experiments, fare in the context of plant ethics.

In animal ethics, one can be guided by similarities between humans and other animal species. For *plant* ethics, this may not work. Arguably, according to our common-sense morality, in most cases there never is anything wrong with harming plants. Instead, many of our considered judgements hold that it is morally acceptable to harm plants by cutting, grafting, harvesting, and eating them. Consider, for example, the cultural habit of putting cut flowers in a vase: having the dying reproductive organs of a plant on display usually does not incite moral concern—on the contrary, it is commonly considered not only morally acceptable, but romantic or beautiful.

The novelty of plant ethics seems to place the burden of proof onto those who challenge the common-sense view and advocate for plant ethics. Methodological challenges might result from this novelty in the form of untrustworthy moral intuitions or inability to recognize plant ethical questions as moral problems. How should one go about rejecting the common-sense view?

Questions in plant ethics can be approached from multiple perspectives. This chapter is mainly interested in approaches that assume that moral considerability can be grounded in plants themselves, i.e., that plants have intrinsic value.[1] Under this assumption two central strategies can be identified. The first strategy focuses on finding possible *morally relevant similarities* between plants and animals, including humans. Based on these similarities, familiar arguments from animal and traditional ethics can be extended to plant ethics (Pouteau 2014, 7–8; Sandler 2018, 43.) This can be called the *extensionist strategy* (cf. Kalhoff 2014; Pellegrino 2018.) The second strategy focuses on highlighting an *ontological* division between plants and animals (Pouteau 2014). According to this strategy, morally relevant properties are different for plants than they are for animals. This will be called the *otherness strategy (for a similar division see Weissglass' chapter (16) in this same volume).* In this section, both strategies will be presented, and the methodological issues related to them will be discussed.

10.2 The Extensionist Strategy

The extensionist strategy plays on the *qualitative similarities* between plants and animals by identifying certain properties thought to be pivotal for moral considerability that are shared by plants and animals alike. There are two complementary ways to realize the extensionist strategy. The *categorical* approach aims to identify categorical properties that are evidently shared between plants and animals, for example being alive (e.g., Goodpaster 1978.) The *scalar* alternative is to identify properties that are shared in differing degrees by plants and animals, placing plants on a continuum with animals with respect to these shared properties—for example

[1] On the axiologies of plants, see Pellegrino (2018).

flourishing (e.g., Attfield 1981), interests (e.g. Johnson 1993), integrity (e.g., Rolston 2002), dignity (e.g., Koechlin 2009; Odparlik 2018), sentience (e.g., Hall 2011), and intelligence (e.g., Hall 2009; Trewavas 2014). These extensionist strategies—*categorical* or *scalar*—do not deny that there are some qualitative differences between plants and animals, but these differences are simply not seen as morally relevant. Just as, in animal ethics, it should not matter morally that bats can echolocate but cows cannot, in the extensionist viewpoint, it does not matter morally that ferns are capable of photosynthesis while dogs are not. What matters morally is that a being fulfills criteria specific to moral considerability. Because the extensionist strategy plays on morally relevant similarities between plants and animals, it does not necessarily involve new philosophical machinery. It rather widens the area of application of the old. In the following section, some extensionist methodological tools are considered.

10.2.1 Conceptualizing Analogousness

Extensionism seems to rest upon an argument *from* analogy: the 'moral circle' (see Singer 2011) can be extended from animals to plants because plants share morally relevant features with animals. The first step is to use conceptual analysis that defines the properties ostensibly shared by plants and animals. The aim is to remove as much vagueness and ambiguity as possible from the definitions, thus clarifying the meaning and extension of the concepts in use (Fosl and Baggini 2020, 34–36.). The analysis phase is followed by a synthesis phase whereby the defined concept is articulated and elaborated upon.

Conceptual analysis is not only descriptive of how a concept is used. It is also prescriptive of how it should be used: what ought to be meant and understood by it. This is typically done by explicating *necessary* and *sufficient* conditions for its application. A necessary condition is a condition which must be met for a given object or property to qualify as an instance of the concept being analyzed (Fosl and Baggini 2020, 219–220). For example, it seems rather uncontroversial that a necessary condition for flourishing is that the organism is viable. For the extensionist strategy, concepts like flourishing need to be analyzed in a way that animal 'flourishing' can be construed as relevantly similar to the flourishing of plants.

Necessary conditions alone are not enough for unambiguous definitions: sufficient conditions must also be satisfied. A sufficient condition is such that meeting it suffices for the object or property to qualify as an instance of the concept being analyzed. A sufficient condition is often not some singular individual condition but a compilation of (often individually necessary) conditions that when fulfilled, together qualify as a sufficient condition (see Fosl and Baggini 2020, 220). For example, sufficient conditions for flourishing may include that an organism is viable throughout its life and is capable of accomplishing its typical life cycle as well as expressing typical characteristics of its species description (Kalhoff 2014, 687).[2]

After clarifying key concepts, the next methodological step is to form an analogy. There are two phases to using analogies: analogical *explanation* and analogical *inference*, or argument *by* analogy (Macagno and Walton 2009; Markman 1997).

10.2.2 Analogical Explanation: Strong and Weak Analogies

An analogical explanation is a claim that two or more cases resemble each other in relevant ways. This is often done to better understand a new or unusual concept or phenomenon through a familiar one (Bailer-Jones 2002; Fosl and Baggini 2020, 59; Juthe 2005, 3). According to André Juthe, analogousness does not convey mere similarity, but rather resemblance. Resemblance means a correspondence between the object of comparison, called the Target-Subject, and the Analogue, namely the object that is compared with the Target-Subject (Juthe 2005, 1, 4). Hence, simply claiming that plants are similar to animals does not lead to an analogical explanation; while a claim that some property or relation p in plants resembles, that is, *corresponds* to a property or relation p* in animals, can be a form of analogical explanation.

The next step is to justify the claim by evaluating whether the correspondence between the properties or relations creates a strong analogy. According to Peter Fosl and Julian Baggini (2020), the strength of an analogy depends on the quantity of qualitatively relevant correspondences. A strong analogy is one where the cases have many similarities and do not have too many dissimilarities concerning their qualitatively relevant correspondences. An analogy may be weak if (a) there are only a few similarities, and/or if (b) the similarities only concern superficial or irrelevant correspondences and/or if (c) there are too many relevant dissimilarities (Fosl and Baggini 2020, 60).

Identifying relevant similarities and differences is a salient problem in the plant ethical context. For example, intelligence has been used both as a separating feature and as a feature that unifies both plants and animals with respect to moral considerability. Plant neurobiological studies suggest that climbing plants seem to exhibit an analogue of intelligence, if 'intelligence' is understood loosely enough. These plants are able to perceive affordances such as 'climbability', which requires controlling the gap between the current state of the vine-support system and the desired goal state of reaching the support (Calvo 2016, 1336). This ability may be interpreted as a sign of 'minimal intelligence'. Yet thus far 'lower' levels of intelligence in animals have been a reason *not* to extend human-level moral considerability to non-human

---

[2] Note that Angela Kallhoff (2014) defends the moral standing of plants based on their capacity to flourish. However, Kallhoff explicitly claims that her approach is not extensionist. Kallhoff conceptualizes flourishing in a way that aims to avoid the pitfalls of extensionist strategies and instead pay tribute to the 'otherness' of plants. Yet, clearly, flourishing can also be used within the extensionist strategy. The key would then be to conceptualize flourishing of animals and plants to be relevantly similar.

animals. Thus, due to its lower levels in comparison to non-human animals, the 'minimal intelligence' expressed in plants could be a reason also *not* to regard plants as morally considerable, even if intelligence *is* a condition of moral considerability. This illustration shows the challenges in defining differences or similarities that make or break an analogy.

10.2.3 Analogical Inference: Argument by Analogy

Following analogical explanation, a common next step is analogical inference, that is, argument *by* analogy. According to it, certain correspondences in analogous cases imply further correspondences between the cases. Take the example of flourishing. Both animals and plants have a need and ability to strive towards homeostasis that is affected by certain preconditions. Some preconditions are conducive to animal well-being and plant flourishing. Arguably, flourishing in plants corresponds to well-being in sentient beings. No wonder then that flourishing is often used as a metaphor to describe the well-being of sentient beings; all organisms wither and die without water, and nutrition—and, in the right conditions, they are said to flourish (Bailer-Jones 2002, 114).

In forming an argument *by* analogy, suggested correspondence between well-being and flourishing is taken to imply further correspondences between plant and animal cases. This correspondence might be in a shared ability to have interests, even though for plants having interests need not mean 'taking an interest' (Sandler 2018, 41). An additional correspondence might involve a possibility to be harmed as opposed to simply being damaged. Here, it is suggested that it is possible to harm plants somewhat similarly if not to the same degree as animals. These correspondences, however, are not enough for arguing that the normative claims that apply to animals also apply to plants. To do so, *parity of reasoning* is often used.

10.2.4 Argument by Parity of Reasoning

Parity of reasoning guides an argument *from* analogy. Unlike the argument *by* analogy, an argument *from* analogy does not argue for analogousness between cases. Rather it argues that due to a pre-established analogousness of two cases, normative conclusions made about one case also apply to the other. Parity of reasoning stipulates that *from similar premises a similar conclusion must follow*. (Fosl and Baggini 2020, 59). It is important to analyze the structural similarities between two arguments, because the similarity that an argument from analogy plays on is not an analogy for example between animal and plant flourishing, but between normative arguments constructed from these cases respectively (Woods and Hudak 1989, 127, 128, 133).

This is illustrated in the following example.

1) Animals have the potential to 'flourish', F, and the possibility to be harmed, H. Hence, they should be treated in some way W.

And

2) As plants have a similar potential to flourish, F*, and supposedly a similar ability to be harmed H*, they should be treated in some way W*.

There is a correspondence between F and F* and H and H*. The 'flourishing' of animals F is analogous to flourishing of plants F* and, through analogical inference, a further similarity between H and H* is established. W and W* are normative conclusions derived from the premises. For example, W could stand for 'one ought to let animals flourish and not harm them unnecessarily', and W* could stand for 'one ought to let plants flourish and not harm them unnecessarily'. Parity of reasoning focuses on W and W*. Here, the argument is no longer about the correspondence between F and F* or H and H*; nor is it about whether F, F*, H and H* are morally relevant properties. The answers to these questions will have been established beforehand, in the formation of the analogy and the subsequent inference. What is argued *from* the analogy, by parity of reasoning, *is that the conclusions W and W* should be similar as well.* Given the similarities between the premises in both cases, a similar conclusion must follow (for want of auxiliary premises) in both cases (Fosl and Baggini 2020, 59; Woods and Hudak 1989).

If there are no morally relevant, divisive factors between plants and animals, the same moral obligations apply in both cases. On the other hand, parity of reasoning can also work the other way: If the arguments have different normative conclusions, then it must be shown that plants and animals do not share (enough) similar relevant properties, and/or that there are also relevant distinguishing properties that justify drawing different normative conclusions. For example, one could add premises about plants' lesser capacity to have interests: this could justify a differing normative conclusion, where interests of plants do not have to be considered similarly as interests of animals. This move would account for our common-sense morality about the treatment of plants (See Sandler 2018). Goodpaster presents another approach by distinguishing moral considerability from moral significance. While moral considerability is about the moral status of an entity, moral significance "aims at governing comparative judgments of moral 'weight' in cases of conflict". (Goodpaster 1978, 311). Here, moral "weight" can be used to explain the common-sense morality that results in different normative conclusions in plant and animal cases.

Arguments from analogy are leading candidates if the aim is to 'bridge the gap' between plant and animal ethics. If moral intuitions are more coherent in animal cases, analogies can be built between animal and plant cases to have more familiar animal cases as a guideline for normative argumentation in plant ethics. Through analogies and parity of reasoning, it is possible to point out inconsistencies and biases in our attitudes toward plants compared to animals. On the other hand, they may also serve as a method to deconstruct *reductio ad absurdum* arguments, where otherwise acceptable premises lead to absurd conclusions, possibly because the analogies are too weak to yield satisfying arguments by parity of reasoning.

10.2.5 Methodological Challenges for the Extensionist Strategy

There are several challenges to the extensionist strategy (Weissglass also discusses challenges to the extensionist strategy in Chapter 16 of this volume). Firstly, analogies are an unstable form of argumentation (Perelman and Olbrechts-Tyteca 2008, 393), meaning that analogical arguments are not truth-conserving. According to some, analogical inference is almost like a form of induction: cases that share several relevant similarities often seem to share other similarities as well. This way analogy-based methods may inherit the problems of induction (Fosl and Baggini 2020, 10–13).

Second, much of the methodology in the extensionist strategy resembles *abductive argumentation*. Abductive argument is an argument to the *best* explanation. However, it fails to guarantee that the selected explanation is the *correct* explanation. Sometimes the 'best' explanation may even serve as an obstacle for acquiring true beliefs (Fosl and Baggini 2020, 50). This may very well be the case in plant ethics, where biases towards human and animal ethics as the best available normative logic may prevent searching and finding other viewpoints that would be more suitable in the context of plant ethics.

Third, it seems clear that plants have some decisive differences compared to animals. Even if those differences are not paramount for moral considerability, they may render analogies and arguments from analogy unsatisfactory and unintuitive. Plants may simply be too alien to humans and animals, making it difficult to extend traditional moral concepts and normative conceptions to them. The *availability error* is an evident risk with the extensionist strategy. Availability error in this context means that because there is limited evidence of the ontology and phenomenology of plants available to humans, inductive reasoning about them may be led astray by incongruous associations with the human/animal case. Humans are simply locked into their own perspective as humans, or animals at most (Fosl and Baggini, 2020).

Many have claimed that considering the interests of plants is absurd (Davion 2006; Kaufman 1994; Sandler 2018; Schmidtz 1998; Thompson 1990; Williams 1995). *Reductio ad absurdum* arguments are arguments aimed at showing that applying some otherwise plausible normative principle to certain contexts would be absurd. Absurdity can also be demonstrated by stretching a contested moral principle in a thought experiment to a context where it produces absurd or counterintuitive conclusions. Some of these arguments suggest that plants cannot have inherent worth, because it is intuitively permissible to treat plants in ways one could not treat humans (Schmidtz 1998); or that taking plants into moral consideration would be absurd since it would also imply that all kinds of artifacts are morally considerable—even though, intuitively, they are not (Kaufman 1994; Thompson 1990).

However, most *reductio ad absurdum* arguments only function when concepts, definitions and conclusions are understood in specific ways. Absurdity of an argument rests on concept definitions or ambiguity of key phrases. Sandler (2018, 49) argues against the absurdity of considering the interests of plants. He argues that due to the different nature of the interests considered, this consideration need not lead to extending to plants the same rights and treatment that we offer to animals. Both animals and plants have interests, but the interests of animals are not to be understood as identical to those of plants. If one does understand them so, then that specific conceptualization of 'having interests' will make the argument absurd; but that understanding, Sandler argues, is wrong.

Intuition presents a further problem to the extensionist strategy. Intuitions are often untrustworthy in complicated or unfamiliar cases—as they may more easily be biased, inconsistent, or unclear (see e.g., Greene 2014). One consequence may be that what seems absurd to some do not seem that way to others. Christopher Stone's *Should Trees Have Standing?* (2010), for example, explores the nature of legal rights and their relation to morality concerning environmental issues. The influential book argues for granting legal rights to the environment, individual plants included. Since 2008 Ecuador has applied this conclusion to their ecocentric constitution, granting legal rights to nature and recognizing the inalienable rights of ecosystems to exist and flourish (Tanasescu 2013). Stone's theorizing and Ecuador's legal practices show that intuitions may vary, and consequently that absurdity is, at least in those cases where it is intuitively established, perspectival.

In conclusion, the extensionist strategy plays on *similarities* between plant and animal life by identifying certain properties pivotal for moral considerability that are shared by both, such as sentience or the capability to flourish. These similarities can be taken as a starting point for expanding the circle of moral considerability, though moral consideration of plants may well differ from that of animals. The challenge with this strategy is to find and recognize relevant shared properties and relations. The extensionist must admit that even though expanding the circle is the only plausible way of taking plants into moral consideration, this strategy might still be flawed or lead to absurd conclusions. This is because in addition to certain shared properties, there is also something fundamentally different between plants and animals, including humans. The next section addresses this matter of an *ontological divide* between plants and animals.

10.3 The Otherness Strategy

According to the *otherness strategy*, plants and animals are qualitatively different. This strategy assumes that the morally relevant qualities in plants and animals are categorically different. Suggestions of potentially morally relevant plant-unique properties include 'non-centredness' and 'unlimited mode of being' (see Pouteau 2014), while not all plant-unique properties need to have moral relevance.

There are at least two approaches that an otherness strategist might take: a deconstructive and/or constructive approach. The *deconstructive approach* to ontological otherness focuses on criticizing the extensionist strategy and draws attention to ontological differences between plants and animals. This approach deconstructs the analogies and subsequent parity arguments without aiming to establish alternative conceptualizations or arguments about plant-unique morally relevant properties. The *constructive approach* attempts to construct alternative concepts and arguments for plant-unique properties. It does not only point out the fundamental difference between plants and animals but aims to characterize said difference. These approaches to the otherness strategy are not mutually exclusive but reinforcing, and in both the focus is on the otherness of plants.

This section focuses on reviewing some methods that could be useful to the otherness strategy, such as conceptual analysis and thought experiments. However, arising methodological challenges cast doubt on the suitability of analytical methods for the otherness approach. Therefore, non-analytical methods are also briefly considered.

10.3.1 Constructive Conceptual Analysis

If the otherness strategy takes a purely deconstructive stance, then a typical disassembling conceptual analysis is clearly useful. But if the aim is to redefine or implement new concepts related to plant-unique properties, a more creative form of conceptual analysis is needed. According to Milos Kosterec (2016, 221–222), a *constructive* conceptual analysis is an analysis that enables the introduction of new concepts initially lacking in a theory. Such activity can also be called conceptual engineering (see e.g., Cappelen 2018; Isaac 2020; Chalmers 2020). Thus, constructive conceptual analysis or conceptual engineering can be used to redefine and provide new ways of understanding plant-related concepts as part of the otherness strategy.

For example, J. L. Arbor (1986, 337–339) introduces new approaches to 'naturalness' and 'integrity' relating to plants. These traits can be tied to self-generated end-states plants seem to have. Naturalness and integrity may be violated when plants are harmed. This is a way to bypass the claim that plants would suffer while still claiming that disvalue can be caused to them. Arbor's argument aims to show that having sensations is not a prerequisite for being an object of moral consideration. Naturalness and integrity may prove as plant-unique properties that create moral questions even when plant interests conflict with non-trivial animal interests such as hunger. Arbor even argues that "sometimes animals, their states and conditions, do not come first" (1986, 339).

Similarly, Vera Bühlmann (2019) provides a non-biochemical analysis of the concept of photosynthesis. Bühlmann defines photosynthesis as economic events of spending, banking, hosting, and servicing the fluid movement of the universe. Bühlmann takes distance from reducing photosynthesis to a chemical process that captures sunlight and converts it to biomass. Instead, her analysis is meant to retain astonishment and awe to plant-unique properties. Bühlmann's analysis suggests that some sources of plants' intrinsic value might have been overlooked and can be acknowledged by substituting the reductive procedural and functional view with one that adds awe to the phenomenon. In both Arbor's and Bühlmann's constructive conceptual analyses, plant-unique properties are evaluated 'in their own right', rather than through the prism of human or animal interests.

10.3.2 Phenomenology: Embracing the Gap Instead of Bridging It

While the extensionist strategy aims to 'bridge the gap' between plants and animals, the methods suitable for the otherness strategy acknowledge and exploit it. Phenomenology is a case in point. Phenomenology is not only a method but a branch of philosophy, which emphasizes the value and meaning of subjective experience (Qutoshi 2018). This approach could serve as a tool to experience and name the gap between plants and animals. Here, the aim is not to get rid of the human perspective. Rather, the purpose of phenomenological approaches can be to experience the very difference between human and plant perspectives, so as to measure the otherness of plants.

With *deconstructive analysis*, perspectives and interpretations that have been concealed or overlooked may be uncovered. Jacques Derrida's deconstruction is a form of philosophical and textual analysis that questions the hierarchical binary oppositions embedded in Western thinking, such as nature-culture, human-animal, man-woman, mind–body (Derrida 2007). Typically, deconstruction aims to present these binaries as rather intertwined and inseparable, but not in a way that would assimilate one to the other. According to Derrida's deconstructionism, the meaning of subjective experience arises in relation to that which the subject is not. In plant ethics, the otherness strategist might turn to classic theories to see the sort of meanings that are attached to non-animal nature or plants. For example, for John Locke, the value of non-animal nature is entirely dependent on its use for humans, suggesting that nature left to its natural states amount to little more than *waste* (Locke 2003, II §42). The search for negations of that view, for example missing plant-centered narratives and contrary interpretations that see plants as independently morally valuable (the opposite of waste), may uncover something new about the value of plants or plant ethics in general.

10.3.3 Methodological Challenges for the Otherness Strategy

Methods of the otherness strategy—like constructing or engineering new concepts, making new distinctions, and searching for missing and contrary interpretations—seem to presuppose a fundamental difference between plants and animals. But it is difficult to consistently explain and communicate the qualitative differences that make plant-otherness (in Chapter 16 of this volume Weissglass also identifies this challenge for the otherness strategy). The otherness strategy also seems to implicitly assume fundamental similarities all animals share, which then in part delineate the 'otherness' of plants.

It seems difficult to approach plant-otherness from a non-animal perspective. One might hope that thought experiments could be helpful in demonstrating these independent values in plants that are not tied to human or animal existence or evaluation. Thought experiments in moral theorizing can be hypothetical or imaginary examples that can be used for defending an argument or exposing moral intuitions in different cases (Jamieson 1993, 682). According to Elke Brendel (2004), thought experiments are mental representations of scientific experiments. These experiments involve planned and controlled manipulation of data to show how variables are dependent on each other. The aim of thought experiments often is to elicit intuitive responses that may then guide further claims or judgments (Brendel 2004, 91–92).

One such thought experiment is Scherer's (1982) *Flora* and *Lifeless*. Picture two far away planets: 'Lifeless', a planet without life but full of beautiful unique landscapes, and 'Flora', a planet full of plant life with no animal and human creatures. Assume that both planets are too distant to be of any use or concern to humans or animals. If you had to destroy one of them, which one would you choose? Arguably, most of us would likely destroy Lifeless and spare Flora. If so, this judgment may show that Flora has more perceived value than Lifeless. Furthermore, the judgment of sparing Flora shows that its perceived value is independent of its usefulness to the perceiver: intuitively, then, Flora has intrinsic value (Scherer 1982). The merit of this thought experiment is that it aims to show the intrinsic value of plants. (Pellegrino 2018.)

However, it is not clear what can be inferred from intuitions elicited from thought experiments like Scherer's, which take place in imaginary worlds. Moral intuitions are formed when one takes the relevant moral features into consideration. If we do not have enough information about the world the thought experiment takes place in (a world where it is possible to destroy whole planets from afar), we might not be aware of the relevant moral features of the events that take place in that world. Thus, there is a risk that intuitions derived from these experiments are untrustworthy (Jamieson 1993, 683).

10.4 Conclusions

This chapter has considered some methodological viewpoints, questions, and challenges in and for plant ethics. It identified two methodologically distinct plant ethical strategies: the extensionist strategy and the otherness strategy. In the extensionist strategy the categorical approach identifies categorical properties that are evidently shared between plants and animals. The scalar approach identifies properties that are shared in differing degrees by plants and animals, placing them on the same continuum. In the otherness strategy, the constructive approach aims to create new ways to understand plant otherness, while the deconstructive approach aims to expose concealed interpretations about plant-unique properties.

Some issues stemming mainly from untrustworthy intuitions concerning plants were addressed with reference to both approaches. The extensionist strategy may seem methodologically friendlier, as it aims at 'bridging the gap' between plants and animals through analogies, and extending ordinary moral concepts, principles, and arguments to plants by parity of reasoning. However, it seems clear that there is something fundamentally different in and about plants that may render the aims and conclusions of the extensionist strategy weak, implausible, or even absurd.

The otherness strategy focuses on and indeed exploits precisely these fundamental differences between plants and animals, including humans. It deconstructs and criticizes the analogies made by extensionists. Otherness strategists may conceptualize plant-uniqueness in creative new ways. The merit of this strategy is that it respects the uniqueness of plants, but it may be methodologically even more challenging to carry out than the extensionist strategy. That these two strategies were explored in this chapter in no way excludes the possibility of other methodological approaches to plant ethical questions. For example, some kind of a hybrid strategy that combines elements from the two might be possible. However, while in a way quite commonsensical, and surely worth pursuing in the future, hybrid strategies are always at risk of inheriting problems from both sides.

References

Arbor, J.L. 1986. Animal Chauvinism, Plant-regarding Ethics and the Torture of Trees. *Australasian Journal of Philosophy* 64 (3): 337–338.

Attfield, Robin. 1981. The Good of Trees. *Journal of Value Inquiry* 15 (1): 35–54.

Bailer-Jones, Daniela. 2002. Models, Metaphors and Analogies. In *The Blackwell Guide to the Philosophy of Science*, ed. Peter Machamer and Michael Silberstein, 108–127. Malden: Blackwell Publishers.

Brendel, Elke. 2004. Intuition Pumps and the Proper Use of Thought Experiments. *Dialectica* 58 (1): 89–108.

Bühlmann, Vera. 2019. Photosynthesis. *Philosophy Today (Celina)* 63 (4): 1037–1050.

Calvo, Paco. 2016. The Philosophy of Plant Neurobiology: A Manifesto. *Synthese* 193: 1323–1343.

Cappelen, Herman. 2018. *Fixing Language: An Essay on Conceptual Engineering*. Oxford: Oxford University Press.

Chalmers, David J. 2020. What is Conceptual Engineering and What Should It Be? *Inquiry*: 1–18.

Davion, Victoria. 2006. Itch Scratching, Patio Building, and Pesky Flies. *Environmental Ethics* 28 (2): 115–128.

Derrida, Jacques. 2007. *Psyche: Inventions of the Other, Volume I* (first published as *Psyché, Invention de l'autre* 1987), ed. P. Kamuf and E. Rottenberg. Stanford, Calif: Stanford University Press.

Fosl, Peter S., and Julian Baggini. 2020. *The Philosopher's Toolkit: Compendium of Philosophical Concepts and Methods*, 3rd ed. Hoboken: Blackwell Publishing.

Goodpaster, Kenneth E. 1978. On Being Morally Considerable. *The Journal of Philosophy* 75 (6): 308–325.

Greene, Joshua. 2014. Beyond Point-and-Shoot Morality: Why Cognitive (Neuro)Science Matters for Ethics. *Ethics* 124: 695–726.

Hall, Matthew. 2009. Plant Autonomy and Human-Plant Ethics. *Environmental Ethics* 31 (2): 169–181.

Hall, Matthew. 2011. *Plants as Persons: A Philosophical Botany*. Albany: State University of New York Press.

Isaac, Manuel. G. 2020. How to Conceptually Engineer Conceptual Engineering? *Inquiry*: 1–24.

Jamieson, Dale. 1993. Method and Moral Theory. In *A Companion to Ethics*, ed. Peter Singer, 476–490. Oxford: Blackwell Publishers.

Johnson, Lawrence. 1993. *A Morally Deep World: An Essay on the Moral Significance and Environmental Ethics*. Cambridge: Cambridge University Press.
Juthe, André. 2005. Argument by Analogy. *Argumentation* 19: 1–27.
Kallhoff, Angela. 2014. Plants in Ethics: Why Flourishing Deserves Moral Respect. *Environmental Values* 23: 685–700.
Kaufman, Frederik. 1994. Machines, Sentience, and the Scope of Morality. *Environmental Ethics* 16 (1): 57–70.
Koechlin, Florianne. 2009. The Dignity of Plants. *Plant Signaling & Behavior* 4 (1): 78–79.
Kosterec, Miloš. 2016. Methods of Conceptual Analysis. *Filozofia* 71 (3): 220–230.
Locke, John. 2003. *Two Treatises of Government: And a Letter Concerning Toleration*, ed. I. Shapiro (originally published in 1689). New Haven: Yale University Press.
Macagno, Fabrizio, and Walton Douglas. 2009. Argument from Analogy in Law, the Classical Tradition, and Recent Theories. *Philosophy & Rhetoric* 42 (2): 154–182.
Markman, Arthur B. 1997. Constraints on Analogical Inference. *Cognitive Science* 21 (4): 373–418.
Odparlik, Sabine. 2018. The Dignity of Plants: An Overview of the Discussion in German-speaking Countries. In *Plant Ethics: Concepts and Applications*, ed. Angela Kallhoff, Marcello Di Paola, Maria Schörgenhumer, 59–69. Schörgenhumer. New York: Routledge.
Pellegrino, Gianfranco. 2018. The Value of Plants: On the Axiologies of Plants. In *Plant Ethics: Concepts and Applications*, ed. Angela Kallhoff, Marcello Di Paola, and Maria Schörgenhumer, 13–29. New York: Routledge.
Perelman, Chaïm, and Lucie Olbrechts-Tyteca. 2008. *The New Rhetoric: A Treatise on Argumentation*, trans. John Wilkinson and Purcell Weaver Indiana: University of Notre Dame Press.
Pouteau, Sylvie. 2014. Beyond 'Second Animals': Making Sense of Plant Ethics. *Journal of Agricultural and Environmental Ethics* 27: 1–25.
Qutoshi, Sadruddin B. 2018. Phenomenology: A Philosophy and Method of Inquiry. *Journal of Education and Educational Development* 5 (1): 215–222.
Rolston, Holmes III. 2002. What Do We Mean by the Intrinsic Value and Integrity of Plants and Animals? In *Genetic Engineering and the Intrinsic Value and Integrity of Animals and Plants. Proceedings of an Ifgene Workshop*, eds. David Heaf and Johannes Wirz, 5–10. Hafan: Ifgene.
Sandler, Ronald. 2018. Is Considering the Interests of Plants Absurd? In *Plant Ethics: Concepts and Applications*, ed. Angela Kallhoff, Marcello Di Paola, and Maria Schörgenhumer, 40–50. New York: Routledge.
Schmidtz, David. 1998. Are All Species Equal? *Journal of Applied Philosophy* 15 (1): 57–67.
Scherer, Donald. 1982. Anthropocentrism, Atomism, and Environmental Ethics. *Environmental Ethics* 4 (2): 115–123.
Singer, Peter. 2011. *The Expanding Circle: Ethics, Evolution, and Moral Progress*. New Jersey: Princeton University Press.
Stone, Christopher D. 2010. *Should Trees Have Standing? Law, Morality, and the Environment*. Oxford: Oxford University Press.
Tanasescu, Mihnea. 2013. The Rights of Nature in Ecuador: The Making of an Idea. *International Journal of Environmental Studies* 70 (6): 846–861.
Thompson, Janna. 1990. A Refutation of Environmental Ethics. *Environmental Ethics* 12 (2): 147–160.
Trewavas, Anthony. 2014. *Plant Behaviour and Intelligence*. Oxford: Oxford University Press.
Williams, Bernard. 1995. Must a Concern for the Environment be Centred on Human Beings? In *Making Sense of Humanity and Other Philosophical Papers 1982–1993*, ed. Bernard Williams, 233–234. Cambridge: Cambridge University Press.
Woods, John, and Brent Hudak. 1989. By Parity of Reasoning. *Informal Logic* 6 (3): 125–139.

Chapter 11
Are Plants Like Patients in Persistent Vegetative States?

Ethan C. Terrill

Abstract Obviously, plants and patients in persistent vegetative states (PPVS) are biologically alive. Plants are capable of circumnutation, growth, and photosynthesis; PPVS breathe, pump blood, and exhibit spontaneous reflexes. The more pressing question is whether they have cognitive capacities. In both cases, there are those who provide a positive and a negative response to the question. This leaves us in a state of indeterminacy. How, then, can we make moral decisions about plants and PPVS based on indeterminacy? This chapter explores the notion of indeterminate moral status in the case of plants and PPVS, which can help us reach conclusions in practical ethics that involve entities with similarly indeterminate cognitive capacities.

11.1 Introduction

Many of us think that we at least owe *patient(s) in persistent vegetative states* (PPVS)—briefly, an individual in "a wakeful unconscious state"—either (i) a good death if we have reason to believe their continued existence is not worth living and there is a negligible probability of recovery, or (ii) life-sustaining resources if we have reason to believe in a non-negligible probability of recovery (Beauchamp and Childress 2019, 11).[1] Many of us also think that—being living organisms—plants are loosely in functionally 'wakeful' unconscious states. Yet many think that we do not owe anything to plants. Suppose your plant begins to wilt due to lack of water: it may be de rigueur to keep your plant from wilting, but many ordinarily think that you need not go out of your way to address the plant's wilting. But people do go out

[1] This says nothing about supervening bioethical principles that may inform or disrupt our reasons to believe either (i) or (ii). Note also that the terminology of PPVS is definitionally unclear. The clinical definition involves a temporal element, i.e., PPVS persist in their state for a month or more, with the prospect of recovery. However, some bioethicists claim that PPVS entails permanency, or a zero probability of recovery, and not merely persistence (DeGrazia et al. 2011, 321–322). While it is unclear how to diagnostically define PPVS, it is unnecessary to make clear for this chapter.

E. C. Terrill (✉)
Washington, D.C., USA
e-mail: ecoleterrill@gmail.com

M. Di Paola (ed.), *The Vegetal Turn*, The International Library of Environmental, Agricultural and Food Ethics 38, https://doi.org/10.1007/978-3-031-68671-9_11

of their way to address the vital needs of PPVS and some would describe care for PPVS as morally obligatory. If the wakeful unconscious states of both entities are prima facie similar, why are there duties towards PPVS and not to plants? And, more importantly, what grounds these duties?

The best conceptual tool to answer these kinds of questions is that of *moral status*, which is about whether some entity has a sake or is "vulnerable to harm or wrongdoing" because that entity has morally relevant interests that matter for its own sake (DeGrazia and Millum 2021, 176; Jaworska and Tannenbaum 2021). Merely being alive is not in itself good or bad for plants and PPVS, which is to say that being alive alone does not confer prudential values along some welfare function, such that moral agents have corresponding moral duties to either plants or PPVS (Višak 2018, 36). A stronger criterion for moral status is the *cognitive capacity account*. Under this account, we have prima facie reasons to believe that plants and PPVS are de minimis unqualified for moral status consideration. There is, however, emerging empirical evidence for the existence of (at least some) cognitive capacities for both plants and PPVS. But does *sufficient* emerging empirical evidence exist to decide whether either/both entities have a specific degree of cognitive capacity to confer (at least some) moral status?

In this chapter, I explore the question of whether plants are like PPVS, in moral terms, or whether there are morally salient differences. In so doing, I do not purport to provide definitive arguments for the conclusion that plants or PPVS have determinate cognitive capacities. On the contrary, I will show the indeterminacy of cognitive capacities for plants and PPVS, borrowing from recent literature discussing the indeterminacy of cognitive capacities in marginal cases (Maher 2021; Birch 2021).[2] Through this lens, I focus primarily on how plants and PPVS figure into ongoing debates in moral status and ethics more broadly; and argue that indeterminacy in cognitive capacities entails an *indeterminate moral status*. So, in this case, plants are like PPVS insofar as they both have indeterminate moral status. And, following Aristotle's *formal equality principle*, we may treat like cases as like ceteris paribus (Gosepath 2021). However, this poses an ethical dilemma, which I call the *spookiness of indeterminate moral status*: without knowing the cognitive capacity variable of an entity with indeterminate moral status, we do not know how to correspondingly act towards, or treat, that entity within our moral calculus. Given this dilemma, I also explore the claim that indeterminate moral status just is a lower degree of partial moral status, and that we should embrace the humility of indeterminacy. Invoking indeterminate moral status, therefore, is sufficient to conclude that plants and PPVS both have a lower degree of partial moral status compared to other entities with determinate rudimentary cognitive capacities and partial moral status—but a degree, nonetheless. To be clear, the claim being made is that while there may not be necessary logical

[2] Maher uses the term *indeterminacy* differently than Birch. For instance, Birch uses indeterminacy to refer to vagueness, uncertainty, and/or an admission of borderline cases in consciousness (Maher 2021, 137). Maher, on the other hand, uses indeterminacy to refer to the predicate "'…is conscious'" (ibid.). For example, Maher writes that "There is currently no fact of the matter as to whether plants belong in "the '…is conscious' predicate extension" (ibid., 140). I favor Birch's use of the term for this chapter.

connections between indeterminate moral status and partial moral status, it is still a morally significant aim to err on the side of the *precautionary principle*. By the end of the chapter, it should be clear that we need to update our philosophical commitments, especially in light of the vegetal turn, to revise neglected minimal duties, since rudimentary cognitive capacities and the lower degrees of moral status are richer than we previously imagined. It would follow that we ought not unnecessarily exploit entities with indeterminate moral status.

11.2 (Cognitive) Capacities and Indeterminacy

Most arguments about the interesting (cognitive) capacities and behaviors of plants and PPVS are often met with incredulity. But of the two, plants are especially vulnerable to assertions about their not having cognitive capacities. Plants are very unlike paradigm adult humans—standard bearers of sophisticated cognitive capacities—as plants lack a wide variety of shared traits, including a head, bilaterally symmetrical body, nervous systems, etc. (Maher 2021, 150). PPVS are more complicated. Unlike plants, PPVS have physical system consistency with humans. The exception, of course, is that PPVS are typically defined by a lack of both awareness of self and their surrounding environment (Stone 2007). Still, it is worth investigating the possibility that both plants and PPVS have functioning cognitive capacities.

Cognitive capacities are neither value neutral nor entirely unambiguous, so I utilize the term broadly. Simultaneously, cognitive capacities exist on a rich and wide spectrum. Cognition can be quite complex and sophisticated on one end, and rudimentary on the other (Terrill 2022, 3). To illustrate this, paradigm adult humans clearly have sophisticated cognitive capacities (e.g., self-awareness, beliefs, desires, language use, and rationality), most non-human animals seem to fall somewhere in the middle between rudimentary and sophisticated cognitive capacities, while material objects such as houses and cars do not have any (Beauchamp and Childress 2019; Harman 2006, 56). Because no one is seriously tempted to accept the claim that plants and PPVS have sophisticated cognitive capacities, we do not need to set the bar too high in our investigation (Terrill 2021, 190). Rudimentary cognitive capacities may involve elementary decision-making, interactions with and adaptions to the environment, and importantly some degree of sentience or the capacity for valenced experiences (e.g., negative affective states like pain) (Barrett and Fischer 2023; Hiernaux 2021; Terrill 2021, 2022).[3]

In this section, I begin with laying out a pithy landscape for both plants and PPVS that reveals the general landscape of capacities before sketching positive and negative claims that (a) give plausible reasons to think that they both have cognitive

[3] Traditionally, evidence for rudimentary cognitive capacities requires possessing neural circuitry, requisite behavioral evidence, etc. However, there is growing awareness that provides nuance to the traditional view, i.e., more nebulous and widespread (Segundo-Ortin and Calvo 2023; Veit and Huebner 2020).

capacities and (b) give plausible reasons to think there is not sufficient evidence in (a). For the purposes of this chapter, I assume that the philosophical disagreements between (a) and (b) are, at least for now, unresolvable.[4] Consequently, plants and PPVS have indeterminate cognitive capacities. This is a case where our current understanding in cognition cannot determinately say where plants and PPVS fall along cognitive capacity dimensions. The upshot of this section, which will be explicated in Sect. 11.3, is that we ought not deny partial moral status to an entity because of its prima facie indeterminate cognitive capacities.[5]

11.2.1 On Plant Capacities

Contemporary plant science suggests that plants are much more "adaptive, flexible, anticipatory, and goal-directed" than we ordinarily think (Segundo-Ortin and Calvo 2019). This claim seems counter intuitive. After all, plants are slow-moving and sessile. And yet, contemporary plant science claims move past *plant awareness disparity*—or the failure to notice plants as complex organisms due to zoocentric attitudes—to get to the root of the fact that plants are unique foragers in their specific ecological niches and remain sensitive to environmental cues (Calvo 2023, 25–29; Calvo and Lawrence 2023; Parsley 2020; Segundo-Ortin and Calvo 2023). Plant responses can be multi-directional, positive, or negative (e.g., roots are positively geotropic, and shoots are positively phototropic). These responses are centralized and subsequently integrated throughout the plant body. For example, roots in the *Arabidopsis thaliana* avoid high saline concentrations in the soil by re-routing to avoid growth constraints (Li and Zhang 2008, 352). Studies suggest that plants are adaptive in behavior to the point that they "can sample and integrate over 20 diverse biotic and abiotic parameters. These parameters are continuously monitored" (Segundo-Ortin and Calvo 2019). Taken together, plant behavior may even be described as intelligent:

> Their roots can avoid obstacles. They can distinguish self from non-self, stranger from kin. If a plant finds itself in a crowd, it will invest resources in vertical growth to remain in light; if nutrients are on the decline, it will opt for root expansion instead. Leaves munched on by insects send electrochemical signals to warn the rest of the foliage, and they're quicker to react to threats if they've encountered them in the past. (Gefter 2023)

The *Pisum sativum*, or the garden pea plant, is one of the go-to behavioral plant models par excellence. In one study that involved a choice of directions in a Y-shaped tube, the roots of the garden pea plant responded strategically in favor of anticipated

---

[4] These disagreements are not merely due to failures of converging shared judgements but rather deeper empirical and theoretical difficulties (Dougherty 2019).

[5] Notice that the amount of evidence (or lack thereof) in this section does not indicate the strength of the argument(s) for or against plants and PPVS having (or not having) cognitive capacities. I make no explicit endorsement of specific methodologies to determine cognitive capacities, nor do I make any definitive endorsements of empirical claims about plants or PPVS.

future nutrient availability against environmental variables (Gagliano et al. 2016). More specifically, the pea plant was exposed to a neutral stimulus (wind) in one arm of the tube and blue light in the other. The two stimuli were switched between the two arms over time, while some plant subjects received both stimuli at the same time. "Most plants showed an adverse conditioned response to the [wind]: they normally grew in the arm exposed to the blue light and grew in the direction of the arm with the [wind] only when this stimulus was coupled with the blue light," and eventually the pea plant made a non-biologically driven association of stimuli between light and wind because it was a good indicator for growth (Nani et al. 2021). The pea plant integrates information within dynamic environments. This suggests that plants make predictions based on anticipations, and "theories on decision-making and optimal behavior developed for [non-human] animals and humans can be applied to plants" (Schmid 2016, 677). From an evolutionary perspective, "plants cannot afford not to be able to anticipate the future," so developmental changes and growth must be adequately assessed over time making integrated and anticipatory behavior valuable (Segundo-Ortin and Calvo 2019).

Another well-known behavioral plant model is the *Mimosa pudica*, or the sensitive plant, known for folding leaves when touched. Gagliano et al. (2014) dropped the plant from specific heights in experiments that would not induce damage to the plant. The study hypothesized that the sensitive plant would learn over time not to fold its leaves when dropped. The results of the experiment revealed evidence of behavioral habituation (i.e., non-associative learning and memory) such that the stimuli from the drop response are remembered and included in the sensitive plant's responses (Calvo et al. 2020, 22). Prior to any talk of plant cognition, what these studies reveal is that we may consider plants as at least having interesting capacities and should take seriously the fact that "we are only scratching the surface" in terms of what plants can do (Ginsburg and Jablonka 2021, 31).

11.2.1.1 Plants Have Cognitive Capacities

Proponents of plants' cognitive capacities begin with the basic problem of measuring cognition, which is that cognitive models are largely based on human standards. This means that we need to rethink our cognitive-theoretic framework and contextualize plants. If plants have rudimentary cognitive capacities, it is because of their functionally equivalent neurobiology—aptly called *plant neurobiology* (Calvo 2016). An empirically informed plant neurobiology is based on the bioelectric properties that plausibly play a significant role in bringing about consciousness and cognitive properties. If plant cognition is based on physiological properties, then the meristematic tissue and vascular system of plants is a helpful starting spot, since the inter-organ system that transports nutrients and metabolic processes also communicates vital information over electric currents (Mediano et al. 2021). There is, then, a strong argument from analogy to be made between the neural correlates of consciousness in human and nonhuman animals with the "vascular correlates of consciousness" in plants (ibid., 86). The vascular structures form a neural-like network of cells interconnected via vascular strands that allow plants flexibility in prioritizing behavioral responses.[6] Take the *Papaya* tree, which has a vascular system that allow it to "coordinate [its] behavior, with electric signaling occurring over long distances through the vascular bundles" (Segundo-Ortin and Calvo 2019). These bundles behave like neuronal synapses with information-processing activities.[7]

11.2.1.2 Plants Do not Have Cognitive Capacities

However, skeptics of plants having cognitive capacities usually make two different arguments against plants having cognitive capacities: that cognition is evolutionarily expensive, and/or plants are simply not equipped with the right features that allow cognitive states (Hamilton and McBrayer 2020; Markel 2020; Višak 2018). For the first argument, there is a standard view which takes cognition to only evolve in non-sessile organisms for the purposes of inter alia avoiding damage and continued survival (Veit and Browning 2023). After all, why would an organism that is not highly mobile need phenomenal consciousness to have interests in avoiding painful valenced states? Cognition would therefore be an evolutionary 'waste' of resources for plants. For the second argument, there is a more obvious argument to be made: plants do not have brains or nervous systems. Some will acknowledge the agential and sapient capacities of plants—even if those capacities are merely adaptive and not intentional—but nonetheless claim that lack of a brain and nervous system makes talk of cognitive capacities inappropriate at best (Ginsburg and Jablonka 2021; Mallatt et al. 2021).

11.2.1.3 Indeterminacy

Because of Sects. 2.1.1 and Sects. 2.1.2, ex hypothesi, plants have indeterminate cognitive capacities.[8]

[6] Similarities between the nervous and the vascular systems are especially pertinent if one holds the *Integrated Information Theory* of cognition to be true, where any system that is "more than the sum of its parts," and has integrated information, is conscious (Mediano et al. 2021, 84). Here, MRI, PET, and electrophotography technologies allow us to trace plant electrophysiology and their networks of integrated information (ibid., 92–93).

[7] Just as anesthesia affects determinately cognitive entities, so too does anesthesia affect plant motor functions. Under anesthesia, the *Pisum sativum*'s circumnutating tendrils cease to function properly, the Venus flytrap's clasp is impeded, and the *Mimosa pudica*'s leaves are prevented from folding when touched. These examples are bolstered by the fact that "targets of anesthesia such as glutamate and GABA are [evolutionarily] shared by animals and plants" (Mediano et al. 2021, 94).

[8] For a recent overview of the back-and-forth literature between proponents and skeptics of plant cognitive capacities, see Terrill and Veit (2024).

11.2.2 On PPVS Capacities

PPVS have sleep–wake cycles seemingly without awareness, usually after suffering some injury or disease (Skene et al. 2009, 247). There are, of course, several real-world cases to draw from, including Terri Schiavo and Karen Ann Quinlan.[9] Historically, PPVS were considered "a kind of human potato, a living body with no mind and no experience of consciousness" (Glenn 2019). In other words, PPVS are organic, living entities without the kind of robust mental life found, for example, in paradigm adult humans. Obviously, this description sounds like how most of us would normally describe plants. From a medical perspective, PPVS are diagnosed with the following criteria: non-awareness, non-voluntary behavior, lack of language comprehension, intermittent wakefulness, incontinence, and variable reflexes (ibid., 337). The question is how we might be able to understand the capacities of PPVS who are unable to verbally respond or report mental states. Notably, PPVS exist in a different kind of state from other conditions that affect cognitive capacities, including minimally conscious states. To distinguish PPVS, some studies indicate that their brain activity is more "comparable to the activity in the brains of healthy individuals under general anesthetic, and quite unlike 'brain dead' patients who have no metabolic neural activity" (Shea and Bayne 2010, 461).

We can consider the study from Adrian Owen et al. to begin parsing out the more complex capacities of PPVS (2006). The study in question used fMRI technology on PPVS to detect brain activity in relation to commands and cues. Researchers tracked brain activity after prompting a series of spatial- and motor-functional representational activities, like playing tennis or walking through rooms in their house (Glenn 2019). The study purportedly revealed that the relevant brain areas corresponded with the questions being asked. Additionally, PPVS responded to autobiographical questions so there is some reason to believe that PPVS can follow minimal instructions. Moreover, there is evidence of image/face recognition via PET scans, where PPVS in the study were shown images of recognizable family members and digitally distorted faces. The scans revealed "activation in the Fusiform Gyrus, which is used for the highly specialized task of face recognition" (ibid., 344).

11.2.2.1 PPVS Have Cognitive Capacities

Proponents of PPVS having cognitive capacities will point to the findings in brain imaging technology as allowing us to draw certain conclusions about such capacities. Ultimately, the question is whether brain imaging technology picks out cognitive capacities. If it does pick out cognitive capacities, then it is plausible to claim that PPVS may have rudimentary cognitive capacities. And, like plants, we now know that one way to possibly understand whether a system is conscious is if it involves the informational flow of electrochemical processing and signals, of which PPVS still have. While the study does not explicitly reveal any relevant details about

[9] See Pierce and Randels (2010) for narrative case studies.

whether PPVS are sentient, the evidence procured from fMRI and PET scans also provides plausible reasons to conclude that PPVS are capable of command-following. If command-following is sufficient for some rudimentary cognitive capacities, then PPVS have some degree of rudimentary cognitive capacities (ibid., 347).

11.2.2.2 PPVS Do not Have Cognitive Capacities

While it is true that absence of evidence itself does not necessarily imply evidence of absence, given current epistemic limitations, skeptics of PPVS having cognitive capacities claim that it is still unclear if responses to verbal cues are "an automatic, non-conscious response" or not (Skene et al. 2009, 248). Additionally, Davies and Levy claim that command-following as an inference for rudimentary cognitive capacities is shaky at best (2016). They argue that PPVS, instead, merely exhibit stimulus-evoked cognition, which lacks endogenous agency—or the kind of agency which is spontaneous and self-initiated—and cannot therefore be considered cognitive in any morally significant sense (Klein 2017, 39). In other words, because PPVS lack flexibility in behavioral and brain activity, then ascriptions of cognitive capacities to PPVS are premature. Additionally, some philosophers who ascribe to the *higher-brain approach* to death believe that if PPVS have irreversibly lost sophisticated cognitive capacities, then the patient is already dead (McMahan 2006). This would mean that talk of rudimentary cognitive capacities would not even get off the ground. Finally, another worry is whether there is any true philosophical significance to fMRI and PET technologies for the purpose of making value judgements (Hacking 1981).

11.2.2.3 Indeterminacy

Because of Sects. 2.2.1 and Sects. 2.2.2, ex hypothesi, PPVS have indeterminate cognitive capacities.

11.3 Moral Status and Indeterminacy

Let us take stock. While there is a growing body of empirical evidence about the interesting and complex behaviors of plants and PPVS, we nevertheless are unable to determinately adjudicate the question of whether plants and PPVS have cognitive capacities. How can moral status address this?

Moral status requires non-arbitrary reasons for conferral and provides us with moral reasons to give weighted priority for some over others (e.g., those with moral status, or a comparatively higher moral status, over those with no moral status). From a normative point of view, the putative belief is that morally relevant interests ground moral status, and cognitive capacities are thought to be a necessary condition for having morally relevant interests (Birch 2021). More narrowly, when dealing with rudimentary cognitive capacities, the *interest model* of moral status generally holds that the possession of morally relevant interests constitutes the conceptual basis for possessing moral status, such that said morally relevant interests inform a moral agent's duties towards the bearer of moral status (DeGrazia 2021, 40). It assumes that moral status is a matter of degree, where the degree is determined by the strength and complexity of cognitive capacities for some entity *E* (DeGrazia 2008). The more rudimentary the cognitive capacities *E* has, the lower its moral status (McMahan 1996, 7). As incremental changes in cognitive capacity are susceptible to non-sharp changes in degrees of moral status, the more cognitively sophisticated the capacities the higher the degree of moral status (McMahan 2007). And the higher the degree of moral status *E* has, the more that *E* matters morally, in its own right and for its own sake. Cognitive capacities, then, both measure and are commensurable with moral status.

If moral status helps us prioritize the well-being of any *E* along the cognitive capacity spectrum, and moral agents have corresponding moral reasons to act in certain positive ways towards any *E* with moral status, then the weight of a moral agent's moral duties towards some *E* is proportional to the degree of moral status that *E* has, with the role and importance of moral status being that of protecting underlying interests. Now suppose the full spectrum of moral status exists on a number line from zero to one. Here, we can disaggregate different degrees of moral status (Terrill 2022). For example, if *E* has no cognitive capacities, and therefore no morally relevant interests, *E* has no concomitant moral status (or MS_0). If *E* has rudimentary cognitive capacities, and therefore some morally relevant interests, *E* has concomitant partial moral status (or $MS_{.01\text{–}0.99}$). If *E* has sophisticated cognitive capacities, and therefore a lot of morally relevant interests, *E* has concomitant full moral status (or MS_1). Based on our working moral status model and the assumptions we make when accounting for cognitive capacities, rocks clearly and determinately fall under MS_0 as they are inanimate and non-living with no cognitive capacities, while a paradigm adult human clearly and determinately falls under MS_1.

To return to the central aim of the chapter, it is enough for my purposes to focus on the lower degrees of partial moral status because no one is seriously tempted to accept the claim that plants and PPVS have sophisticated cognitive capacities. Below, I discuss how we can coherently extend partial moral status to marginal cases at the cusp of our moral circle by expanding the parameters of moral status through indeterminate moral status. I will argue that while plants and PPVS have indeterminate cognitive capacities, and thus not the MS_1 conferred to paradigm adult humans, given an expansive and wide enough scope of moral status, indeterminate moral status may ground a lower degree of de facto partial moral status.

11.3.1 Indeterminate Moral Status

Any relevant model of moral status needs to address how we should respond to moral indeterminacy. Just as fuzzy areas exist between cognitive categories, so too do we need to update our concept of moral status to allow for the possibility of fuzzy areas where we fall short in our ability to delineate determinate moral status cases.

If *E* has indeterminate cognitive capacities, then ipso facto *E* would also have indeterminate moral status. Drawing from Julian and Wilkinson (2019), indeterminate moral status is defined as follows: if there is a non-zero chance that *E* has indeterminate cognitive capacities, then *E* has indeterminate moral status (King 2019). Hereafter, call *E* with indeterminate moral status *E**.

The worry with *E** is the "risk that such beings are treated wrongfully"—this is the spookiness of indeterminate moral status (Douglas et al. 2013, 694). This worry emerges from epistemic incompleteness about cognitive capacities, whereby we cannot attach determinate, corresponding moral status to *E**. But this could well be taken to counsel the adoption of reflective equilibrium in the face of indeterminacy and its constitutive spookiness, which should lead us to produce only tentative and revisable moral judgements about *E**. We may also factor in beliefs about the probabilities that *E** may fall somewhere along the moral status spectrum, without having a specific and exact cognitive capacity value to determine its placement on the spectrum (MacAskill et al. 2020, 196). The thinking is that *E** neither has supra-moral status, existing beyond the realms of our moral status spectrum, nor sub-moral status or MS_0.

Analogically, in law, only judicial discretion to clarify indeterminacy is required, and existing laws may benefit from indeterminate cases in the form of new laws (see Hart 1961). For example, in borderline cases of child neglect, it may be indeterminate whether borderline case *B* determinately counts as child neglect. But *B* allows us to answer normative questions about whether *B* should count as child neglect, thus yielding the possibility of fruitful progress in law (Wodak 2021, 779). Similarly, indeterminate moral status provides moral agents the opportunity to reflect on, and plausibly update, our moral priors and reasons to generate particular moral decisions about *E**, including the conferral of lower degrees of de facto partial moral status (Leiter 1995).

Because plants and PPVS both have indeterminate moral status, and ignoring an alternative option that is epistemically open-minded and assigns non-zero probability to plants and PPVS having full moral status (i.e., MS_1), we have three options given indeterminate moral status:

1. We are epistemically open-minded and assign zero probability to plants and PPVS having any moral status (i.e., MS_0).
2. We are epistemically open-minded and assign non-zero probability to plants and PPVS having a degree of partial moral status (i.e., $MS_{.01\text{-}0.99}$).[10]

[10] A reviewer raises the question of whether we will need to probabilistically represent the many possible degrees of partial moral status that plants and PPVS may in fact have under (2). While a point d'appui, probabilistically representing the many possible degrees of partial moral status using a kind of probability distribution is outside the scope of this chapter. Nevertheless, probabilistically representing the many possible degrees of partial moral status is important to build a more robust conceptual framework of moral status in toto, beyond indeterminacy cases. In my view, the probabilistic representation of the many possible degrees of partial moral status depends upon the quality

3. We are epistemically closed-minded and morally indifferent to plants and PPVS having any moral status.

11.3.1.1 Response to (1)

While epistemically open-minded, advocates of (1) might argue that we would not want to unnecessarily make a moral error of *commission*, where we falsely conclude some E^* has determinate moral status when it does not (Shea and Bayne 2010, 463). However, unless and until we can rule out alternative hypotheses by improving our epistemic position on E^*, the burden of proof for claiming that E^* has MS_0 shifts to the critics, to determinately reveal that entities such as plants and PPVS do not have cognitive capacities. This is because in the asymmetry of risk between treating E^* in such a way as if they are cognitive when they may not be and treating E^* in such a way as if they are not cognitive when they may be, the latter reveals a failure of a duty of due care under moral indeterminacy.[11] Another way to put this is that an error of *omission*—where we falsely conclude E^* does not have determinate moral status when it does—is morally riskier than an error of commission. Even if it turns out that plants and PPVS determinately have MS_0, it is still morally optimal to lean towards the more liberal error given the lower degree of partial moral status that such error will require us to confer.

11.3.1.2 Response to (3)

Advocates of (3) fail to account for the precautionary implications of indeterminacy. In such cases as plants and PPVS, we have reason to apply the precautionary principle so that we treat entities with indeterminate cognitive capacities *as if* they have cognitive capacities (Birch 2021). Indeterminate moral status also leaves room for moral status adjustments across time, given the ever-present possibility of relevant epistemic changes. In other words, if we gain evidential confidence in plants and PPVS having determinate cognitive capacities, it makes less sense to remain epistemically closed-minded and indifferent (Veit 2023, 95). Suppose that I make a judgement call C^1 at time t^1, but at t^2 I gather more epistemic evidence and make judgement call C^2. At t^2 I rationally believe that judgement C^1 is morally impermissible. We ought not unnecessarily constrain our future judgement calls, particularly if that involves the risk of morally troublesome outcomes and is done on epistemically inconclusive

of cognitive capacity empirical data and theoretical assessments. I imagine that this project would, hopefully, be able to estimate the likelihood of different degrees of moral status given any E or E^* and thereby assign corresponding moral weight (for an example of a project along these lines, see Fischer [2022]).

[11] Broadly, I take a duty of due care to mean having an obligation to take reasonable precautions that prevent unnecessary harms when φing. A moral agent thus violates a duty of due care when she negligently fails to φ, thus causing incidental harm to others (Schwenkenbecher 2014).

grounds. An implication of the precautionary principle, then, is that "if it is indeterminate whether an [E^*] is conscious or not, then it is decision-obligatory to treat it in all respects as if it were conscious in any context where it may be caused suffering" (Birch 2021, 16–17). To undermine this verdict, the skeptic would need an epistemically superior verdict. Because of indeterminacy, the skeptic cannot provide such a verdict. Therefore, we should invoke the precautionary principle to confer partial moral status. The skeptic who chooses indifference, then, also makes a decision that constitutes a violation of some duty of due care.

11.3.1.3 Preference for (2)

As can be seen, (1) and (3) are premised upon moral recklessness and violate some duty of due care. Imagine a case where you stand at a door knowing that someone is on the other side. If you were to, say, fire a weapon through the door without knowing who exactly was on the other side, we would say that that action was reckless rather than reasonable behavior and thus blameworthy. Like this imagined case, (1) and (3) disregard probable risks. Contra (1) and (3), (2) makes the claim that it would be at least somewhat wrong to harm plants and PPVS even if the risk of harm is extremely low. It seems to me that (2) is the best posture to adopt given our current epistemic limitations: if, given the evidence at our disposal, the probability that we are ready to assign to the possibility of cognitive capacities in plants and PPVS increases, so too does the justifiability of conferring a degree of partial moral status and motivating at least some moral considerability. In other words, indeterminate moral status makes a moderate claim about hedging our moral bets: we need not make a determinate decision one way or the other about the cognitive capacities of E^* to apply the precautionary principle and adopt proportionate measures by assigning partial moral status to indeterminate moral status (Birch and Browning 2021, 56–57).

11.3.2 Morally Salient Differences?

If plants and PPVS both have indeterminate moral status, and if we buy (2) such that they also both have a lower degree of partial moral status, then we may treat like cases as like ceteris paribus. Why is this so? More formally, because I have not set out to specify precise lower degrees of partial moral status for plants and PPVS, we may assume *equal consideration*. This entails that "we ought to attribute roughly equal moral weight or importance to X's and Y's comparable interests" (DeGrazia 2007, 316). Because of indeterminacy, equal consideration is pertinent to comparability cases of E^*. Even so, when we parse out the ceteris paribus clause, it is important to trace the potential myriad of morally salient differences between plants and PPVS.

First, PPVS' cognitive capacities were at one point determinate. An adult human patient at t^1 is a person with MS_1 in virtue of having sophisticated cognitive capacities, but at t^2 the patient suffers a brain injury and enters a coma, and at t^3 is diagnosed as

PPVS—where she now has indeterminate cognitive capacities (Suchy-Dicey 2009, 129). On the other hand, at any point in time plants only have indeterminate moral status. And, according to principles of respect for autonomy, we may be required to respect prior wishes or advance directives provided by the human patient before being diagnosed as PPVS.[12]

Second, there is a lot of evidence that suggests humans have *status quo biases* that inform our moral judgements in favor of prioritizing other humans (Earp et al. 2021). In the case of PPVS, "some familial and professional caregivers find cultural, religious, symbolic, or emotional barriers to withholding or withdrawing" life-sustaining treatment (Beauchamp and Childress 2019, 166). For this reason, even if it turns out that PPVS, who have lost their previous sophisticated cognitive capacities, have a determinately lower degree of moral status than plants, there are still supervening reasons to treat PPVS differently than plants, including social cohesion and stability (DeGrazia 2021).

Extrinsic reasons aside, should we nevertheless give equal consideration to plants as we do PPVS? Suppose you—a moral agent—find yourself in a moral emergency and must choose between saving the life of a plant, a PPVS, a parrot, or that of a paradigm adult human. Because of the widely shared normative claim that moral agents have stronger moral duties towards *E* with determinate sophisticated cognitive capacities than they do towards *E* with determinate rudimentary cognitive capacities, it is obvious that you should save the life of the paradigm adult human because paradigm adult humans have MS_1 given their sophisticated cognitive capacities, while plants, PPVS, and parrots do not. This becomes more complicated if we move beyond individuals to aggregates, especially in cases where the aggregate interests of plants may outweigh the aggregate interests of PPVS (Sebo 2023). Continuing to parse the nuances of indeterminate moral status may help us accommodate these thorny questions in applied ethics, since it helps to focus our scope of moral reasoning and decision-making, including the possibility of a stronger move towards the acceptance of (some) rights in the vegetal turn (Polk 2024, this collection). For now, we may tentatively hold onto the following moral status commitments: plants and PPVS similarly do not have MS_0, but do have some lower degree of de facto partial moral status. The upshot is that it elevates the moral floor.

11.4 Conclusion

I began the chapter by discussing the indeterminacy of plant and PPVS cognitive capacities. Indeterminacy is a classic case of risk management in our decision-making, and in the moral case a negligible risk of harm can be enough to make some actions (at least somewhat) wrong. I argued that to avoid what may later turn out to be moral errors, a viable concept of moral status should accommodate indeterminacy,

[12] Merely referring to principles in bioethics does not eo ipso resolve complicated matters within bioethics itself.

such that we have some morally modest reasons to think that indeterminate moral status is in correspondence with partial moral status. Indeterminate moral status, and by extension partial moral status, will not generate overly demanding moral duties but will afford moral agents the opportunity to expand the moral circle. It seems, then, that we ought to (at least slightly) update our bioethics and environmental ethics in the process. For environmental ethics, there are some pragmatic implications that may yield some high-impact actions for the environment if plants have indeterminate moral status. For starters, plants having indeterminate moral status provides us with another moral reason to adopt a more conservationist attitude and ecologically sustainable practice toward individual plants in nature than we would have otherwise. For bioethics, the pragmatic implication for PPVS having indeterminate moral status is that we may "succeed in avoiding a misdiagnosis and its potentially grave consequences" (Glenn 2019, 349).

Accordingly, as our moral circle expands, what we need is a bottom-up approach to moral status. By starting with marginal cases with indeterminate moral status, we may have a better sense of moral status and its relationship to cognitive capacities. Still, moral agents may worry that indeterminate moral status is difficult to digest, especially if it turns out that indeterminacy is a feature, instead of a bug, in our moral theories. If it is a feature in our moral theories, then that gives us more reason to be both flexible and inclusive in the face of indeterminacy. It is also prudent to claim that more empirical and theoretical research is required. Furthermore, we need to reassess our methodological tools for studying plants and PPVS to meet them where they are at, since it would be a mistake to think that any plausible and possible *E* with (some) cognitive capacities and moral status must mirror the sophisticated cognitive capacities of paradigm adult humans. Perhaps we ought not rely on neuron counts alone, or we should broaden our understanding of nervous systems (Miguel-Tomé and Llinás 2021; Shriver 2022). Cognition, and therefore moral status, has the potential to be far more expansive in scope; and that allows moral agents to consider moral reform. While many believe that "the adjective 'vegetative' expresses that degradation and loss of dignity," we might instead think about how vegetal states can help us flesh out the gradations of indeterminate cognitive capacities and thereby imagine a wider range of moral status. Indeterminate moral status in vegetal states may create better conditions for both humans and non-humans in such states (Glenn 2019, p. 349). After all, indeterminacy highlights the complexity and uncertainty of the world we live in and the need for ethical decision-making to be informed by careful reflection, ongoing inquiry, and a willingness to acknowledge and learn from our limitations.

References

Barrett, Meghan, and Bob Fischer. 2023. Challenges in Farmed Insect Welfare: Beyond the Question of Sentience. *Animal Welfare* 32 (e4): 1–5. https://doi.org/10.1017/awf.2022.5.

Beauchamp, Tom L., and James F. Childress. 2019. *Principles of Biomedical Ethics*, 8th ed. New York: Oxford University Press.

Birch, Jonathan. 2021. Materialism and the Moral Status of Animals. *The Philosophical Quarterly* 72 (4). https://doi.org/10.1093/pq/pqab072.

Birch, J., and Browning, H., 2021. Neural organoids and the precautionary principle. *The American journal of bioethics*, 21 (1): 56–58.

Calvo, Paco. 2016. The Philosophy of Plant Neurobiology: A Manifesto. *Synthese* 193 (5): 1323–1343. https://doi.org/10.1007/s11229-016-1040-1.

Calvo, Paco, Monica Gagliano, Gustavo M. Souza, and Anthony Trewavas. 2020. Plants are Intelligent, Here's How. *Annals of Botany* 125 (1): 11–28. https://doi.org/10.1093/aob/mcz155.

Calvo, Paco, and Natalie Lawrence. 2023. *Planta Sapiens: The New Science of Plant Intelligence.* New York: WW Norton & Co.

Davies, Will, and Neil Levy. 2016. Persistent Vegetative State, Akinetic Mutism and Consciousness. In *Finding Consciousness: The Neuroscience, Ethics, and Law of Severe Brain Damage*, ed. Walter Sinnott-Armstrong, 122–136. Oxford: Oxford University Press.

DeGrazia, David. 2007. Human–animal Chimeras: Human Dignity, Moral Status, and Species Prejudice. *Metaphilosophy* 38 (2–3): 309–329.

DeGrazia, David. 2008. Moral Status as a Matter of Degree? *The Southern Journal of Philosophy* 46 (2): 181–198. https://doi.org/10.1111/j.2041-6962.2008.tb00075.x.

DeGrazia, David. 2021. An Interest-Based Model of Moral Status. In *Rethinking Moral Status*, ed. Steve Clark, Hazem Zohny, and Julian Savulescu, 40–56. Oxford: Oxford University Press.

DeGrazia, David, Thomas A. Mappes, and Jeffrey Brand-Ballard. 2011. *Biomedical Ethics.* New York: McGraw-Hill Higher Education.

DeGrazia, David, and Joseph Millum. 2021. *A Theory of Bioethics.* New York: Cambridge University Press.

Dougherty, Tom. 2019. Vagueness and Indeterminacy in Ethics. In *The Routledge Handbook of Metaethics*, eds. Tristram McPherson and David Plunkett, 185–193. New York: Routledge, Taylor & Francis Group.

Douglas, Thomas, Rachell Powell, and Julian Savulescu. 2013. Is the Creation of Artificial Life Morally Significant? *Studies in History and Philosophy of Science Part c: Studies in History and Philosophy of Biological and Biomedical Sciences* 44 (4): 688–696. https://doi.org/10.1016/j.shpsc.2013.05.016.

Earp, Brian D., Killian L. McLoughlin, Joshua T. Monrad, Margaret S. Clark, and Molly J. Crockett. 2021. How Social Relationships Shape Moral Wrongness Judgments. *Nature Communications* 12 (1). https://doi.org/10.1038/s41467-021-26067-4.

Fischer, Bob. 2022. An Introduction to the Moral Weight Project. *Rethink Priorities.* October 31. https://rethinkpriorities.org/publications/an-introduction-to-the-moral-weight-project.

Gagliano, Monica, Michael Renton, Martial Depczynski, and Stefano Mancuso. 2014. Experience Teaches Plants to Learn Faster and Forget Slower in Environments Where It Matters. *Oecologia* 175 (1): 63–72. https://doi.org/10.1007/s00442-013-2873-7.

Gagliano, Monica, Vladyslav V. Vyazovskiy, Alexander A. Borbély, Mavra Grimonprez, and Martial Depczynski. 2016. Learning by Association in Plants. *Scientific Reports* 6 (1). https://doi.org/10.1038/srep38427.

Gefter, Amanda. 2023. What Plants Are Saying About Us: Your brain is not the root of cognition. *Nautilus* 48. http://nautil.us/.

Ginsburg, Simona, and Eva Jablonka. 2021. Sentience in Plants: A Green Red Herring? *Journal of Consciousness Studies* 28 (1–2): 17–33.

Glenn R. Butterton, 2019. How Neuroscience Technology is Changing Our Understanding of Brain Injury, Vegetative States and the Law, 20 N.C. J.L. & Tech. 331.

Gosepath, Stefan. 2021. Equality. *Stanford Encyclopedia of Philosophy*. https://plato.stanford.edu/entries/equality/.

Hacking, Ian. 1981. Do We See through a Microscope? *Pacific Philosophical Quarterly* 62 (4): 305–322. https://doi.org/10.1111/j.1468-0114.1981.tb00070.x.

Hamilton, Adam, and Justin McBrayer. 2020. Do Plants Feel Pain? *Disputatio* 12 (56): 71–98. https://doi.org/10.2478/disp-2020-0003.

Harman, Elizabeth. 2006. Sacred Mountains and Beloved Fetuses: Can Loving or Worshipping Something Give It Moral Status? *Philosophical Studies* 133 (1): 55–81. https://doi.org/10.1007/s11098-006-9006-7.

Hart, H.L.A. 1961. *The Concept of Law*. Oxford: Oxford University Press.

Hiernaux, Quentin. 2021. Differentiating Behaviour, Cognition, and Consciousness in Plants. *Journal of Consciousness Studies* 28 (1–2): 106–135.

Jaworska, Agnieszka, and Julie Tannenbaum. 2021. The Grounds of Moral Status. *Stanford Encyclopedia of Philosophy*. https://plato.stanford.edu/entries/grounds-moral-status/.

Julian, Koplin, and Dominic Wilkinson. 2019. *Journal of Medical Ethics*. 45: 457–458 Published Online First: 11 Jul 2019. https://doi.org/10.1136/medethics-2019-105644

King, Mike. 2019. Implications of Moral Uncertainty: Implausible or Just Unpalatable? *Journal of Medical Ethics* 45 (7): 451–452. https://doi.org/10.1136/medethics-2019-105588.

Klein, Colin. 2017. Consciousness, Intention, and Command-Following in the Vegetative State. *The British Journal for the Philosophy of Science* 68 (1): 27–54.

Leiter, B. 1995. Legal Indeterminacy. *Legal Theory* 1 (4): 481–492. https://doi.org/10.1017/S1352325200000227.

Li, X., and W.S. Zhang. 2008. Salt-Avoidance Tropism in Arabidopsis Thaliana. *Plant Signaling & Behavior* 3 (5): 351–353. https://doi.org/10.4161/psb.3.5.5371.

MacAskill, William, Krister Bykvist, and Toby Ord. 2020. *Moral Uncertainty*. Oxford: Oxford University Press.

Maher, Chauncey. 2021. The Indeterminacy of Plant Consciousness. *Journal of Consciousness Studies* 28 (1–2): 136–154.

Mallatt, J., M.R. Blatt, A. Draguhn, D.G. Robinson, and L. Taiz. 2021. Debunking a Myth: Plant Consciousness. *Protoplasma* 258 (3): 459–476. https://doi.org/10.1007/s00709-020-01579-w.

Markel, Kasey. 2020. Lack of Evidence for Associative Learning in Pea Plants. *ELife* 9 (June). https://doi.org/10.7554/elife.57614.

McMahan, Jeff. 1996. Cognitive Disability, Misfortune, and Justice. *Philosophy & Public Affairs* 25 (1): 3–35. https://doi.org/10.1111/j.1088-4963.1996.tb00074.x.

McMahan, Jeff. 2006. An Alternative to Brain Death. *Journal of Law, Medicine & Ethics* 34 (1): 44–48. https://doi.org/10.1111/j.1748-720x.2006.00007.x.

McMahan, Jeff. 2007. Challenges to Human Equality. *The Journal of Ethics* 12 (1): 81–104. https://doi.org/10.1007/s10892-007-9020-9.

Mediano, Pedro A.M.., Anthony Trewavas, and Paco Calvo. 2021. Information and Integration in Plants: Towards a Quantitative Search for Plant Sentience. *Journal of Consciousness Studies* 28 (1–2): 80–105.

Miguel-Tomé, Sergio, and Rodolfo R. Llinás. 2021. Broadening the Definition of a Nervous System to Better Understand the Evolution of Plants and Animals. *Plant Signaling & Behavior* 16 (10): 1927562. https://doi.org/10.1080/15592324.2021.1927562.

Nani, Andrea, Gabriele Volpara, and Andrea Faggio. 2021. Sentience With or Without Consciousness. *Journal of Consciousness Studies* 28 (1–2): 60–79.

Owen, A.M., M.R. Coleman, M. Boly, M.H. Davis, S. Laureys, and J.D. Pickard. 2006. Detecting Awareness in the Vegetative State. *Science* 313 (5792): 1402. https://doi.org/10.1126/science.1130197.

Parsley, K.M. 2020. Plant Awareness Disparity: A Case for Renaming Plant Blindness. *Plants People Planet* 2 (6): 598–601. https://doi.org/10.1002/ppp3.10153.

Pierce, Jessica, and George Randels. 2010. *Contemporary Bioethics: A Reader with Cases*. New York: Oxford University Press.

Polk, Amber. 2024. In Defense of (Some) Plant Rights. In *The Vegetal Turn*, ed. Marcello Di Paola. New York: Springer.
Schmid, Bernhard. 2016. Decision-Making: Are Plants More Rational than Animals? *Current Biology* 26 (14): R675–R678. https://doi.org/10.1016/j.cub.2016.05.073.
Schwenkenbecher, Anne. 2014. Collateral Damage and the Principle of Due Care. *Journal of Military Ethics* 13 (1): 94–105. https://doi.org/10.1080/15027570.2014.910015.
Sebo, Jeff. 2023. The Rebugnant Conclusion: Utilitarianism, Insects, Microbes, and AI Systems. *Ethics, Policy & Environment* 26 (2): 249–264. https://doi.org/10.1080/21550085.2023.2200724.
Segundo-Ortin, Miguel, and Paco Calvo. 2019. Are Plants Cognitive? A Reply to Adams. *Studies in History and Philosophy of Science Part A* 73 (February): 64–71. https://doi.org/10.1016/j.shpsa.2018.12.001.
Segundo-Ortin, Miguel and Paco Calvo. 2023. "Plant sentience? Between romanticism and denial: Science." *Animal Sentience* 33 (1). https://doi.org/10.51291/2377-7478.1772.
Shea, Nicholas, and Tim Bayne. 2010. The Vegetative State and the Science of Consciousness. *The British Journal for the Philosophy of Science* 61 (3): 459–484. https://doi.org/10.1093/bjps/axp046.
Shriver, Adam. 2022. Why Neuron Counts Shouldn't Be Used as Proxies for Moral Weight. *Rethink Priorities*. November 28. https://rethinkpriorities.org/publications/why-neuron-counts-shouldnt-be-used-as-proxies-for-moral-weight.
Skene, L., D. Wilkinson, G. Kahane, and J. Savulescu. 2009. Neuroimaging and the Withdrawal of Life-Sustaining Treatment from Patients in Vegetative State. *Medical Law Review* 17 (2): 245–261. https://doi.org/10.1093/medlaw/fwp002.
Stone, Jim. 2007. Pascal's Wager and the Persistent Vegetative State. *Bioethics* 21 (2): 84–92. https://doi.org/10.1111/j.1467-8519.2007.00528.x.
Suchy-Dicey, Carolyn. 2009. It Takes Two: Ethical Dualism in the Vegetative State. *Neuroethics* 2 (3): 125–136. https://doi.org/10.1007/s12152-009-9039-7.
Terrill, Ethan C. 2021. Plants, Partial Moral Status, and Practical Ethics. *Journal of Consciousness Studies* 28 (1–2): 184–209.
Terrill, Ethan. 2022. How I Learned to Stop Worrying and Respect Post-Persons. *Journal of Ethics and Emerging Technologies* 31 (1): 1–14. https://doi.org/10.55613/jeet.v31i1.84.
Terrill, Ethan C., and Walter Veit. 2024. Better to be a Pig Dissatisfied than a Plant Satisfied. *Journal of Agricultural and Environmental Ethics* 37 (4). https://doi.org/10.1007/s10806-024-09922-4.
Veit, Walter. 2023. Confidence Levels or Degrees of Sentience? *Asian Bioethics Review* 15: 93–97. https://doi.org/10.1007/s41649-022-00230-5.
Veit, Walter and Bryce Huebner. 2020. Drawing the Boundaries of Animal Sentience. *Animal Sentience* 29 (13). https://doi.org/10.51291/2377-7478.1595.
Veit, Walter and Heather Browning. 2023. *Defending Sentientism*. Preprint. http://philsci-archive.pitt.edu/21817/.
Višak, Tatjana. 2018. Utilitarian Plant Ethics. In *Plant Ethics: Concepts and Applications*, ed. Angela Kallhoff, Marcello Di Paola, and Maria Schörgenhumer, 30–39. Abingdon, Oxon: Routledge.
Wodak, Daniel. 2021. The Nature and Value of Vagueness in Law. *Ethics* 131 (4): 777–781.

Chapter 12
In Defense of (Some) Plant Rights

Amber Polk

Abstract In this chapter, I present a novel argument defending the claim that certain, exceptional plants possess moral rights. First, I set the stage for developing an agent-centered, plant rights view, which grounds moral rights on the nature of practical reason rather than the natural properties of moral subjects. Then, I draw on the concept of the sublime from aesthetics to ground our duties of non-harm (and positive assistance) to exceptional plants. I discuss the importance of rights in a plant ethic because they are action-guiding for us in how we treat plants, which is what matters most to plants. I also note the importance of plant rights as boundaries on the self-obsessive end of ever increasing human preference satisfaction. Finally, I concede that the scope of plant rights must be limited because of the pressure that rights extensionism in the non-human sphere places on the possibility of human freedom for self-determination.

12.1 Introduction

Rights have often been a first step in welcoming new individuals to the realm of moral inclusion (Nash 1989). This chapter explores whether there are any grounds for recognizing plant rights and whether plant ethics should consider rights as part of its moral machinery. In Sect. 12.2, I briefly map out the background of non-anthropocentric western ethics and locate plant ethics in that space. In Sect. 12.3, I argue that some plant rights can be normatively grounded via an aesthetic argument that certain exceptional plants are right holders. In Sect. 12.4, I discuss the importance of plant rights in plant ethics. However, such importance should not be taken to entail that the entirety of plant ethics is or should be synonymous with plant rights. Finally, in Sect. 12.5, I conclude with a brief comment on the limits of rights in plant ethics.

A. Polk (✉)
University of Alabama School of Law, Tuscaloosa, AL, USA
e-mail: apolk@law.ua.edu

M. Di Paola (ed.), *The Vegetal Turn*, The International Library of Environmental, Agricultural and Food Ethics 38, https://doi.org/10.1007/978-3-031-68671-9_12

12.2 Non-Anthropocentric Ethics and the Emergence of Plant Ethics

The relationship between humans and plants has not historically been the focal point of non-anthropocentric ethics. At the beginning of the modern environmental movement in the 1970s, the extension of moral consideration to others beyond human beings was initially split between arguments extending ethics to include animals and arguments extending ethics to include everything else. Animal ethicists argue that many (if not all) of our deontic moral theories include animals as moral patients, i.e., subjects of right or wrong actions.[1] The animal welfarists appeal to variations on consequentialist moral theories (Bentham 1948; Singer 1975), and the animal rights theorists appeal to variations on deontological moral theories, such as Kantianism and contractarianism (Korsgaard 2018; Regan 1983; Rowlands 2009). Because of this tradition, animal ethics generally speaks in the language of deontic moral theories. Its goal has always been directly answering the question of what humans owe to animals (and why).

Environmental ethics, however, struggled to invoke the notion of right and wrong action, which is the focal point of deontic ethics and reflected in animal ethics. This is because it has no unified target for moral inclusion. Environmental ethics contends with the possibility that plants, mountains, rivers, species, and whole ecosystems might be natural objects worthy of moral inclusion and respect. The challenge of environmental collectivities produced a contingency of environmental ethicists (the biocentrists) that looked to ecology, the science of ecological wholes and systems, for new moral principles (Callicott 1989, 1999; Leopold 1949). This resulted in many years of discourse describing and attempting to reconcile the tension between animal ethics and biocentrism (Callicott 1980, 1988; Sagoff 1984; Warren 1983). Other environmental ethicists imported deontic concepts (e.g., intrinsic value and moral considerability) to bear on various natural objects without importing the rest of a deontic ethical framework (e.g., rights and duties and their moral significance) (Goodpaster 1978; Rolston III 1975). Plants were never wholly ignored during this time (e.g., Rolston III 1991), but a plant ethic was not a focal point in environmental ethics until recently.

In this recent and uniquely plant ethics, three types of approaches can be identified. First, the traditionalist approaches borrow from environmental ethics the concepts of intrinsic value and moral considerability and apply them to plants specifically (e.g., Sandler 2018). Second, the extensionist approaches focus on extending various established moral frameworks to include plants as members of the moral community. For

---

[1] Western ethics can be divided into the aretaic (virtue ethics) and the deontic (consequentialism and deontology). Deontic ethics' core concept is that of duty (or obligation). Morality is a matter of right and wrong action—that is, 'thin' ethical concepts. The aretaic, in contrast, is not solely focused on right and wrong action but also considers an agent's character, dispositions, emotional states, and other ancillary 'thick' ethical concepts. Modern moral philosophy was primarily concerned with deontic ethics through the mid-twentieth century, with a revival of the aretaic occurring in the second half of the twentieth century (Anscombe 1958; Williams 1985).

example, Matthew Hall (2009, 2011, 2019) has argued that plants are autonomous beings and, therefore, persons. Angela Kallhoff (2014, 2018) and Maria Schörgenhumer (2018) have advocated for rethinking the human-plant relationship through the lens of virtue ethics, and Tatjana Visak (2018) has suggested that even if plants are not welfare subjects under utilitarianism, we still have reason to protect them. Third, the revisionist approaches challenge the implicit metaphysical assumptions in our moral theorizing, arguing that a plant ethic will be necessarily intertwined with developing a new plant ontology. For example, Michael Marder (2013) has defended a new plant ontology, "plant-thinking," with the goal of reimagining our ethical relationship with plants. Sylvie Pouteau (2014, 2018) has suggested that a plant ethic needs to recognize and appreciate the 'openness' and 'infinite' nature of plants (as opposed to the 'closedness' and 'finite' nature of animals) and not treat plants as second animals.

Arguing for (some) plant rights falls into the category of extensionist approaches, and I turn to that next.

12.3 Normatively Grounding (Some) Plant Rights

To take the idea of plant rights seriously, there must be some reason for thinking that at least some plants are the holders of rights. Traditionally, there have been two ways in which rights are normatively grounded. They are grounded on the nature of the moral patient (patient-centered rights views), or they are grounded on the nature of the moral agent (agent-centered rights views). In this section, I will first suggest that a plant rights view should be agent-centered, not patient-centered. Then, I will provide an agent-centered, aesthetic argument that exceptional plants possess rights.

12.3.1 Patient- Versus Agent-Centered Rights Views

Most of the discussions in plant ethics have been patient-centered—that is, they have focused on the nature of plants themselves, whether plants have a good or can be harmed, whether plants are autonomous and therefore persons, whether plants have welfare interests, etc. In such views, the focus is on whether plants possess the correct property for grounding the right kind of value for moral inclusion (Pellegrino 2018).

Extending these approaches to arguments for plant rights could easily parallel Tom Regan's (1983) argument for animal rights. The logic of that argument contains two steps: properties to value and then value to right holding. For Regan, the type of value that entails right holding is inherent value. This is because to be inherently valuable is to be an end-in-oneself and, therefore, a (Kantian) moral patient (a right holder, someone to whom moral agents owe duties). For Regan (1983, 243–248), the property sufficient for being inherently valuable is being the 'subject-of-a-life.' Scientific evidence suggests that mature mammals (and some birds) have a rich enough mental life to be considered subjects-of-a-life. Therefore, having identified properties to value and value to right holding, fully mature mammals (and some birds) possess rights, correlative to moral agents' duties to such animals not to kill them, experiment on them, or trap them (Regan 1983, 330–99). A patient-centered plant rights approach may adopt this logic, arguing that subject-of-a-life is too narrow and that the correct property for moral inclusion is perhaps personhood. Adopting such an approach and finding convincing, e.g., Hall's (2011) work on plant personhood, it would follow that plants are inherently valuable and, therefore, right holders.

There are two significant challenges that cut against adopting a patient-centered approach to plant rights. The first challenge is that patient-centered views focus entirely on which property is the correct property for selecting all and only those individuals who possess inherent value. The second challenge is that it is not immediately apparent that inherent value, or any other kind of value, is what grounds rights. I discuss these two challenges in turn.

With regards to the first challenge, in general, patient-centered rights movements have focused intensely on the natural-property-to-value question. For example, we abandoned race as a natural property to delineate the realm of moral inclusion by ending slavery. We abandoned sex as a natural property to delineate the realm of moral inclusion in the women's rights movements. Animal rights theorists continue to push for abandoning the natural property homo sapiens to delineate the realm of moral inclusion. Some extensionist plant ethicists have suggested plants possess certain natural properties, such as personhood (Hall 2011), the capability of flourishing (Kallhoff 2014, 2018), or even just being alive (Goodpaster 1978), which are sufficient for value and, therefore, moral inclusion.

A couple of problems with patient-centered views arise from this focus on grounding moral value on natural properties. First, there is no necessary connection between the natural and good (or right). To maintain a patient-centered view of plant rights, one would need to reject the naturalistic fallacy in some meaningful way, which is a tough hurdle to overcome. Second, whether implicitly or explicitly, patient-centered approaches rely heavily on human empathy and identification with 'others' to expand the realm of moral inclusion. This reliance pushes the debates over which natural property is the dividing line for moral inclusion into deadlock. At a certain point, people are either convinced that sentience, capacity to suffer, capacity to flourish, etc., is sufficient for moral inclusion, or they are not. Arguing for an agent-centered approach to plant rights can overcome some of this deadlock, as I discuss in the next section.

The second challenge for patient-centered rights views goes less to practical considerations in moving the debate forward and more to what it is to be a rights view. If we reexamine Regan's animal rights approach, the order of 'property to value to rights' seems to be inverted. Subjects-of-a-life are not inherently valuable on the grounds of the natural property they possess. They are inherently valuable because the nature of practical reason compels moral agents to treat them as inherently valuable and, therefore, as right holders. More generally, this mistake of moving from value to rights fails to recognize that rights (and their correlative duties) are derived from the structure of practical reason. They are not supervenient upon or derived from value in the world that moral agents are supposed to just recognize. One facet of modern deontological morality was to derive principles of action without having to adopt and defend some theory of value to ground our principles of action. The goal was to move beyond debates regarding 'true' or 'correct' sources of external value, upon which moral agents rarely entirely agree.

There are two traditional deontological moralities, both of which reflect reason's priority over value in grounding principles of action: Kantianism and contractarianism. While Kant's (2012) categorical imperative is often invoked as the guiding principle for determining what constitutes moral action, it is less emphasized in those discussions just how the categorical imperative itself is derived. The categorical imperative is derived from the nature of practical reason itself. Because moral agents can reflect on the means they take to achieve their ends, they can also reflect on the ends themselves and question whether the end is worth pursuing. When a moral agent acts in pursuit of an end, she is not only willing the means to the end but also the end itself. Because practical reason is universal, to will an end is to will it universally for all moral agents. This limits what ends a moral agent can rationally pursue because not all ends can be willed universally. To will some ends universally, such as killing other moral agents for fun or lying indiscriminately, undermines willing those ends individually, so reason would command a moral agent not to pursue those ends. In this way, moral duties are owed to right holders, not because the moral agent recognizes some value in the right holder, but because those moral duties are inherent in the nature of the moral agent's practical reason—that is, they are agent-centered.

This same agent-centeredness is seen in the grounding of rights and their correlative duties in contractarianism. In contractarianism, the idea is that moral agents can reflect upon and recognize the untenably of a world where moral agents have no restrictions on what they may do. In such a world, because all actions are permissible, moral agents have no reason *not* to chronically aim to dispossess one another of life and resources, thus creating a "nasty, brutish, and short" Hobbesian (1996) state of nature. The rational solution to this problem is that each moral agent should be willing to give up some of their permissions to act in certain ways in exchange for that same restraint from other moral agents. These reciprocal promises constitute the duties that moral agents owe each other and the rights that moral agents can claim against each other. In this way, contractarian-derived rights and their correlative duties are also agent-centered. They emerge from the command of practical reason possessed by moral agents who are faced with living in the state of nature absent some mutually agreed upon and reciprocal restrictions on their conduct.

Such agent-centered approaches have been argued for in animal rights theory more recently (Korsgaard 2018; Rowlands 2009). They solve the two challenges for patient-centered rights views that I discussed above. First, both Korsgaard's and Rowlands' animal rights views move beyond the stagnant debate that tries to ground rights and duties on natural properties (via the intermediary concept of value). Second, our obligations to animals (and their rights) in these views originate from the nature of practical reason in moral agents. They do not require a moral agent to recognize and appreciate something external to the moral agent in deciding what she ought to do. Given this landscape, I think attempting an agent-centered argument for some plant rights can move the needle beyond stagnant, patient-centered debates and reinvigorate the conceptual origins of rights views. I turn to such an attempt next.

12.3.2 The Case for (Some) Agent-Centered Plant Rights

One agent-centered argument for plant rights could take up Christine Korsgaard's (2018) agent-centered approach for animal rights and extend it to plants. The gist of the argument for animals is that when we, as moral agents, exercise our practical reason, we will not only our choice universally but also that which we take as our end. Because we are natural creatures with natural ends which we pursue through practical reason, we will our natural ends. Because practical reason demands that we will our ends universally, it follows that we are rationally required to respect other creatures who pursue their own natural ends. This is something that animals do, so we are rationally required to recognize that animals are part of the realm of moral inclusion. We owe animals duties not to treat them as mere means for our ends. The extension to plants requires only a showing that plants pursue their own natural ends. This extension is plausible, and Korsgaard (2004, n. 69; 2018) recognizes the implications of her animal rights argument for the moral status of plants. I do not defend this extension here because the view does not provide a clear limitation on generating plant rights, which I believe is necessary for the reason I discuss in Sect. 12.5.

Instead, I aim to provide a narrower agent-centered argument for recognizing *some* plant rights. The view invokes aesthetic concepts and relies on the claim that certain exceptional plants evoke the feeling of the sublime for humans. Because of the relationship between the sublime and morality for moral agents, humans have duties to these exceptional plants. Most importantly, humans have duties not to harm and duties to assist such plants, particularly when humans are the source of the threat of harm. There are three parts to defending this view. First is the concept of the sublime, what it is and is not. Second is the question of which plants evoke the feeling of the sublime in human moral agents and why. Third, it must be shown that the sublime is related to human morality in such a way that our duties to exceptional plants (and their correlative rights) follow from the nature of human practical reason. This third part maintains the argument's agent-centeredness.

To begin, sublimity is not a property of exceptional plants. It is not a property of external things at all. It is an aesthetic feeling that human moral agents experience in certain situations, and it does not require that humans empathize or identify with 'others' in the way that patient-centered views typically need in order to expand the moral realm to new individuals. Because the sublime, as I intend here, is not a property of plants, it diverges from nineteenth-century Romantic views of the sublimity of nature (e.g., Hitt 1999). Those discussions consider nature sublime because of its otherness and power over humans. This is not the concept of sublimity I wish to invoke here. I intend sublimity in a somewhat Kantian way (Matthews 1996). It is a simultaneous feeling of both attraction and repulsion (or pleasure and displeasure) to a sensible object on account of something about that object being inaccessible to human reason. In judgments of the sublime, we cannot fully represent a sensible object to our faculty of reason. "The sublime is a segment of a circle, grasped by imagination, with the rest of the circle demanded and as it were dreamt of by reason, but not given" (Murdoch 1959, 49). The sublime exists for humans because our animal nature is insufficient to match the nature of reason. This shortcoming evokes a feeling of frustration (because we cannot fully represent the sensible object for ourselves) simultaneously with a feeling of pleasure (admiration or respect) (because we are made aware that we possess the faculty of reason that makes us supersensible beings) (Matthews 1996).

The next question is which plants evoke this feeling for human moral agents and why. The plants that evoke feelings of the sublime for human moral agents include very old plants, such as the ancient redwoods and sequoias (including the General Sherman Tree and President Tree in the Giant Forest Grove) and the ancient bristlecone pines (including Methuselah), as well as individuals born from ancient clonal root systems (many of which push 10,000 years old). The age of these living things far surpasses the lifespan of a human being. Their lives proceed on a completely different timescale than a human life (Craciun 2024). Such ancient living things force us to confront the limitation of our sensibilities through our possession and exercise of reason. They produce a feeling of attraction (and admiration) (because we can conceptualize the age of such living plants through reason) and a feeling of repulsion (because our sensible experience of life is limited by the timescale of our lifespans and so we fail to fully sensibly represent for ourselves a life thousands of years long).[2]

It is possible that other exceptional plants evoke the feeling of sublimity in the absence of great age. Such plants include those that are rare, i.e., belonging to an endangered species, or very unusual in some regard, such as having a unique propagation mechanism or a very limited habitat range. Rarity, precariousness, or very unusual features pose challenges to harmonizing sensibility and reason similar to greatness of age, and they can evoke the feeling of frustration at our inadequacy in representing sensible objects and the admiration of possessing and using reason as a supersensible faculty. For example, the endangered and bizarre plant known as the corpse flower is one such plant that evokes the sublime in human moral agents. In such a plant's presence, its membership in an endangered species and its long timescale propagation strategy fail to be sensibly represented to match reason's understanding of the plant's nature. That failure of representation is both frustrating in our inadequacy and admirable as we are made aware of our supersensible faculty of reason.[3]

[2] For Kant, the sublime was evoked by things that were of great size (how large, in the mathematically sublime) or great strength (how powerful, in the dynamically sublime) (Ferguson 2012). Very old plants seem to be a relative of sorts to the mathematically sublime.

[3] I recognize that this view may imply that many other non-plant natural objects evoke the feeling of the sublime in human moral agents and we may have obligations to such natural objects (and they may have rights against us). I do not defend or refute this implication here because it is beyond the

Finally, having argued that exceptional plants evoke the feeling of the sublime in human moral agents, I must establish that sublimity generates duties on human moral agents not to destroy (and to positively assist) the objects which evoke the sublime. These duties arise because of the similarity between the sublime and moral *feeling*. Moral feeling also consists of simultaneous sensations of displeasure and respect caused by the regular conflict between desire and the moral law.[4] This means that our feeling of the sublime is similar to the feeling we get when we reflect on what ought we do and choose to act morally. Because reason in and of itself is not (indisputably) motivating for human moral agents, acting morally in real life requires cultivating moral feeling and acquiescence to the moral law in situations of conflict between desire and reason. Because the sublime evokes feeling similar to moral feeling, experiencing the sublime helps us cultivate moral feeling and the motivation required to act morally. Given our animal nature, the experience of the sublime can help us become better moral actors.

For this reason, we ought not to harm (and we ought to assist) those living things which evoke the feeling of the sublime, including the exceptional plants I mentioned above. We owe to those exceptional plants duties of non-harm and assistance because they are a part of the external world that is crucial for human agents to cultivate moral motivation. Such duties are grounded in our nature as human moral agents, not in some value that the exceptional plants possess and is supposed to be recognized by us. As such, this argument is agent-centered and represents an attempt to move beyond the patient-centered stalemates in non-anthropocentric ethics.

12.4 Why Rights in Plant Ethics

To this point, I have presumed that rights and their correlative duties are desirable moral concepts in plant ethics without providing any defense for this position. Accordingly, I will provide a brief defense here. First, we attach importance to being recognized as a right holder. In large, heterogeneous, and anonymous societies, individuals do not take the position, "I'd rather not have a right. I'd rather be morally considerable in another way that isn't being a right holder." Individuals fight for the recognition of their rights. This is not to say that individuals never *waive* their rights, but waiving a right is not the same as not being recognized as possessing the right in the first place. Waiving a right is a right holder's choice not to claim a right that the individual is already recognized to possess.

So, why do individuals fight to be recognized as possessing certain rights? Rights have a special force about them that other moral concepts do not. Rights place their

---

scope of this chapter, but I do not think it implausible that we have duties to much more of nature than we currently recognize.

[4] "Moral feeling includes a negative feeling of pain as a result of striking down particular sensible desires and the general principle of self-conceit. The feeling of striking down self-conceit is humiliation. But this same humiliation awakens respect for the moral law" (Matthews 1996, 169).

holders in a position of entitlement against moral agents, and this entitlement is not generally overridable or discounted to further other aims. Rights carry categorical force in their recognition, something no other moral concept can claim. The exact nature of this categorical force has been debated. It has been called a "trump" (Dworkin 1977), an "all-things-considered ought" (Thompson 1990), a "side constraint" (Nozick 1974), and a "prima facie" (Searle 1978) or "pro tanto" reason (Frederick 2014). Regardless of one's preferred theory of the force of rights, rights place boundaries on consequentialist reasoning and its imperative to maximize "The Good." While The Good is not necessarily, or entirely, an anthropocentric concept (Singer 1975; Bentham 1948), The Good can almost always be increased by disregarding non-human life forms to further human preference satisfaction. A right, therefore, bestows on the right holder a claim against being treated as a mere instrumentality for the preference satisfaction of humans. Possessing such a claim represents inclusion in a very important part of the moral realm for animals or plants. Recognizing animals or plants as right holders would prevent humans from disregarding them for the self-obsessive end of ever increasing human preference satisfaction.

Further, rights are part of deontological morality, a branch of deontic ethics. Because deontic ethics focuses on the duties of moral agents, it necessarily focuses on moral action rather than moral emotions, disposition, or character. While moral emotions, disposition, and character are important in a complete and plausible human ethic (Anscombe 1958; Williams 1985), they are significantly less important in a plant ethic. As living, teleological beings, what matters most for plants is how humans treat them. Whether or not we act with the correct disposition, empathy, or care is not particularly relevant to the plant because plants do not share in human emotions like other humans and (some) animals do. For example, we may have an obligation to feed our pet dogs, and our dogs may have a correlative right to be fed by us. A complete human–dog ethic would encourage the cultivation of the correct disposition around feeding our pet dogs, e.g., that it is rewarding and should be pleasant to take care of our canine companions who depend on us for so much. This emotional component of the relationship between humans and dogs would arise because dogs can share in human emotional life. If we have similar obligations to our houseplants to nourish them (which could include playing music for them), whether we do so with care, love, and affection rather than grudgingly makes little difference *to the plant*. Plants do not share in our emotional lives. So, I think a plant ethic must include and emphasize how a moral agent ought to treat plants, independent of a moral agent's character and feelings about plants. The value that rights bring to plant ethics lies in this categorical force commanded by rights claims regarding certain treatment. It is not captured by any other moral or ethical concept.

Talk of plant rights in a plant ethic also presents some advantages compared to the traditionalist and revisionist approaches simply because the language of rights is well-established and relatively concrete. One of the biggest problems that emerges from the traditionalist approaches to plant ethics is that such approaches do not tell moral agents the practical importance of extending the moral realm to include plants. One could grant that plants are intrinsically valuable or morally considerable. Such recognition, however, does not make clear how those value concepts translate into (categorical) principles of moral action for moral agents, which are what matters most to plants. For example, Rolston III (1991) appeals to the notion of moral respect for living things, but that only gets him to conclude that we should not mutilate majestic sequoias for amusement (78). It says nothing about whether we are obligated to protect them from wildfires (or human-caused climate change more generally).

Similar problems exist for the revisionist approaches in plant ethics. However, revisionist approaches also face the challenge of creating and motivating a revolution in thought in metaphysics, ethics, and the relationship between metaphysics and ethics. There is a reason that deontic ethics has been the dominant position for conversations regarding morality (as well as law and politics) since the Enlightenment. Enlightenment thought centered on developing an understanding of how individuals existing in essentially anonymous relationships with one another ought to treat one another. Homogeneity of lived experience and homogeneity of values no longer provided common ground from which norms of behavior could be derived. Regardless of what we think of this move from shared values and community to individualism, the current world consists of extremely heterogenous and almost exclusively anonymous relationships among human beings. The language of rights and duties makes clear what is required of moral agents and when moral agents are free to act as they wish. The revisionist plant ethics abandons this framework and leaves moral agents with even less understanding of how they ought to treat plants than the traditionalist approaches. For example, Sylvie Pouteau's (2014, 2018) work on the metaphysics of plants as open and infinite beings (contrasted with animals as closed and finite beings) is one new plant ontology. As is Thomas Nail's (2024) suggestion that plants (indeed all Earth life) are cosmic terrestrial patterns of motion rather than the categorically distinct entities suggested by modern biological taxonomy. However, accepting such metaphysical views of plants does not readily translate into answers to the question, "How ought I treat this plant?" Even if western ethics accommodates a new ontology of plants, western ethics presumes particular questions that it is trying to answer. One of those questions is "How should I treat X?" So, the revisionist plant ethicists have a two-fold challenge: (1) justifying the new plant ontology and (2) deriving a plant ethic that provides action guidance for moral agents that depends on the new plant ontology. This is a tall order.

Among the extensionist approaches to the plant ethic, one may wonder what advantages rights and their correlative duties have over some other established moral concept, such as interests, personhood, or flourishing. One advantage of appealing to rights and duties is being able to argue for a legal right by grounding it on the moral right. Suppose the moral rights of plants become legal rights. In that case, the power of the state can be invoked to remedy rights violations, enjoin rights violations before they occur, and compel assistance to plants when it is morally required. One might think that personhood is sufficient for achieving this outcome. However, legal personhood is not a sufficient condition for bringing a legal claim. Legal personhood is merely a necessary condition for bringing a legal claim. One must also possess a legal right with identifiable content against an identifiable, obligated party. A moral right can provide this identifiable content against an identifiable, obligated party and, therefore, the basis for bringing a legal claim in a way legal personhood cannot.

At this point, a keen reader may wonder, if rights are so important to plant ethics, why have I only argued for rights for a handful of exceptional plants? In the argument I advanced above, most plants are not right holders. In response to this concern, I want to emphasize that the recognition of rights for some exceptional plants does not eliminate other, less categorical, ethical concepts from articulating ways in which the relationship between humans and plants should change. It may be that the genetically modified cornstalk has no right not to be cultivated for food. It may even be that the genetically modified cornstalk has no right that I not kill it for no reason. This does not imply that it is becoming of my character to wantonly strike it down. Killing living things for no other reason than enjoyment at extinguishing the life of another is and has been a sure sign of a deficient character in humans.

The deeper reason my argument above is limited to exceptional plants comes from the real challenge for a complete plant ethic, which are the cases of unremarkable plants that either interfere with or are critical resources for satisfying legitimate human ends in some fashion. For example, I landscape my property with several bushes. One of them overgrows its lot, and I trim it. Eventually, I change my mind about the aesthetic of my lawn and decide to remove the bush, killing it. What does plant ethics say about my actions here? What if the bush could not survive in the climate where I live without me providing yearly fertilizer and regular irrigation? Would that affect whether I am permitted, under any part of a complete plant ethic, to kill the bush or let it die and then remove it? The variety of situations where plant interests conflict with human interests is substantial. Consider an oak tree that has overgrown the roof of my house and poses a threat to damaging my roof in a storm or a local grove of tall pine trees that stand in the way of the power company's efforts to build a transmission line that will be more resilient in the face of climate change. While there are good reasons to recognize some plant rights, there must be a limit to plant rights, and it is to that which I will now turn.

12.5 The Limits of Plant Rights

To advocate for the recognition of some plant rights is to advocate for some amount of rights extensionism. Despite the importance and value of rights, widespread rights extensionism will face challenges to its own project at some point. This is because recognizing new rights (and their correlative duties) necessarily reduces the sphere of optional action for moral agents. In other words, when I have an obligation to someone not to do something, I am no longer at liberty to do or not do that thing. I am obligated not to do it. Thus, my sphere of optional actions is smaller than if I had no obligation not to do that thing. This kind of freedom is crucial to living a self-directed life as a human individual.[5] If, at every juncture in the road of practical

[5] Here, I am thinking of Bernard Williams' (1982) concept of S projects, projects of subjective satisfaction, which individuals pursue from their internal, subjective motivations to act (101–113).

reason, what I ought to do is determined, I have no effective freedom. I become a moral robot devoid of creativity and individual personality.

One might interpret this concern as an argument against recognizing new rights because to do so would produce too onerous a morality on moral agents. While I generally agree that the onerousness of morality is not an argument against it, in very broad rights extensionism, especially to the nonhuman world, the issue of onerousness takes on a different meaning. When morality only concerns human affairs, even if I had no moral discretion in my actions regarding other human beings, I would not be a moral robot devoid of creativity and individual personality. I would have the moral freedom to pursue my self-directed ends in the sphere of non-human affairs, perhaps to the detriment of the interests of non-human beings. If morality commanded me to act (or refrain from acting) in certain ways throughout the entire non-human world, I would lose the space to realize myself as an individual. Everything I do would be morally obligated, and I would become merely a conduit through which moral action is realized in the world. Where the boundary lies between my sphere of moral obligations and my sphere of autonomy for self-actualization, I cannot say or defend here. I posit only that plant rights in plant ethics cannot possibly extend to all plants without having to confront the threat of collapsing human freedom for self-determination into nothingness.[6]

12.6 Conclusion

In this chapter, I have argued that plant ethics should incorporate rights into its moral machinery and that we should recognize exceptional plants as right holders. I argued that those plant rights can be normatively grounded on an agent-centered rights argument. Exceptional plants evoke the sublime in human moral agents. Because the sublime is similar to moral feeling, human moral agents can (and should) cultivate moral motivation by experiencing the sublime. So, we have duties of non-harm (and positive assistance) to exceptional plants. I discussed the importance of rights in a plant ethic because they bring a categorical force to bear on how we are required to treat certain plants, something no other ethical concept does. Finally, I noted that plant rights must be limited because of the pressure that rights extensionism places on the possibility of human freedom for self-determination.

---

[6] I make no commitment regarding the onerousness of other ethical concepts in plant ethics. My intuition is that because rights carry categorical force, something no other ethical concept invokes, the onerousness of rights extensionism is higher than extensions of other ethical concepts.

References

Anscombe, G.E.M. 1958. Modern Moral Philosophy. *Philosophy* 33 (134): 1–19.

Bentham, Jeremy. 1948. *A Fragment on Government with An Introduction to the Principles of Morals and Legislation.* Oxford: Basil Blackwell.

Callicott, J. Baird. 1980. Animal Liberation: A Triangular Affair. *Environmental Ethics* 2 (4) (Winter): 311–38.

Callicott, J. Baird. 1988. Animal Liberation and Environmental Ethics: Back Together Again. *Between the Species* 4 (3): 163–169.

Callicott, J. Baird. 1989. *In Defense of the Land Ethic: Essays in Environmental Philosophy.* Albany: State University of New York Press.

Callicott, J. Baird. 1999. *Beyond the Land Ethic: More Essays in Environmental Philosophy.* Albany: State University of New York Press.

Craciun, Adriana. 2024. The Vegetal Return: Plant Time and the Curious Engines of Enlightened Knowledge. In *The Vegetal Turn*, ed. Marcello Di Paola. New York: Springer.

Dworkin, Ronald. 1977. *Taking Rights Seriously.* Cambridge: Harvard University Press.

Ferguson, Frances. 2012. Reflections on Burke, Kant, and *Solitude and the Sublime. European Romantic Review* 23 (3): 313–317.

Frederick, Danny. 2014. Pro-Tanto versus Absolute Rights. *Philosophical Forum* 45 (4): 375–394.

Goodpaster, Kenneth. 1978. On Being Morally Considerable. *Journal of Philosophy* 75(6) (June): 308–25.

Hall, Matthew. 2009. Plant Autonomy and Human-Plant Ethics. *Environmental Ethics* 31(2) (Summer): 169–181.

Hall, Matthew. 2011. *Plants as Persons: A Philosophical Botany.* Albany: State University of New York Press.

Hall, Matthew. 2019. In Defence of Plant Personhood. *Religions* 10 (5): 317.

Hitt, Christopher. 1999. Toward an Ecological Sublime. *New Literary History* 30 (5) (Summer): 603–623.

Hobbes, Thomas. 1996. *Leviathan.* Cambridge: Cambridge University Press.

Kallhoff, Angela. 2014. Plants in Ethics: Why Flourishing Deserves Moral Respect. *Environmental Values* 23 (6) (December): 685–700.

Kallhoff, Angela. 2018. The Flourishing of Plants: A Neo-Aristotelian Approach to Plant Ethics. In *Plant Ethics: Concepts and Applications*, ed. Angela Kallhoff, Marcello Di Paola, and Maria Schörgenhumer, 40–50. New York: Routledge.

Kant, Immanuel. 2012. *Groundwork of the Metaphysics of Morals*, trans. and ed. Mary Gregor and Jens Timmermann. Cambridge: Cambridge University Press.

Korsgaard, Christine. 2004. Fellow Creatures: Kantian Ethics and Our Duties to Animals. *Tanner Lectures on Human Values* 24: 77–110.

Korsgaard, Christine. 2018. *Fellow Creatures: Our Obligations to Other Animals.* Oxford: Oxford University Press.

Leopold, Aldo. 1949. *A Sand County Almanac.* Oxford: Oxford University Press.

Marder, Michael. 2013. *Plant-thinking: A Philosophy of Vegetal Life.* New York: Columbia University Press.

Matthews, Patricia. 1996. Kant's Sublime: A Form of Pure Aesthetic Reflective Judgment. *The Journal of Aesthetics and Art Criticism* 54 (2) (Spring): 165–180.

Murdoch, Iris. 1959. The Sublime and the Good. *Chicago Review* 13 (3) (Autumn): 42–55.

Nail, Thomas. 2024. On the Geology of Plants. In *The Vegetal Turn*, ed. Marcello Di Paola. New York: Springer

Nash, Roderick Frazier. 1989. *The Rights of Nature: A History of Environmental Ethics.* Madison: University of Wisconsin Press.

Nozick, Robert. 1974. *Anarchy, the State, and Utopia.* New York: Basic Books.

Pellegrino, Gianfranco. 2018. The Value of Plants: On the Axiologies of Plants. In *Plant Ethics: Concepts and Applications*, ed. Angela Kallhoff, Marcello Di Paola, and Maria Schörgenhumer, 13–27. New York: Routledge.

Pouteau, Sylvie. 2014. Beyond 'Second Animals': Making Sense of Plant Ethics. *Journal of Agricultural and Environmental Ethics* 27 (1): 1–25.

Pouteau, Sylvie. 2018. Plants as Open Beings: From Aesthetics to Plant-Human Ethics. In *Plant Ethics: Concepts and Applications*, ed. Angela Kallhoff, Marcello Di Paola, and Maria Schörgenhumer, 82–97. New York: Routledge.

Regan, Tom. 1983. *The Case for Animal Rights*. Oakland: California University Press.

Rolston III, Holmes. 1975. Is There an Ecological Ethic? *Ethics* 85 (2) (January): 93–109.

Rolston III, Holmes. 1991. Environmental Ethics: Values in and Duties to the Natural World. In *Ecology, Economics, Ethics: The Broken Circle*, eds. F. Herbert Bormann and Steven R. Kellert, 73–96. New Haven: Yale University Press.

Rowlands, Mark. 2009. *Animal Rights: Moral Theory and Practice*, 2d ed. New York: Palgrave Macmillan.

Sagoff, Mark. 1984. Animal Liberation and Environmental Ethics: Bad Marriage, Quick Divorce. *Osgoode Hall Law Journal* 22 (2) (Summer): 297–307.

Sandler, Ronald. 2018. Is Considering the Interests of Plants Absurd? In *Plant Ethics: Concepts and Applications*, ed. Angela Kallhoff, Marcello Di Paola, and Maria Schörgenhumer, 40–50. New York: Routledge.

Schörgenhumer, Maria. 2018. Caring for Plants: Cultivating Relational Virtues. In *Plant Ethics: Concepts and Applications*, ed. Angela Kallhoff, Marcello Di Paola, and Maria Schörgenhumer, 110–118. New York: Routledge.

Searle, John. 1978. *Prima Facie* Obligations. In *Practical Reasoning*, ed. Joseph Raz, 81–91. Oxford: Oxford University Press.

Singer, Peter. 1975. *Animal Liberation: A New Ethics for Our Treatment of Animals*. New York: New York Review Books.

Thompson, Judith. 1990. *The Realm of Rights*. Cambridge: Harvard University Press.

Visak, Tatjana. 2018. Utilitarian Plant Ethics. In *Plant Ethics: Concepts and Applications*, ed. Angela Kallhoff, Marcello Di Paola, and Maria Schörgenhumer, 30–39. New York: Routledge.

Warren, Mary Anne. 1983. The Rights of the Nonhuman World. In *Environmental Philosophy*, ed. Robert Elliot and Arran Gare, 109–134. Queensland: University of Queensland Press.

Williams, Bernard. 1985. *Ethics and the Limits of Philosophy*. Cambridge: Harvard University Press.

Williams, Bernard. 1982. *Moral Luck*. Cambridge: Cambridge University Press.

Chapter 13
Plants as Subjects in Botanic Gardens: Toward Environmental Engagement

Thomas Heyd

Abstract This chapter begins with a brief exploration of what it means to view plants as subjects. This is followed by an account of gardens, and their role as sites of human-plant collaboration. The historical development of botanic gardens, and their special functions as recreational displays and sites for plant conservation are discussed next. This is followed by the proposal that botanic gardens, even more than other gardens, exhibit human-plant collaboration, and the claim that, as such, these gardens offer outstanding opportunities for viewing plants as subjects in botanic gardens. It is concluded that recognition of plants as subjects may contribute in an important way toward greater inclusion of the non-human world in our consideration.

13.1 Introduction

Plants have until recently only had a marginal place in our moral consciousness despite the fact that all animal life, human life included, is directly or indirectly dependent on plants for food and that, in contrast to other features of our environment on which we depend, such as air, water or table salt, plants are alive, in many ways responsive to interventions in their proximate space, and intelligent in their own specific ways (see, e.g., Calvo 2016; Calvo et al. 2017; Calvo and Lawrence 2023; Mancuso and Viola 2015; Pollan 2001; Pouteau 2014; Trewavas 2014, 2017).

As of late plants have, nonetheless, become of increasing interest from natural scientific as well as philosophical perspectives (e.g., see Hall 2009; Heyd 2006; Heyd 2007; Heyd 2012; Hiernaux 2021; Kallhoff 2002; Kallhoff et al. 2018; Kallhoff et al. 2018; Koechlin 2009; Pouteau 2014; Odparlik 2018; Sandler 2018; van Bueren and Struik 2005; Warber et al. 2003). In the following it is proposed that there are good reasons to view plants as subjects, that gardens require human-plant collaboration,

T. Heyd (✉)
Department of Philosophy and School of Environmental Studies, and Centre for Global Studies, University of Victoria, Victoria, Canada
e-mail: heydt@uvic.ca

M. Di Paola (ed.), *The Vegetal Turn*, The International Library of Environmental, Agricultural and Food Ethics 38, https://doi.org/10.1007/978-3-031-68671-9_13

which, furthermore, makes evident their subjecthood, and that the special conditions present in botanic gardens are particularly well suited to this task.

13.2 Plants as Subjects: Recognition of Their Otherness

To recognize plants for what they are is not a straightforward matter because of their great differences from us humans. Here it is suggested to take a Levinasian approach according to which to truly engage other beings precisely is to recognize them *in their otherness* (Lévinas 1989; Perpich 2008). Recognizing others *as others* in face-to-face relations also means that those others *not* be considered as mere background but, rather, *for themselves*.[1] On this view, recognition of others *as others* is a pre-condition for respect and consideration, such as may be proposed by a plant ethics.[2]

Contrary to extensionist approaches in ethics, which argue from features held in common between other beings and ourselves, the Levinasian perspective does *not* rely on extending consideration on the basis of *similarities* shared with those others, be those similarities in terms of membership in our species, sentience, or even goal-directedness. So, while animal ethics generally seeks its grounding in the supposition that non-human animals share with us a concern for the avoidance of pain and suffering (Singer 1989, 1990, 1993), or having a care for one's life that justifies ascription to them of rights (Regan and Singer 1989), the Levinasian approach is grounded on a distinctly different approach.

From this latter point of view, to truly be present to plants requires engaging them as our counterparts—*no matter* the gulf that separates our selves from them, and *in virtue* of a recognition of their otherness. Notably, however, once we are embarked on recognition of other beings in face to face relations (Lévinas 1989; Perpich 2008), this also implies recognition of their subjecthood, at least in one sense of the term.

As argued by Davies (2019) regarding ecological subjecthood, the notion of "a subject" makes reference to at least two distinct meanings.[3] It is rather common to directly associate this notion with the supposition that subjects are autonomous in the Kantian sense, that is, capable of moral choices, grounded in their specific subjectivity, thereby providing us with a grasp of who they are in relation to the rest of beings in their world and of our moral obligations to them. Understood this way, the set of subjects constitutes a relatively limited group, as it implies that even among humans there will be some who do not qualify as such because of a lack of moral maturation or missing capacities for subjectivity (cp. Ingensiep 2007).

---

[1] This approach echoes a point also emphasized by Michel Serres (1990), who argues that we have reached a point in time in which we treat nature as mere background at our peril.

[2] Regarding proposals for plant ethics (see, e.g., Arbor 1986; Hall 2009; Heyd 2012; Hiernaux 2021; Kallhoff 2018; Marder 2013; Pouteau 2014).

[3] Also see Bernstein (1992) on an expanded moral community that may include plants among its subjects. Neal (2012) has a similar approach as Davies on subjecthood, though limited to the context of human dignity.

This, however, is only one way of viewing subjecthood, since it is directly modelled on what it means to be a full member of the (human) moral community, thereby constituting subjects as both moral patients and moral agents. As argued by Davies, though, another way to conceiving of subjecthood, going at least as far back as Aristotle, is available insofar as we may recognize the capacity for subjecthood, understood as agency, without necessarily ascribing full, morally relevant, autonomy. Aristotle attributed such capacity for agency to all living beings, thereby including plants as well as animals (Aristotle *De Anima* 2.4.415b18 (Aristotle 1961); also see Bos 2010; Bernstein 1992; Sprague 1991).[4]

So, from this second perspective on subjecthood it would be mistaken to suppose that, in order for plants to be subjects one would have to attribute human---like subjectivity to plants. Nor does recognition of plants as subjects in this sense mean that we view them as autonomous in a Kantian sense, i.e., as capable of envisioning their behavior within a universe of possible behaviors, freely choosing from among them, and being morally responsible for their choices. Nor, moreover, does it mean that we suppose them as capable of recognizing themselves as autonomous subjects.

Of course, we may ask what this alternative notion of subjecthood amounts to in a positive sense. In answer to this question we may note that recognition of an other in face-to-face relations means that we come to see the being in question as the type of being that it is, at least in part, as the result of internal organizing principles. That is, that it is under the direction of its own (either "inbuilt" or autonomously developed) logic or mode being (cp. Pouteau 2014; also see Bernstein 1992; Sprague 1991). It is for this reason that, when we recognize an entity as "an other" we see it as our counterpart who, moreover, calls for a (possibly ethical) response from us.

Strikingly, as in the case of animals, the recognition of otherness of plants thereby confronts us with a whole other evolutionary history of adaptation to environments than our own (also see Singer 1989). In the case of plants this evolutionary history has led to adaptations that may appear especially alien to us, since they are based on different kinds of perceptual capacities and intelligence than those common among "the higher" animals (e.g., vertebrates) (see, e.g., Calvo and Lawrence 2023; Mancuso and Viola 2015; Trewavas 2014, 2017). Consequently, in acknowledging subject status to plants we are opening ourselves up to becoming present to the very substantial differences between our human mode of being in the world and theirs.[5]

This opening up to plants, of course, does not mean that we can *understand* them in a full sense any more than we can fully understand non-human animals (cp. Nagel 1974). Recognizing plants as subjects means, rather, a readiness to acknowledge that there are many significant differences from the type of being that we are, to accept those differences as part and parcel of their otherness, and even to pursue those differences as worthy of exploration. Incidentally, the comparison of ways of

[4] In contemporary times a number of writers (including Bernstein 1992; Davies 2019; Heyd 2007; Neal 2012) have argued in various ways that a notion of subjecthood involving agency may be extended to an even wider set of beings than those standardly considered "living". Regarding agency also see Latour 2014.

[5] Also see Marder (2013) on the idea of opening ourselves up to plants, which, he argues, should lead to respect for them.

being that this entails probably constitutes at least part of the pleasure experienced by gardeners, arborists, botanists and nature enthusiasts in their interactions with plants.

13.3 Gardens: Sites for Human-Plant Collaboration

Since much of human behavior degrades, or even destroys, the habitat of plants, the dependence of plants on adequate natural spaces to live in should lead us to reflect on the places where they may flourish and be appreciated as subjects. Here I propose that, while their wild habitats of origin certainly should be our primary focus when it comes to protection of spaces where they may flourish, we may also consider gardens, and especially botanic gardens, as suitable places for their flourishing and for being appreciated for what they are.

The concept, history, and function of gardens have been a topic of discussion for some time (see, e.g., Dixon Hunt 1988; Miller 1993; O'Brien 2010; Ross 1988; Turner 2005). Gardens have also received a certain degree of philosophical attention (e.g., Cooper 2006/2008; di Paola 2018a; di Paola 2018b; Hall 2010; Hall 2011; Leddy 1988; Reichard 2011; Reichard and Raven 2011; Vogt 2017), though there remain a number of open questions such as, for example, what exactly makes for a garden. Notably, while gardens commonly are associated with plants, there are some traditions that consider certain sites as gardens even if only populated with sand, pebbles and rocks. This is the case of the Japanese rock or dry landscape garden (*kare sansui*), often found in Zen gardens, such as at Ryōan-ji, Kyoto, Japan (see Heyd 2002).

In any case, when planted, gardens distinguish themselves from the rest of vegetated areas by the fact that the plants contained in them usually are physically separated (at least to some degree) from the remainder of plants (wild plants) beyond its boundaries, and that their selection, distribution on the land reserved for them, and care, are under the direction of humans. Nonetheless, the distinction between gardens and other agricultural endeavors, such as seeded fields, remains somewhat fluid, since what distinguishes gardens from those other planted areas mostly seems to come down to scale of occupation of land.

Plants in gardens are in a similar position as plants in other human-managed spaces insofar as they may have *instrumental* functions in relation to human goals. As such, these garden plants have a similar role in human endeavors as plants in larger scale plantations. However, when considered more closely, it is evident that plants in gardens tend to be beneficiaries of greater individualized attention in terms of time, effort and physical resources dedicated to the flourishing of each plant than their counterparts in open fields, where the focus generally is on quantity rather than quality. This difference in the amount of care per plant is especially true when gardens have decorative or artistic functions (as botanic gardens often have, among other things).

Common to both gardens and larger agricultural endeavors is that, in contrast to plants in the wild, plant flourishing turns out to be *a collaborative endeavor*. For plants to be productive in humanly confined areas they have to be treated as subjects capable of entering into collaborative relationships with humans. Gardeners (and agriculturalists) can only succeed in their own gardening (and agricultural) aims by being attentive to the aims and needs of plants. In other words, plants can only contribute to the goals of gardeners to the degree that their nature, including their innate goal-orientation (see Kallhoff 2018; also see Aristotle, Physics 199a 23–30; 199a 29 (Aristotle 1936)), is activated to cooperate with gardeners' goals. Practically this means that they require sufficient quantities of water, adequate levels of light, adequate temperatures, soil quality, and so on.

It may perhaps be objected that if gardens serve the economic or aesthetic goals of humans, such as the production of food stuffs or the embellishment of property, it is unclear in what sense humans actually *collaborate* with, and not simply *exploit*, plants in gardens. In response it may be granted that, insofar as plants are forced to produce through the application of techniques such as genetic manipulation or the abundant use of fertilizers, the proper description of the relation between humans to plants may well be in terms of exploitation. However, even in those cases, in order for gardens to exist at all requires attentiveness to, and care by, gardeners to plant needs to enable the latter to contribute through their genetically set potentials.

To illustrate, we may contrast the interaction of humans with plants in gardens to the actions required in other extractive activities, such as the procurement of water from water seams, or of iron from hematite, for example, which are *not* dependent on any level of collaboration from non-human parts since these activities do not require the alignment of human behavior with the agency of other beings. Given that the contribution of plants is an indispensable part of what makes gardens possible, and, noting that such agency entails that plants function as subjects, this suggests that the experience of gardens offers one important avenue for the recognition of plants as subjects.

13.4 The Many Functions of Botanic Gardens

Botanic gardens share with ordinary, standard, gardens the characteristic of being planted and tended by humans, and that their plants receive more individualized attention than those in larger scale agricultural settings (for a definition of botanic gardens, see, e.g., Sim 2001). While, as noted, the more general topic of gardens has already been approached somewhat from the point of view of philosophy, very little attention has been paid from that disciplinary perspective so far to this special case of gardens (but see Heyd 2006, 2007, 2012).

Here it is proposed that, while botanic gardens share with other gardens (as well as other agricultural endeavours) that they exist because of the collaboration between humans and plants, they are special insofar as that the focus in these spaces is the flourishing of *plants as such*, and only secondarily any benefits that humans may receive from them. As such, botanic gardens offer outstanding opportunities to experience plants as active makers of their own existences, and, for this reason, are exceptionally well suited to exhibit plants *as subjects*.

13.4.1 Historical Perspective on the Roles of Botanic Gardens

Historically, today's botanic gardens can be traced to a diversity of planted sites with various sorts of aims (Dixon Hunt 1988). Initially, they emerge out of Mediaeval "physick gardens", which were designed to supply monasteries with herbs and medicinal plants. John Prest (1981) argues that there is another origin, which can be traced to the discoveries in the late Renaissance of plants from far-away lands, formerly unknown in Europe. This suggested to Europeans that they may be encountering specimens that once populated the original Garden of Eden, described in the Hebraic texts (Genesis 2:4–6), which inspired them to recreate that mythic garden.

Gardens populated with a great diversity of exotic plants also were built and maintained as living warehouses to supply the expanding colonial powers, particularly Britain, with commercially valuable plants (such as cocoa and tea) for dissemination in the territories that they had occupied around the world (Drayton 2000; Brockway 1979; Maldonado Polo 2000; but also see Endersby 2000). Gardens with useful plants, moreover, were set up at crucial stopover places (for example, on the southern tip of Southern Africa), to function as *en route* supply centres of vital fruits, such as scurvy---preventing citrus (Drayton 2000).

Some gardens in subtropical locations, such as the Jardín de aclimatación de La Orotava, Tenerife, furthermore, served as intermediate acclimatization points of exotic plants, such as potatoes and bananas, intended for ultimate cultivation in European countries (Thacker 1979), or even as keystones of eco---nationalist projects (Ginn 2008). In modern times the primary aim of botanic gardens has been research and education even if, eventually, the turn toward laboratory dependent research techniques, made living plant collections relatively irrelevant for botanical research (Seibert 1956).

13.4.2 Leisure Recreation in Botanic Gardens

Gardens support human beings in a variety of ways, among which the supply of fresh vegetables in kitchen gardens and of fruits in orchards may be of fundamental importance. This has always been the case in many settled societies with deep roots in the land. Gardens, however, may also become sources of essential food supplies in metropolitan areas that have long lost touch with surrounding rural areas, as is evident in the increasing popularity of urban gardening.[6] Furthermore, in countries with a relatively high standard of living, those who can afford a space in which they can set up gardens often find that the activity of gardening offers them physical and mental health benefits.

To all these benefits to humans one may add the sheer aesthetic pleasure of being surrounded by plants. That pleasure can be considered under at least two points of view. On the one hand, such aesthetic pleasures may be considered as an appreciation of the *intrinsic* value of plants, insofar as plants are recognized as sophisticated forms of life that strike us in their differences from ourselves as humans. On the other hand, such aesthetic pleasures may be considered in terms of their *instrumental* value, insofar as plants provide us with a number of appreciable benefits, such as pleasing shade, contributions to landscape diversity, as well as the promotion of psychological health (as recently discussed regarding the benefits of "forest bathing", for example).

Though botanic gardens generally do not supply the public with fresh vegetables or fruits, nor allow for participation in their maintenance by laypersons as gardeners, they generally do offer well-tended aesthetic displays. The attractiveness of these displays is increased, furthermore, by the fact that these types of gardens tend to feature plants from geographic and climatic regions beyond the local, thereby adding a certain exoticism to the experiences that they facilitate.[7] As such, botanic gardens offer valuable benefits in terms of leisure recreation for urban populations. This is especially so in large metropolitan areas, where access to natural spaces, relegated to the periphery, away from places of habitation, generally is difficult (see O'Malley 1992; Robinson 1996; Seibert 1956).[8]

While leisure recreation in botanic gardens may vary in the degree to which their plants are appreciated for their instrumental versus their intrinsic value, from this perspective the fundamental orientation toward the plants exhibited remains anthropocentric. Viewed this way, the predominant relation between human beings and plants in these gardens is one of instrumental use. This, however, need not mean that those plants cannot be viewed as subjects in their own right, as we will see below.

13.4.3 Plant Conservation in Botanic Gardens

As is well known by now, land use changes around the world are provoking ever more concerning decreases in biodiversity in wild spaces, both in relation to plant and

---

[6] A case in point, in which community gardens became the only way to address food shortages, is Havana, Cuba, after 1989, as a consequence of the collapse of the Soviet Union and the enduring US American embargo (Plonska and Saramifar 2019).

[7] On the added aesthetic value arising in cases of rarity see, e.g., Russow (1981), who makes a case with regard to rare animals, but which similarly applies to plants. Also see James (2023).

[8] Botanic garden advocates have long recognized this value of botanic gardens and have suggested ways in which these spaces could heighten their contribution to recreational goals. For example, by actively developing courses for the broad public in topics ranging from horticultural techniques to plant species recognition (Avery Jr. 1957).

animal species.[9] Presently, the losses in species diversity are in an accelerating trend, due to increasing demands by corporations and nation-states to exploit all available resources in order to feed the economic growth demanded by financial markets. The global loss of species, moreover, is further exacerbated by anthropogenic climate change.

Of course, the primary means of maintaining biodiversity is by active protection of the ecosystems that are the home space of the species in question. Given the pressure on such places, however, remedial measures gain in importance. For example, with regard to plants, through seed saving at regional, national and international levels. (One such project, with global aspirations, is the much discussed Svalbard Global Seed Vault.)

It is in this context that botanic gardens are seen to play an important role in the *ex situ* conservation of plants, as proposed, for example, by the International Union for the Conservation of Nature (IUCN) and the World Wildlife Fund (WWF) in their *Botanic Gardens Conservation Strategy* (Heywood 1989).[10] This method of supporting the conservation of plants is not without its problems though, since it is argued that only a relatively small gene pool can be preserved this way and that, if the ecosystems in which the plants were originary have disappeared, such conservation efforts may only serve as museal displays (Folsom 1996; but see Maunder et al. 2001).

In any case, when botanic gardens are understood as sites for plant conservation, the aim in view is *not* to benefit humans in some way (at least not immediately or directly). Rather, viewed this way, plants, as representatives of certain species, are being considered as worthy of protection and care, largely for their inherent or intrinsic value, that is, for a value that human beings recognize in them *as such* and not merely because they are useful for the satisfaction of human interests.

13.5 Human-Plant Collaboration in Botanic Gardens

The relationship of humans with plants in gardens is not without conflict, since humans normally seek to control the selection of plants in gardens by supporting the flourishing of a certain set (either because useful in some practical way, or pleasant to look at), while suppressing others (treated as "weeds"). Even when gardens are seen as spaces for plant conservation, it is human choices that determine the state of the garden, though the control that humans exert comes in degrees, both in terms of intensity and frequency of interventions, and in terms of species composition allowed on the parcel of land occupied by gardens. So, some gardens are subject to very thorough regimentation, including regular pruning, treatments with herbicides and pesticides and weeding, while others are basically left to themselves once a certain species assortment has been set up.

[9] Also see Sarkar (2005) with regard to the transformative value of biodiversity.

[10] Regarding *ex situ* conservation, also see Mounce et al. (2017).

Despite human control, however, gardens, as noted, only exist because of the collaboration between gardeners and plants, in which gardeners support the flourishing of plants which participate with their distinct, species-specific capabilities. The fact that humans and plants in gardens have to be closely in tune with each other has often been recognized in the literature (see, for example, Dixon Hunt 1988; Crawford 1983). Dixon Hunt, for instance, speaks of the garden as "a site of conflict and dialogue" in which there has to be "accommodation of both nature and culture" (Dixon Hunt 1988, 272).

In botanic gardens this need for accommodation is magnified because of the *prima facie* commitment in these sites to the flourishing of plants in such a way that they can clearly exhibit their species-specific characteristics. That is, while in other gardens plants may be pruned, fertilized, or otherwise manipulated for purely aesthetic or food productive reasons, in botanic gardens interventions tend to be focussed on those actions that will allow plants to exist in their most authentic manner.

To make this possible, up to a point the designers of botanic gardens seek to create small-scale replicas of the originary environments of the plants selected and, when these gardens contain exotic specimens, designers apply further manipulations to their environments, so that the ground, sun exposure, humidity levels, and so on, are adequate. So, somewhat ironically, in order to help plants flourish in an authentic manner in botanic gardens requires a high degree of manipulation of environmental factors (for example, in greenhouses), thereby involving the creation of something like simulacra (see Baudrillard 1994). However, since in botanic gardens the focus is plant flourishing as such, the point of all of these manipulations is to allow plants much freer expression than they would have in other human-controlled spaces.

In short, while botanic gardens are artifactual spaces, involving a high degree of design and, subsequently, human intervention in their maintenance, this artifactuality precisely is put in place in order to allow plants to exhibit their natures. In this sense botanic gardens represent outstanding examples of collaboration between humans and plants. Furthermore, insofar as these gardens allow plants as much spontaneity as is possible in artificial (humanly-manipulated) environments, the human-plant relationship exhibited there facilitates viewing it as a relation of subjects to subjects.

13.6 Plants as Subjects in Botanic Gardens

As already noted, nothing can replace the kind of possibilities for the full and free expression that plants can find in their environments of origin, since there are climatic, edaphic, and ecosystemic subtleties present in those places that may be hard to reproduce in humanly-created spaces. Consequently, for people who seek to experience plants in their authenticity, witnessing the expression of plants "in the wild" certainly should to be preferable. Nonetheless, to city-dwelling people of today, botanic gardens offer several advantages over their wild counterparts.

First, botanic gardens offer accessibility. Most urban centers only have limited spaces with plant cover and, where available, they tend to be populated with a very restricted set of species (for example, the few herbaceous plant species occupying the ubiquitous front yard lawns). Many people, moreover, do not have the financial means and time to frequent wild vegetated places on the periphery of their cities, and even less to travel to the many places from which the plants cultivated in botanic gardens come from.

Botanic gardens, furthermore, have the advantage that they usually feature visitor-friendly introductions to the plants on site through labels and, occasionally, through tours. For laypersons these gardens thereby facilitate deeper cognitive insight into the mode of existence of plants than would be obtained by simply traversing non-cultivated natural areas (even if in-person visits to forests, steppes, or wetlands, may generate significant affective connections).

Because of their intentional design, botanic gardens, moreover, may offer aesthetic experiences that ordinary people would not have had even in the places of provenance of the gardens' contents. For example, in the Jardí Botànic of Barcelona visitors are guided through landscapes with a palimpsest of Mediterranean types of vegetation from around the planet, including from places such as the Chilean coastal region, California, South Africa's Western Cape, and West Australia. Such aesthetic experiences from assemblages of plants not found together in the wild may then facilitate a deeper understanding of the plants on view by way of comparison in one space of parallel evolutionary paths leading to similar solutions for living in certain climatic conditions.

Recognition of others as subjects, as noted, implies that they not be treated as mere background or as mere resources to be exploited. Rather, recognition as subjects calls for consideration of their flourishing and acknowledgment of the difference between our own mode of being and theirs. Insofar as botanic gardens are spaces of collaboration, these conditions are satisfied to a large degree. Plants that are cultivated in them are foregrounded, as their flourishing is the guide to the management of these spaces, and the display of the difference between the mode of existence of plants and ours is their primary aim.

As such, botanic gardens offer opportunities for rethinking the place of plants in relation to humans and for the recognition of plants as subjects. In this way botanic gardens may, moreover, support the possibility of the integration of plants in the moral community, and thereby facilitate the eventual development of a plant ethic.

13.7 Conclusion

Plants certainly are the most abundant visible life form on the planet. They fulfil a great number of vital functions for animals, human beings included, ranging from the provision of oxygen and the production of nourishment, to the furnishing of materials for shelter and heating materials. Nonetheless, for the most part, plants are seldom seen as considerable for themselves.

Gardens, however, require close attention to the needs of plants if they are to flourish. Outstanding Botanic gardens are created and maintained to serve a multiplicity of functions, ranging from research, education, and recreational displays, to ex situ preservation of species in danger of extinction. Moreover, insofar as botanic gardens aim at the authentic display of plants, they offer especially good opportunities to observe the collaboration between humans and plants typical of garden spaces.

Insofar as in these gardens *the ends* of plants, namely, the realization of their potentialities as the kind of plants that they are, precisely is *the means* for the achievement of the other ends that these gardens are designed for, these gardens direct visitors toward an appreciation of plants as others, that is, as subjects and not as mere objects. This, in turn, may enable people to more easily develop the kind of inclusiveness of the non-human that is needed to counteract the self-defeating and destructive over-exploitation of the natural environment prevalent today. From this perspective, seeing plants as subjects acquires special urgency, and botanic gardens, as sites in which plants may be recognized as subjects, may play a crucial role.[11]

Finally, it may be added that botanic gardens could even more explicitly integrate into their educational programs accounts of how each plant species fits into certain niches in the environment, while also contributing to the functioning of the ecosystems that they are part of (also see Huckle 2016). It may, furthermore, be valuable for botanic garden designers to dedicate exhibition space to illustrate how each type of plant fits into their ecological niches of original provenance in contrast with how their selection in industrial agriculture may undermine those place-specific adaptations.

Moreover, viewing plants as subjects through visits to botanic gardens may transform gardening in other spaces, as in people's front or back yards, and may lead to greater cooperation with the natural tendencies of plants. For example, leaving more room in gardens for local indigenous species and applying less aggressive ways of obtaining instrumental value from them (as, for example, by avoiding pesticides, herbicides, fungicides, and adopting methods such as permaculture).

References

Arbor, J.L. 1986. Animal Chauvinism, Plant-Reading Ethics and the Torture of Trees. *Australasian Journal of Philosophy* 64 (3): 335–339.

Aristotle. 1936. Physics. In *Aristotle's Physics*, ed. W.D. Ross. Oxford: Clarendon Press.

Aristotle. 1961. De Anima. In *Aristotelis: De Anima*, ed. W.D. Ross. Oxford: Clarendon Press.

Avery, G.S., Jr. 1957. Botanic Gardens—What Role Today? An 'Operation Bootstraps' Opportunity for Botanists. *American Journal of Botany* 44: 268–271.

Baudrillard, J. 1994. *Simulacra and Simulation.* Ann Arbor: The University of Michigan Press.

Bernstein, M. 1992. Towards a More Expansive Moral Community. *Journal of Applied Philosophy* 9 (1): 45–52.

[11] Section III of this text summarises certain key ideas from Chapter 12, Heyd (2007). I am indebted to the botanist Joan Pedrola, designer of the plant displays at the Jardí Botànic of Barcelona, for provoking my initial reflections on botanic gardens.

Bible (English Standard Version). 2009. *Genesis.* Oxford: Oxford University Press.

Bos, A.P. 2010. Aristotle on the Difference Between Plants, Animals, and Human Beings and on the Elements as Instruments of the Soul (De Anima 2.4.415b18). *Review of Metaphysics* 63 (4): 821–841.

Brockway, H. 1979. Science and Colonial Expansion: The Role of the British Royal Botanic Garden. *American Ethnologist* 6 (3): 449–465.

Calvo, P. 2016. The Philosophy of Plant Neurobiology: A Manifesto. *Synthese* 193 (5): 1323–1343.

Calvo, P., V.P. Sahi, and A. Trewavas. 2017. Are Plants Sentient? *Plant Cell Environment* 40: 2858–2869. https://doi.org/10.1111/pce.13065. Accessed 18 March 2023.

Calvo, P., and N. Lawrence. 2023. *Planta Sapiens: Unmasking Plant Intelligence.* New York, NY: WW Norton & Co.

Cooper, D.E. 2006/2008. *A Philosophy of Gardens.* Oxford: Oxford University Press.

Crawford, D. 1983. Nature and Art: Some Dialectical Relationships. *The Journal of Aesthetics and Art Criticism,* 42 (1): 49–58. https://doi.org/10.2307/429946

Davies, E.S. 2019. *Ecological Subjecthood* (Ph.D. thesis). Canberra: The Australian National University.

Di Paola, M. 2018a. *Ethics and Politics of the Built Environment.* Cham: Springer International Publishing. https://doi.org/10.1007/978-3-319-71166-9_5.Accessed28March2023.

Di Paola, M., ed., 2018b. Gardens and Ethics. In *Ethics and Politics of the Built Environment Gardens of the Anthropocene.* Cham, Switzerland: Springer International Publishing.

Dixon Hunt, J. 1988. Gardens: Historical Overview. In *Encyclopedia of Aesthetics*, ed. M. Kelly, vol. 2, 271–274. Oxford: Oxford University Press.

Drayton, R. 2000. *Nature's Government: Science, Imperial Britain, and the Improvement of the World.* New Haven CT: Yale University Press.

Endersby, J. 2000. A Garden Enclosed: Botanical Barter in Sydney 1818–1839. *British Journal for the History of Science* 33 (118): 313–334.

Folsom, J.P. 1996. The Issues and Ethics of Plant Collections. *Public Garden* 11 (4): 24–29.

Ginn, F. 2008. Extension, Subversion, Containment: Eco-nationalism and (Post)colonial Nature in Aotearoa New Zealand. *Transactions of the Institute of British Geographers* 33 (3): 335–353.

Hall, M. 2009. Plant Autonomy and Human—Plant Ethics. *Environmental Ethics* 31 (2): 179–180.

Hall, M. 2010. Escaping Eden: Plants in a Gardener's World. In *Gardening Philosophy for Everyone*, ed. D. O'Brien, 38–47. Oxford: Wiley-Blackwell.

Hall, M. 2011. *Plants as Persons: A Philosophical Botany.* Buffalo, NY: SUNY.

Heyd, T. 2002. Nature Restoration Without Dissimulation: Learning from Japanese Gardens and Earthworks. *Essays in Philosophy* 3 (1), Environmental Aesthetics. http://commons.pacificu.edu/cgi/viewcontent.cgi?article=1036&context=eip.

Heyd, T. 2006. Thinking through Botanic Gardens. *Environmental Values* 15: 197–212.

Heyd, T. 2007. *Encountering Nature: Toward an Environmental Culture.* Aldershot UK: Ashgate.

Heyd, T. 2012. Plant Ethics and Botanic Gardens. PAN: Philosophy. *PAN: Philosophy, Activism, Nature* 9: 37–47.

Heywood, V.H. 1989. *Botanic Gardens Conservation Strategy.* Richmond, UK: IUCN Botanic Gardens Secretariat, Kew.

Hiernaux, Q. 2021. The Ethics of Plant Flourishing and Agricultural Ethics: Theoretical Distinctions and Concrete Recommendations in Light of the Environmental Crisis. *Philosophies* 6: 91. https://doi.org/10.3390/philosophies6040091.

Huckle, J. 2016. Reconstructing & Representing Nature: The Challenge for Botanic Garden Educators. In *The Power for Change*, eds. L.A. Sutherland, K.T. Abraham, and J. Thomas. Richmond, Surrey, UK: Botanic Gardens Conservation International.

Ingensiep, H.W. 2007. Consciousness and its Place in a 'Natural Hierarchy': Considerations Concerning the Role of Consciousness in Modern Philosophy and Ethics. *Synthesis Philosophica* 4 (2): 301–317.

James, S.P. 2023. Rarity and Endangerment: Why do they Matter? *Environmental Values.* https://doi.org/10.1177/09632719231171836.

Kallhoff, A. 2002. *Prinzipien der Pflanzenethik. Die Bewertung pflanzlichen Lebens in Biologie und Philosophie.* Frankfurt/New York: Campus Verlag.

Kallhoff, A. 2018. The Flourishing of Plants. In *Plant Ethics: Concepts and Applications*, eds. J. Kallhoff, M. di Paola, and M. Schörgenhumer. New York: Routledge.

Kallhoff, A., M. di Paola, and M. Schörgenhumer, eds. 2018. *Plant Ethics: Concepts and Applications.* New York: Routledge.

Koechlin, F. 2009. The Dignity of Plants. *Plant Signal Behavior* 4 (1): 78–79.

Latour, B. 2014. Agency at the time of the Anthropocene. *New Literary History* 45: 1–18.

Leddy, T. 1988. Gardens in an Expanded Field. *British Journal of Aesthetics* 28 (4): 327–340.

Lévinas, E. 1989. *The Levinas Reader.* Oxford UK: Blackwell.

Maldonado Polo, J.L. 2000. La expedición botánica a Nueva España, 1786–1803:El Jardín Botánico y la cátedra de botánica. *Historia Mexicana* 50 (1): 5–56.

Mancuso, S., and V. Viola. 2015. *Brilliant Green: The Surprising History and Science of Plant Intelligence.* Washington DC: Island Press.

Marder, M. 2013. *Epoché* 18 (1): 247–265.

Maunder, M., S. Higgens, and A. Culham. 2001. The Effectiveness of Botanic Garden Collections in Supporting Plant Conservation: A European Case Study. *Biodiversity and Conservation* 10 (3): 383–401.

Miller, M. 1993. *The Garden as an Art.* Albany NY: State University of New York Press.

Mounce, R., P. Smith, and S. Brockington. 2017. Ex Situ Conservation of Plant Diversity in the World's Botanic Gardens. *Nature Plants* 3: 795–802. https://doi.org/10.1038/s41477-017-0019-3.

Nagel, T. 1974. What is it Like to be a Bat? *Philosophical Review* 83 (4): 435–450. https://doi.org/10.2307/2183914.

Neal, M. 2012. Not Gods but Animals: Human Dignity and Vulnerable Subjecthood. *Liverpool Law Review* 23 (3): 177–200.

O'Brien, D. 2010. *Gardening Philosophy for Everyone.* Oxford: Wiley-Blackwell.

Odparlik, S. 2018. The Dignity of Plants. In *Plant Ethics: Concepts and Applications*, eds. J. Kallhoff, M. di Paola, and M. Schörgenhumer. New York: Routledge.

O'Malley, T. 1992. Art and Science in the Design of Botanic Gardens, 1730–1830. *Garden History: Issues, Approaches, Methods* 279 (302): 64–87.

Perpich, D. 2008. *The Ethics of Emmanuel Levinas.* Stanford CA: Stanford University Press.

Plonska, O., and Y. Saramifar. 2019. *The Urban Gardens of Havana.* Cham: Springer International Publishing. https://doi.org/10.1007/978-3-030-12657-5_3.

Pollan, M. 2001. *The Botany of Desire: A Plant's-eye View of the World.* Random House.

Pouteau, S. 2014. Beyond "Second Animals": Making Sense of Plant Ethics. *Journal of Agricultural Environmental Ethics* 27: 1–25. https://doi.org/10.1007/s10806 013-9439-x.Accessed28March 2023.

Prest, J. 1981. *The Garden of Eden: The Botanic Garden and the Re-Creation of Paradise.* New Haven, CT: Yale University Press.

Regan, T., and P. Singer, eds. 1989. *Animal Rights and Human Obligations.* Oxford: Oxford University Press.

Reichard, S.H. 2011. *The Conscientious Gardener: Cultivating a Garden Ethic.* Berkeley CA: University of California Press.

Reichard S., and P. Raven. 2011. *Toward a Garden Ethic The Conscientious Gardener: Cultivating a Garden Ethic.* Berkeley, CA: University of California Press.

Robinson, F. 1996. The People-Plant Connection. Public. *Garden* 11: 18–20.

Ross, S. 1988. *What Gardens Mean.* Chicago IL: University of Chicago Press.

Russow, M. 1981. Why Do Species Matter? *Environmental Ethics* 3: 101–112.

Sandler, R. 2018. Is Considering the Interests of Plants Absurd? In *Plant Ethics: Concepts and Applications*, eds. J. Kallhoff, M. di Paola, and M. Schörgenhumer. New York: Routledge.

Sarkar, S. 2005. *Biodiversity and Environmental Philosophy: An Introduction.* New York: Cambridge University Press.

Seibert, R.J. 1956. Arboreta and Botanical Gardens in the Field of Plant Sciences and Human Welfare. *American Journal of Botany* 43: 736–738.

Serres, M. 1990. *The Natural Contract*. Ann Arbor Mich: University of Michigan Press.

Sim, J. 2001. *Botanic Garden. Chicago Botanic Garden Encyclopedia of Gardens History and Design*, vol. 1, 172–175. Chicago, IL: Fitzroy Dearborn.

Singer, P. 1989. All Animals are Equal. In *Animal Rights and Human Obligations*, eds. T. Regan, and P. Singer, 215–226. Oxford: Oxford University Press.

Singer, P. 1990. *Animal Liberation*. New York: Avon Books.

Singer, P. 1993. *Practical Ethics*. Cambridge: Cambridge University Press.

Sprague, R.K. 1991. Plants as Aristotelian Substances. *Illinois Classical Studies* 16 (1/2): 221–229.

Thacker, C. 1979. *The History of Gardens*. Berkeley CA: University of California Press.

Trewavas, A.J. 2014. *Plant Behaviour and Intelligence*. Oxford University Press.

Trewavas, A.J. 2017. The Foundations of Plant Intelligence. *Interface Focus* 7 (3). https://doi.org/10.1098/rsfs.2016.0098. Accessed 28 March 2023.

Turner, T. 2005. *Garden History, Philosophy and Design 2000 BC to 2000 AD*. Spon Press London.

Van Bueren, E.T.L., and P.C. Struik. 2005. Integrity and Rights of Plants: Ethical Notions in Organic Plant Breeding and Propagation. *Journal of Agricultural and Environmental Ethics* 18: 479–493. https://doi.org/10.1007/s10806-005-0903-0. Accessed 28 March 2023.

Vogt, B. 2017. *A New Garden Ethic: Cultivating Defiant Compassion for an Uncertain Future*. New Society Publishers, Ottawa ON. https://canadacommons.ca/artifacts/1880860/a-new-garden-ethic/2630180/. Accessed 28 March 2023.

Warber, S.L., M.D. Fetters, and P.B. Kaufman. 2003. Finding a Moral Compass for Human Plant Interaction. *Alternative Therapies in Health and Medicine* 9 (2): 100–105.

Chapter 14
The Land Ethic: The "Key-Log" for the Vegetal Turn?

Casey Rentmeester

Abstract Recent work in the vegetal turn movement has helped to cement the idea that plants should be seen less as nearly inanimate objects and more as highly complex agents that actively interact with the natural world. Various traditions in environmental philosophy are amenable as theoretical frameworks for the vegetal turn, including biocentrism, deep ecology, and the land ethic. Upon chronicling the vegetal turn and its intersection with these traditions, the author argues that the land ethic is best suited to serve this purpose. Published in 1949, Aldo Leopold's *A Sand County Almanac* outlines his land ethic that attempts to move away from the understanding of human beings as masters over nature and towards viewing humans as plain members and citizens of the biotic community. Not only is the land ethic a fitting framework for the vegetal turn, the combination of the two may serve as a key-log to initiating a paradigm shift in our understanding of the human-nature relationship, which, the author claims, is desperately needed to confront the environmental crisis.

14.1 Introduction

The presuppositions—or *presumptions*—of human exceptionalism that have dominated the first three metaphysical eras of Western civilization are finally coming to bear on the human species with the onset of anthropogenic climate change, which threatens human civilizations as we have come to know them. From Aristotle in the ancient era, we learned that humans are the only rational animals, and that other animals and plants have lower souls (Aristotle 1907). From St. Thomas Aquinas in the medieval era, we learned of the *Scala Naturae*, that is, the Great Chain of Being, in which humans are atop the hierarchy of all physical beings, uniquely sharing the spirit world with angels and God (Davies 2016). From René Descartes in the modern era, we learned that humans are uniquely ensouled subjects who just happen to have objective bodies, while plants and animals are merely mechanically automated objects with no souls whatsoever (Descartes 1641/2008). In our current postmodern

C. Rentmeester (✉)
Bellin College, Green Bay, WI, USA
e-mail: casey.rentmeester@bellincollege.edu

M. Di Paola (ed.), *The Vegetal Turn*, The International Library of Environmental, Agricultural and Food Ethics 38, https://doi.org/10.1007/978-3-031-68671-9_14

era, we simply don't seem to know where to turn in our conceptualization of human beings (Taylor 1992), and as we face the mounting and increasingly urgent crisis of anthropogenic climate change, we seem to be doubling down on the very narratives of human exceptionalism that have gotten us into these troubled and troubling circumstances (Rentmeester 2021). The proposed solutions to climate change in the realms of engineering and technology, wherein we geoengineer the planet's climate through carbon dioxide removal or solar radiation management, are examples of entrenchments of the human exceptionalism mentality. The very word "geo-engineer," with "geo" meaning earth, should give us pause, and alarm us as to the level of human hubris this mentality has bred.

Where can we turn to right the civilizational ship? While there is likely no magic bullet to answer this question, surely one of the aspects of this answer needs to be knocking humans off our collective pedestal by recognizing our proper place in the natural world. Two examples of attempts to do just that are the animal liberation movement of the 1970s, which urged us to expand moral consideration to nonhuman animals by showing our similarities with them (Singer 1975, 2009; Regan 1983), and the more recent and nascent vegetal turn as exemplified in this volume, which attempts to enlarge moral consideration even further, and indeed invites a more general rethinking of the morality categories we live by. Michael Marder and Gary Francione's recent debate highlights some of the obvious differentiating features between the animal rights movement and the vegetal turn, most pointedly, their disagreement as to whether plants are sentient or merely have "nonconscious intentionality" (Marder 2016, 176–187). In the animal rights movement, plants are not deemed as sentient and thus have no preferences, thereby leaving them out of the realm of moral consideration, while intellectuals in the vegetal turn movement do ascribe preferences and thus moral consideration to plants. In terms of the history of environmental ethics, the usual place to look for a theoretical context that is hospitable to plants is biocentrism (Taylor 1981, 1986), which has garnered considerable criticism (Watson 1983; Plumwood 1991), and thus fallen out of favor among most environmental philosophers (Basl 2019). The other obvious theory that would be amenable to the vegetal turn would be deep ecology, but that also has serious limitations that I'll outline, as corroborated by others (Sylvan 1985; Grey 1986). In this chapter, I suggest that Aldo Leopold's land ethic provides a more solid basis for the vegetal turn, which not only gives plants their due respect but does so in the context of a generally more appropriate understanding of the human relationship with the rest of the natural world.

14.2 The Vegetal Turn and Philosophy

In order to understand the land ethic as a helpful conceptual basis for the vegetal turn, we need some clarity as to what is meant by that latter phrase. What, precisely, do we mean by the vegetal turn? While we can disagree as to the details and complexities of the vegetal turn, at its heart lies a shift in the understanding on the part of humans of what it means *to be* a plant. Recent research in the existential features of plant behavior offers the possibility of a paradigm shift in our understanding of vegetal life in which plants are seen less as nearly inanimate objects and more as highly complex agents that actively interact with the world, albeit on a timescale imperceptible to the human eye in real time.[1] This paradigm shift lies at the heart of the vegetal turn, and the reason this should be understood as such is linked to our primary relationship to plants throughout Western history. Michael Marder, one of the most important thinkers in the vegetal turn movement, sums up this relationship succinctly as follows: "although—akin to us—they are living creatures, we fail to detect the slightest resemblances to our life in them and, as a consequence of this failure, routinely pass a negative judgment of their worth" (2013a, 3).

Our inability to see any affinity of plant life to our own is linked with what Leopold referred to as the Abrahamic conception of the land (Leopold 1949, viii) in which humans are on a pedestal far removed from all other living beings in the so-called "Great Chain of Being." The vegetal turn posits that "it is necessary to rethink the status of plants not only as objects to be protected but also as subjects to be respected" (Marder 2013b, 47). For Marder, plants exhibit "the agency or the capacity to actively mold the world" (2013b, 47). Indeed, recent research in plant behavior has demonstrated that plants are perhaps sentient beings in that they process information, defend themselves from harm in highly sophisticated ways, share resources, and communicate with one another (Mancuso and Viola 2015). In fact, as Stefano Mancuso and Alessandra Viola note, "plants have all five senses that humans do: sight, hearing, touch, taste, and smell—each developed in a 'plant' way, of course, but no less real" (2015, 3–4). In reorienting our understanding of plant life by chronicling the ways in which plants are similar to humans, Mancuso and Viola help us to rethink the ways in which we've separated humans from the rest of nature.

In thinking through proper the role of philosophers in the vegetal turn, it seems clear that one task is to chronicle the aspects of our traditions that have relegated plants to the status of mere resources, while recognizing with Marder that "we cannot rid ourselves of such preconceptions overnight" (2016, 46). In other words, we need to come to terms with how deeply entrenched our Abrahamic conception of the land is tied to our intellectual history in general, which will then give us greater clarity as to how plants have been relegated to the status of mere resources. If Marder (2013a, 6) is right that "the ideological roots of both the deepening environmental crisis and the exploitation of plants are buried in the treatises by some of the most emblematic representatives of [Western metaphysics]," then it is incumbent upon us to highlight which concepts have sanctioned wanton environmental destruction so we can get clarity about them and attempt to leave those destructive concepts to the wayside as we chart forth a more appropriate worldview that diminishes our sense of human exceptionalism. To do so, we need to chronicle, for instance, "the violence Aristotle's

[1] The term "paradigm shift" comes from Thomas Kuhn, who argued that "the transition from a paradigm in crisis to a new one … is far from a cumulative process, one achieved by an articulation or extension of the old paradigm. Rather it is a reconstruction of the field from new fundamentals, a reconstruction that changes some of the field's most elementary theoretical generalizations as well as many of its paradigm methods and applications" (1970, 84–85).

thought has unleashed against plants" (Marder 2013a, 21) and how this bias has been taken up by us, given his towering influence on all subsequent Western metaphysics, to the detriment of ourselves and our planet. We also need to chronicle the ways in which the myth of the medieval Great Chain of Being and the Cartesian claim that nonhuman animals and plants are merely mechanical automata have gotten us into major civilizational trouble whereby we still arrogantly view the planet as a machine to be fixed through geoengineering.

A critique of our tradition, though, is not enough. The primary task of philosophers in the vegetal turn is not merely to chronicle how our relationship to the land has been formed through our intellectual history but rather to mine concepts and perspectives from our traditions—many of which are embedded in minority perspectives—that provide a conceptual basis for the vegetal turn. The land ethic, I argue, is one of those minority traditions that can serve this role.

14.3 Contextualizing the Land Ethic

In *A Sand County Almanac* (Leopold 1949, 224), the canonical text of the land ethic movement, Aldo Leopold utilizes the analogy of a "key-log" to discuss the shift that needs to occur away from thinking of the land merely as a repository of resources and towards understanding the land as a biotic community consisting of interdependently related parts. This analogy refers to cases in which logs that are being transported down a river get stuck, causing a logjam that can only be freed if the key-log is identified and removed, thus allowing the flow to resume. According to Leopold, "The 'key-log' which must be moved to release the evolutionary process for an ethic is simply this: quit thinking about decent land-use as solely an economic problem" (1949, 224). The purely economic understanding of the land results from viewing land as a mere commodity, which he called the "Abrahamic concept of land" (Leopold 1949, viii)—a conception that he attempted to rectify by revealing "land as a community to which we belong" (Leopold 1949, viii).[2] I argue here that whereas most of the Western philosophical tradition viewed plants as mere resources for human consumption (if they were mentioned at all), thereby being much in line with the Abrahamic conception Leopold was criticizing, the land ethic provides a fitting conceptual basis and quite possibly the key log for the vegetal turn, such that plants are properly understood as active members of the biotic community that deserve respect.[3]

[2] Regarding the Abrahamic concept of the land, Leopold states, "Abraham knew exactly what the land was for: it was to drip milk and honey into Abraham's mouth" (1949, 205). For an argument regarding the significance of the Judeo-Christian worldview and its effect on the understanding of the human-nature relationship, see White (1967).

[3] Michael Marder (2013a, 2) rightly notes that "if animals have suffered marginalization through the history of Western thought, then non-human, non-animal living beings, such as plants, have populated the margin of the margin, the zone of absolute obscurity undetectable on the radars of our

In order to understand the relationship between the vegetal turn and the land ethic, we need some context as to the current status of the land ethic among intellectuals. J. Baird Callicott, the most famous champion of the land ethic since Leopold, states that "of all the environmental ethics so far devised, the land ethic, first sketched by Aldo Leopold (1887–1948), is most popular among professional conversationists and least popular among professional philosophers" (2001, 204). This is likely due to the fact that environmental philosophy was framed as a version of applied ethics at its inception, a domain whose mainstream philosophical postures are generally at odds with the holistic approach found in Leopold.[4] In *Animal Liberation* (1975/2009)—the founding canonical text for the animal liberation movement—Peter Singer applied utilitarianism to the question of whether animals should be morally considered, and that encouraged framing the entire discipline of environmental philosophy as applied ethics. Professional philosophers trained in ethics who had an interest in environmental concerns typically took an environmental issue and applied ethical theories to it.[5] So while Singer and the other towering figure in animal ethics, deontologist Tom Regan, surely disagreed as to *why* animals deserve moral consideration—for Singer it is because animals are sentient and for Regan it is because they are subjects of a life—from a methodological perspective, their approaches were similar in that they both extended existing theories to a novel sphere of moral consideration.

Callicott explains why the land ethic is so different from these models of applied ethics. Leopold's training was not in philosophy but in the more practical field of wildlife and forest management.[6] Moreover, his intellectual influences were not moral theorists like John Stuart Mill or Immanuel Kant but rather evolutionists like Charles Darwin and ecologists like Charles Elton. In *Descent of Man*, Darwin (1871) argued that moral feelings are just as embedded in the process of natural selection as physical characteristics, which proved influential in Leopold's understanding (Callicott 1987, 191). This is why Leopold speaks of the key-log as "releasing the evolutionary process for an ethic," as noted above: he views a shift towards the land ethic as a process of civil evolution (1949, 203). From Elton, Leopold learned that "ecological relationships determine the nature of organisms rather than the other way around" (Callicott 1987, 200). Elton's community model of ecology, wherein each living being has a "place in its community [according to] its relations to food and enemies" (Elton 1927/2001: 50) is evident in Leopold's understanding of the land as a biotic community. Indeed, Leopold met Elton in 1931, and that profoundly influenced his conception of the land (cf. Simberloff 2012).

---

conceptualities." For an attempt to reconsider Western philosophical history from the perspective of plant life, see Marder (2014).

[4] For a more robust and historical commentary on environmental philosophy as merely applied ethics, see the preface to my *Heidegger and the Environment* (Rentmeester 2016).

[5] As Martin Schönfeld (2016, 194) notes, while applied ethics is "timely and important, … it is also, and equally, limited. It fails to contribute insight about the existential *meaning* of the [environmental] crisis."

[6] Leopold's 1933 work, *Game Management*, is still regarded as essential reading in the realm of wildlife and forest management. Cf. Leopold (1933/1986).

The astute observatory manner of Leopold's writings is also at a stark contrast to the more abstract and theory-loaded writing typically found in applied ethics. Believing that "we can be ethical only in relation to something we can see, feel, understand, love, or otherwise have faith in" (Leopold 1949, 214), in *A Sand County Almanac* (1949), his most widely read work, Leopold begins by chronicling the seasonal changes he observed near his farm in Wisconsin, thus contextualizing and setting the emotional tone for the final chapter, "The Land Ethic," where he outlines an "ethic dealing with man's relation to land and to the animals and plants which grow upon it" (1949, 203).

Having beautifully described the ways in which the soil, plants, and animals depend upon each other, Leopold conceptualizes what these interdependent relationships look like in an ecosystem through his land pyramid, a concept he borrowed from Elton (1927/2001). Noting that the pyramid's "functioning depends on the cooperation and competition of its diverse parts" (Leopold 1949, 215), Leopold argues for three primary propositions about the land as a result of his land pyramid: (1) land "is not merely soil; it is a fountain of energy flowing through a circuit of soils, plants, and animals" (1949, 216); (2) "the native plants and animals kept the energy circuit open; others may or may not" (1949, 218); and (3) "man-made changes are of a different order than evolutionary changes, and have effects more comprehensive than is intended or foreseen" (1949, 218). These three propositions speak of the interdependence of eco-systemic entities, of how to sustain the health of ecosystems, and of what is at stake when humans—with their Abrahamic conception of the land—situate themselves as "conqueror(s) of the land-community" rather than as "plain member(s) and citizen(s) of it" (1949, 204). Leopold eventually makes a normative precept explicit, in the rule-like language that is preferred in applied ethics: "A thing is right when it tends to preserve the integrity, stability and beauty of the biotic community. It is wrong when it tends otherwise" (1949, 224–225).[7] This precept captures the tripartite emphases of the land ethic well: integrity aligns with interdependence, stability with sustainability, and the non-anthropocentric emphasis on the biotic community as a whole aligns with human humility. Nonetheless, that oft-quoted precept is really more of a beginning than it is a conclusion, from Leopold's perspective. He believes that formulas alone will achieve little: that one must have a

[7] Famously, Tom Regan has argued that the land ethic's prioritization of the integrity of the biotic community would lead to "environmental fascism" in that, if carried to an extreme, it could sanction the killing of a human being in favor of preserving a rare wildflower (Regan 1983, 362). This worry may be especially urgent in the Anthropocene (Crutzen and Stoermer 2000), when the Abrahamic conception of the land seems to have won the planetary scene (Rentmeester 2021) and humans have become the most destructive beings on the planet. However, Callicott (2001, 211) has argued, persuasively in my opinion, "the land ethic is an accretion—that is, an addition—to our several accumulated social ethics, not something that is supposed to replace them". Thus, it is not as if eco-systemic integrity trumps all other values; nor is it necessarily and always the case that eco-systemic integrity just obliterates the importance of relationships with specific entities, including living beings. In fact, Leopold gives multiple examples of individual entities that deserve moral consideration (the oak tree itself being one, as noted below (1949, 9), but also the pine trees on his property, which he especially esteemed (1949, 70).

lived relationship with the land to overcome something as deeply rooted as the Abrahamic conception and those notions of human exceptionalism that justify it. It cannot merely be a matter of taking a theory, principle, or rule and applying it to some new domain, case, or context. Rather, it must be a matter of *living amongst* and indeed *living closely with* nonhuman beings, as a community, living our interdependences, articulating sustainability, and exercising well-founded human humility.

Leopold states that "the evolution of a land ethic is an intellectual as well as emotional process" (1949, 225). He makes it very clear that it is inconceivable "that an ethical relation to land can exist without love, respect, and admiration for land, and a high regard for its value" (Leopold 1949, 223). Theories and principles can help there but, for Leopold, the core need is for humans to spend more time in and with the biotic community—for "your true modern is separated from the land by many middlemen, and by innumerable physical gadgets. He has no vital relationship to it; to him it is the space between cities on which crops grow" (1949, 223–224). Go there, Leopold urges—for the farmer who cultivates the land has a stronger appreciation for the harvest than the city dweller who obtains food from the grocery store, just as the person who actually chops the wood has a greater appreciation for the heat that it provides.

14.4 The Proper Place of Plants

Plants play a special role in Leopold's discussion. For instance, in the chapter "Good Oak," Leopold reflects after chopping up a lightning-struck oak tree on his property, and states: "If one has cut, split, hauled, and piled his own good oak, and let his mind work the while, he will remember much about where the heat comes from, and with a wealth of detail denied to those who spend the weekend in town astride a radiator" (1949, 6). For Leopold, humility and respect for nature are matters of experiential training, generated by engaging and working with nature reflectively (i.e., letting one's mind "work the while"). This sort of reflective experience, or even experiential knowledge, simply cannot be obtained via theory alone, however respectful of nature the theory's mandates may be. Later in the book, Leopold juxtaposes this humble relationship with nature to its Abrahamic alternative, with an explicit reference to plant life. We often talk of "love of the land," he says—think, for instance, of Woody Guthrie's song "This Land is Your Land," which has been a cultural staple in the United States since the 1940s—but Leopold poses the question as to what and whom we actually love. His answer regarding plant life is unequivocal: "Certainly not the plants, of which we exterminate whole communities without batting an eye" (1949, 204).

What is the proper relationship to plants from the perspective of the land ethic? The land ethic does not mandate a hands-off form of respect for the nonhuman (Leopold was, after all, a professor of game *management*). In fact, Leopold explicitly says that "a land ethic of course cannot prevent the alteration, management, and use of these 'resources,' but it does affirm their right to continued existence" (1949, 204). Plants, in this conception, can be used as resources, but that in no way establishes that their existence is meant solely to serve human interests. In any case, as Leopold notes, "the difficulty, however, is not so much in proving this principle in the abstract as in getting people to see and respect its applications" (1953/1993, 172). That requires a Gestalt shift away from the Abrahamic conception of the land but not all the way to hands-off reverence—a shift that will allow us to make sense, for instance, of Leopold's response to the loss of the lightning-struck oak tree: "we mourned the loss of the old tree, but knew that a dozen of its progeny standing straight and stalwart on the sands had already taken over its job of wood-making" (1949, 9).[8] There is respect for the dead old tree, but the community relationships that the land ethic wants us to explore and nurture are not only with specific plants, animals, or landscapes that have some forms of special meaning to ourselves, but with the ecosystem as a whole and its complex, extended, generative and regenerative workings.

Leopold makes the case that genuine education in the moral status of trees requires a new way of seeing them, rather than an application of theoretical concepts regarding them. He states, for instance, "the autobiography of an old board is a kind of literature not yet taught on campuses" (1949, 25). Although Leopold was of course a professor in an academic setting, he maintained strongly that certain sorts of knowledge can only be developed experientially. He talks through a farmer whose land includes a veteran bur oak as someone who "owns a historical library, and a reserved seat in the theater of evolution" (Leopold 1949, 30). The nature-attuned farmer sees and feels the changes in the environment in ways that simply can't be replicated in the college classroom. Genuine love of nature in the context of the land ethic requires an experiential connection with nature in his account.

14.5 Why the Land Ethic Over Other Alternatives?

Having chronicled the land ethic and its potential for serving as a foundation for the vegetal turn, an appropriate question to pose is why the land ethic is more fitting than biocentrism or deep ecology, which also ascribe moral consideration to plants. The main theorist of biocentrism is Paul Taylor, whose *Respect for Nature* (1986) is considered the founding philosophical representation of this movement.[9] Taylor (1986, 71) argues, "To have the attitude of respect for nature is to regard wild plants and animals of the Earth's natural ecosystems as possessing inherent worth." For Taylor, this notion of inherent worth is egalitarian in nature: every entity that has inherent worth has it equally. This leads his view to be akin to "a life-centered theory of environmental ethics that gives impartial consideration to every species" (Taylor

[8] I should point out that framing the vegetal turn as a paradigm shift is linked with this notion of a Gestalt shift, since Kuhn's conception of paradigms was influenced by Gestalt psychology, as he informs us in the preface of *The Structure of Scientific Revolutions* (Kuhn 1970, vi).

[9] I use the term "founding *philosophical* representation" since the founder is truly the Christian thinker, Albert Schweitzer. Cf. Schweitzer's explication of his reverence for life theory (1923).

1986, 306). While Taylor does recognize that biological competition is inevitable, all living beings deserve *prima facie* equal and impartial consideration on his account, thus making Taylor's theory a version of biocentric egalitarianism.

Taylor (1986, 122) argues that each organism is a "teleological center of life," which he defines as a "unified, coherently ordered system of goal-oriented activities that has a constant tendency to protect and maintain the organism's existence." Due to the fact that we are moral agents, humans can choose to respect teleological centers of life by recognizing their inherent worth or they can choose to disregard them and treat them as mere resources with simply economic worth. Taylor explicitly references plant life in his discussion as to how to apply his theory:

> I take it that trees, for example, have no knowledge or desires or feelings. Yet it is undoubtedly the case that trees can be harmed or benefited by our actions. We can crush their roots by running a bulldozer too close to them. We can see to it that they get adequate nourishment and moisture by fertilizing and watering the soil around them. Thus we can help or hinder them in the realization of their own good. (1981, 200)

To help a tree achieve "the full development of its biological powers" (Taylor 1981, 199) is to do good, while to thwart it is to do bad, in Taylor's account, even if the tree has no feelings or desires from his perspective.

One can see how biocentrism is superior to the human exceptionalism sort of mentality that has reigned in Western metaphysics, at least from the perspective of respecting plant life, thus making it a rival candidate to serve as the theoretical basis for the vegetal turn. However, Taylor's explicit denial of plant sentience is clearly problematic, now knowing what we know regarding plant life. Above and beyond this, there are three primary reasons as to why the land ethic is superior to biocentrism in its ability to serve as a matrix for the vegetal turn: (1) it doesn't rely upon the notion of inherent worth; (2) its land pyramid retains some notion of a hierarchy and doesn't succumb to egalitarianism; (3) it is based in a lived relationship with the land, rather than a theoretical, applied ethics. Regarding the notion of inherent worth, Mary Anne Warren (1987, 435) has famously noted that this concept "appears as a mysterious non-natural property which we must take on faith." Rather than putting faith in a mysterious notion, Leopold connects the worth of living entities with their connection to the health of the ecosystem as a whole. Thus, it is perfectly reasonable to exterminate non-native invasive species of plants, for example, in Leopold's account if they threaten the health of the whole ecosystem. One need not consider the "inherent worth" of the species before doing so. Secondly, and related to ecosystemic health, rather than an espousal of egalitarianism, Leopold's land ethic is built upon a pyramid in which "health is [regarded as] the capacity of the land for self-renewal" (Leopold 1949, 221) of its energy circuits. For Leopold (1949, 178), we need to be mindful of the unique ways in which the "soil–plant-animal-man food chain" plays out in each ecosystem, which will lead to differing levels of worth for species, depending on context. For example, Leopold understood certain species like wolves in the Apache National Forest and Carson National Forest to be what we would refer to as "keystone" species such that they play a vital role in the overall health of the ecosystem (Flader 1994). From this perspective, the moral status of the wolves in these contexts may reasonably be higher than, for instance, a less integral species to the ecosystem such as the mountain lions that infrequently roam across these terrains. Finally, regarding the third point, Val Plumwood (1991, 6) has shown how Taylor's biocentrism "becomes an essentially *cognitive* matter (that of a person believing something to have 'inherent worth' and then acting from an understanding of ethical principles as universal)." As we have seen, while Leopold's land ethic does include cognitive elements, he is insistent that one must have an experientially-based relationship to the land in order to form the proper ethical intuitions towards it, which will then guide the appropriate behaviors toward it.

What about deep ecology, which also espouses moral consideration for plants and shares in the land ethic's holistic approach? Is it a fitting alternative to the land ethic? The founder of deep ecology is Arne Naess, who juxtaposes shallow ecology with deep ecology. While shallow ecology's central objective is the health and affluence of persons in developed countries, a commitment to deep ecology entails an ontological shift in our understanding of what it means for a natural entity to be *away from* seeing things as isolated individuals and *towards* conceiving "organisms as knots in the biospherical net or fields of intrinsic relations" (1973, 95). Naess shows us that shallow ecology is pragmatic in nature, while deep ecology is fundamentally ontological. At first glance, Naess and Leopold's conceptions have a lot in common. After all, Leopold would agree that all natural entities are intrinsically related: the health of one species is contingent upon the health of the others that compete and cooperate with each other. They also agree as to the means of bringing about the appropriate perspective in that an ecologically conscious ethic needs to be cultivated through living closely with nature. Naess, who called his approach "ecosophy," argued that "a philosophy, as articulated wisdom, has to be a synthesis of theory and practice" (1988, 258) and explicitly stated that "ecological knowledge and the life-style of the ecological field-worker have *suggested, inspired, and fortified* the perspectives of the Deep Ecology movement" (1973, 98). Just as Leopold cut his teeth in the vast natural surroundings of the United States, Naess cultivated his ethic in the mountains of Norway, calling his own philosophy "Ecosophy T" after his favorite mountain, Tvergastein. Thus, Naess and Leopold are, at the surface level, two peas in a pod.[10]

At a deeper level, though, deep ecology and the land ethic have their differences, and the primary difference has to do with deep ecology's belief in inherent value. Naess explicitly insists "upon the value in themselves, the autotelic value, of every living being" (1988, 265–266), the same trait Warren called a "mysterious non-natural property" when criticizing biocentrism that was noted above. When posing the question as to "why has *any* living being autotelic value?" Naess replies, "Faced with the ever returning question of 'why?,' we have to stop somewhere" (1988, 266). Naess, it seems, agrees with Wittgenstein's famous claim that "if I have exhausted the justifications I have reached bedrock, and my spade is turned" (1968, §217), thus noting that at a certain point normative justifications must end. Leopold, on

[10] Christian Diehm has pointed out a further similarity between the two in that Arne Naess understood deep ecology through the lens of Gestalt ontology (2006), a narrative that is consistent with Leopold's land ethic.

the other hand, doesn't need to appeal to non-natural properties when he speaks of ethics as maintaining the integrity and stability of the ecosystem, aspects that can be assessed using biological markers.[11] This is why Leopold can, without any hesitation whatsoever, speak of "which tree, if any, needs felling for the good of the land" (1949, 68), upon surveying his land after the leaves had fallen in autumn. For Leopold, it is contextual, eco-systemic thinking that ultimately guides the land ethic, while for deep ecologists, the fundamental shift is one towards imbuing all natural beings with inherent worth and erring on the side of non-interference.[12]

14.6 Concluding Thoughts

On the whole, the Leopold's call is not merely for a shift in intellectual emphases, but rather a shift in perspective and ultimately in our relationship to the rest of the natural world. Instead of the Abrahamic conception of seeing nature as a repository of commodities to be exploited, he calls us to recognize the interdependent nature of the biotic community, the ways in which to maintain its health, and the need to adopt an attitude of humble stewardship towards the natural world. This matrix is fitting for the vegetal turn, which embraces a similar call to knock human beings off their collective pedestals. Indeed, the shift away from the Abrahamic conception of land may very well be the key log to initiate the paradigm of the vegetal turn to a broader audience. In the end, as Leopold notes, "the hope of the future lies not in curbing the influence of human occupancy—it is already too late for that—but in creating a better understanding of the extent of that influence and a new ethic for its governance" (1986, 21).

[11] I should point out, though, that there are versions of deep ecology that end up being a combination of Naess's view and Leopold's that could be viable. Warwick Fox, the philosopher responsible for popularizing deep ecology in Australia, argues that "the deep ecologist's conception … need not imply that, at any given moment, all 'knots' (i.e., organisms) in the 'biospherical net' are constituted of equally complex relations" (1984, 204). He concludes that not all organisms have equal inherent worth.

[12] Part of this has to do with the fact that Gandhi had a monumental impact on Naess's thought. As Bill Devall and George Sessions note, "[Naess's] work on the philosophy of ecology, or *ecosophy*, developed out of his work on Spinoza and Gandhi and his relationship with the mountains of Norway" (1985, 225). Gandhi's commitment to *ahimsa*, that is, to non-violence, is defined by Naess as "direct influence in the direction of an *increase* of level of actual self-realization" (1974, 53), which is juxtaposed with *himsa*, that is, "avoidable direct influence in the direction of a *decrease* of level of actual self-realization" (1974, 53). Since all living beings have autotelic value in Naess's account, we should avoid *himsa* in the very least and ideally practice *ahimsa*.

References

Aristotle. 1907. *De Anima*. Trans. R. D. Hicks. Cambridge: Cambridge University Press.

Basl, John. 2019. *The Death of the Ethic of Life*. Oxford and New York: Oxford University Press.

Callicott, J. Baird. 1987. *Companion to a Sand County Almanac: Interpretive and Critical Essays*. Madison, WI: University of Wisconsin Press.

Callicott, J. Baird. 2001. The Land Ethic. In *A Companion to Environmental Philosophy*, ed. Dale Jamieson, 204–217. Malden, MA and Oxford: Blackwell.

Crutzen, Paul J., and Eugene F. Stoermer. 2000. The "Anthropocene." *The International Geosphere-Biosphere Programme (IGBP) Newsletter* 41: 17–18.

Darwin, Charles. 1871. *The Descent of Man and Selection in Relation to Sex*, vol. I. London: John Murray.

Davies, Brian. 2016. *Thomas Aquinas's Summa Contra Gentiles: A Guide and Commentary*. New York and Oxford: Oxford University Press.

Descartes, René. 2008. *Meditations on First Philosophy*. Trans. Michael Moriarty. Oxford and New York: Oxford University Press.

Devall, Bill, and George Sessions. 1985. *Deep Ecology: Living as if Nature Mattered*. Layton, UT: Gibbs Smith.

Diehm, Christian. 2006. Arne Naess and the Task of Gestalt Ontology. *Environmental Ethics* 28 (1): 21–35.

Elton, Charles. 2001. *Animal Ecology*. Chicago and London: The University of Chicago Press.

Flader, Susan L. 1994. *Thinking Like a Mountain: Aldo Leopold and the Evolution of an Ecological Attitude toward Deer, Wolves, and Forests*, 2nd ed. Madison, WI: University of Wisconsin Press.

Fox, Warwick. 1984. Deep Ecology: A New Philosophy for Our Time. *The Ecologist* 14: 194–204.

Grey, William. 1986. A Critique of Deep Ecology. *Journal of Applied Philosophy* 3 (2): 211–216.

Kuhn, Thomas. 1970. *The Structure of Scientific Revolutions*, 2nd ed. Chicago: The University of Chicago Press.

Leopold, Aldo. 1949. *A Sand County Almanac and Sketches Here and There*. London, Oxford, and New York: Oxford University Press.

Leopold, Aldo. 1986. *Game Management*. Madison, WI: The University of Wisconsin Press.

Leopold, Aldo. 1993. *Round River: From the Journals of Aldo Leopold*, ed. Luna B. Leopold. New York and Oxford: Oxford University Press.

Mancusco, Stefano, and Alessandra Viola. 2015. *Brilliant Green: The Surprising History and Science of Plant Intelligence*. Trans. Joan Benham. Washington, D.C.: Island Press.

Marder, Michael. 2013a. *Plant Thinking: A Philosophy of Vegetal Life*. New York: Columbia University Press.

Marder, Michael. 2013b. Should Plants Have Rights? *The Philosophers' Magazine* 62: 46–50.

Marder, Michael. 2014. *The Philosopher's Plant: An Intellectual Herbarium*. New York: Columbia University Press.

Marder, Michael. 2016. *Grafts: Writings on Plants*. Minneapolis, MN: Univocal.

Naess, Arne. 1973. The Shallow and the Deep, Long-Range Ecological Movement. *Inquiry* 16: 95–100.

Naess, Arne. 1974. *Gandhi and Group Conflict: An Exploration of Satyagraha, Theoretical Background*. Oslo-Bergen-Tromsø: Universitetsforlaget.

Naess, Arne. 1988. Identification as a Source of Deep Ecological Attitudes. In *Deep Ecology*, 2nd ed., ed. Michael Tobias, 256–270. San Marcos, CA: Avant Books.

Plumwood, Val. 1991. Nature, Self, and Gender: Feminism, Environmental Philosophy, and the Critique of Rationalism. *Hypatia* 6 (1): 3–27.

Regan, Tom. 1983. *The Case for Animal Rights*. Berkeley and Los Angeles, CA: University of California Press.

Rentmeester, Casey. 2016. *Heidegger and the Environment*. London and New York: Rowman & Littlefield International.

Rentmeester, Casey. 2021. How Could We Drink Up the Sea? From Technological Nihilism to Dwelling in the Anthropocene. *Das Questões* 13: 12–29.
Schönfeld, Martin. 2016. Future-Oriented Philosophy and Wisdom East and West. In *Wisdom and Philosophy: Contemporary and Comparative Approaches*, eds. Hans-Georg Moeller and Andrew K. Whitehead. London and New York: Bloomsbury Academic.
Schweitzer, Albert. 1923. *Civilization and Ethics*, trans. A. Naish. London: Black.
Simberloff, Daniel. 2012. Integrity, Stability, and Beauty: Aldo Leopold's Evolving View of Nonnative Species. *Environmental History* 17: 487–511.
Singer, Peter. 2009. *Animal Liberation*. New York: HarperCollins.
Sylvan, Richard. 1985. A Critique of Deep Ecology. *Radical Philosophy* 41: 10–22.
Taylor, Charles. 1992. *Sources of the Self: The Making of Modern Identity*. Cambridge: Cambridge University Press.
Taylor, Paul. 1981. The Ethics of Respect for Nature. *Environmental Ethics* 3 (3): 197–218.
Taylor, Paul. 1986. *Respect for Nature*. Princeton, NJ: Princeton University Press.
Warren, Mary Anne. 1987. Difficulties with the Strong Rights Position. *Between the Species* 2 (4): 433–441.
Watson, Richard A. 1983. A Critique of Anti-Anthropocentric Biocentrism. *Environmental Ethics* 5 (3): 245–256.
Wittgenstein, Ludwig. 1968. *Philosophical Investigations*, 3rd ed., trans. G. E. M. Anscombe. New York: MacMillan.
White, Jr., Lynn. 1967. The Historical Roots of our Ecologic Crisis. *Science* 155: 1203–1207.

Chapter 15
Arboreality and the Ethics of Standing in Place

Laura Pustarfi

Abstract Trees and plants are receiving renewed attention as work in both the sciences and humanities is affirming their intelligence, agency, and relationality. If trees are more than merely instrumental beings, what is our ethical responsibility toward them? Taking a phenomenological approach, trees themselves are grounded in a particular place. Edward S. Casey's phenomenological work on place gives insight into one approach to arboreal ethics by considering how trees create place both within and outside of the forest. While trees may not have a right to life, they may have a right to place, and an ethic of standing in place considers both individual and surrounding, acknowledging ecological interrelationships. In this chapter, I take Casey's phenomenology of place as a starting point for an arboreal ethic which recognizes trees as non-human others that can be considered for some rights and also acknowledges that our lives depend on taking theirs. I will also consider vegetal ethics, particularly in Michael Marder's philosophy, and other place-based approaches to environmental ethics. Rather than propose a fully developed ethic, I uncover some of the unique issues trees pose to Western ethics and point towards potential respectful relationships with trees and plants.

15.1 Introduction

Trees are grounded in place. As comparatively long-lived and immobile beings, trees are often backgrounded as part of the environment or nature rather than addressed on their own terms. Trees exhibit an alterity to human being that the current Western philosophical paradigm has employed to cast trees as only instrumental, either as resource for human consumption or as beautiful scenery. However, research in the fields of plant signaling and behavior as well as critical plant studies question this view and its implications. Scientific studies are showing that trees and plants communicate, learn, remember, and share resources among other capacities (Karban et al. 2013; Gagliano et al. 2014; Simard 2017) and this work is beginning to show that trees

L. Pustarfi (✉)
California Institute of Integral Studies, San Francisco, US
e-mail: lpustarfi@ciis.edu

M. Di Paola (ed.), *The Vegetal Turn*, The International Library of Environmental, Agricultural and Food Ethics 38, https://doi.org/10.1007/978-3-031-68671-9_15

are intelligent, agential, and relational. Using this new understanding, scholars in the humanities are questioning Western beliefs about plants as inferior, stemming from Platonic and Aristotelian roots, and relations between human beings and plant beings (Hall 2011). This emergent focus on plants, sometimes called the Vegetal Turn, is present in philosophical discussions, with Michael Marder (2013a), Matthew Hall (2011), and Luce Irigaray (Irigaray and Marder 2016) leading the philosophical conversation on plant being and ethics. Though conversations on plants necessarily include trees, these massive, woody-stemmed beings deserve further attention due to their unique status in vegetal realms, including such superlatives as largest, tallest, and oldest.

As trees are reconsidered, our human relationships to trees must also be reconsidered. Taking into account that humans must take plant lives to maintain our own, how can we articulate an arboreal ethic that both allows for human sustenance and respects trees? Unless moved, trees stand situated for the duration of their lives. Ethical considerations toward trees must take arboreal existence into account, rather than imposing an anthropocentric view. If we accept such capacities as intelligence, agency, and relationality in trees, arboreal ethics must be revisioned. This task is complicated by our human need to take trees and plants for sustenance, and trees present a particularly challenging ethical case. Contrary to animal ethics, where it can be claimed that humans should avoid eating or using any animal products, humans must use trees and plants for our food, medicines, and shelter. In Alfred North Whitehead's words, "life is robbery," though "the robber requires justification" (1978, 105). Insights from the phenomenological tradition, particularly philosopher Edward S. Casey's work on place, bolsters one approach to arboreal ethics by considering how trees create place both within and outside of the forest (2001, 2009). Because trees themselves are grounded in place, arboreal rights may include rights to place. Casey's work on place, including work on trees in place with Marder, provides a structure for examining the unique challenge of articulating an arboreal ethic. Further, Marder's discussion of rights for plants as well as other place-based approaches to ethics can also extend to trees (2013b). In this paper, I show how Casey's phenomenological conception of place provides a framework for thinking through arboreal ethics, and I propose arboreal place-based rights as one potential step toward more respectful relationships with trees. While trees may not have a right to life, they may have a right to place in an arboreal ethic.

15.2 Arboreal Implacement

Trees are both larger and older than humans with a dramatically different expression of livingness as sessile, modular beings. Our arboreal neighbors' lives are intertwined with our own through exchange of breath and our use of them in innumerable ways.[1] Casey's statement, "place is what takes place between body and landscape" expands to include the arboreal bodies grounded both locally and in forests globally. As he reminds us, to be is to be in place. Place is the condition as well as the limit of existence (2009, 15). Casey states, "animals, perhaps even plants, possess their own equivalents of embodiment and implacement" (2001, 693). He differentiates arboreal implacement from human perception of place, stating, "a tree stands in its own place. Its life is sedentary. It is a life in one place, a life without anxiety. Not only is a tree in its place; it actively contributes to its place, filling it up with its own organic substance. It knows no menacing void, even though to move from its own place is to risk the death of the organism" (Casey 2009, xi–xii). Humans and animals experience movement from place to place, bodily encountering contrast as they are continually implaced and displaced. Trees, unless forcibly uprooted and replanted, remain standing in one place throughout their lives. How does this embodiment inform ethics as it regards trees?

Michael Marder's thought on plants, from which his work on trees derives, is based around his vegetal anti-metaphysics or what he calls phytophenomenology. This affirms plants' intentionality and subjectivity but does not include "autonomy, unity, individuality, personhood, or will" (Marder 2012, 1371). Critiquing both Aristotelian and post-Kantian metaphysics, Marder claims, "vegetation accomplishes a *living* reversal of metaphysical values and points toward the collapse of hierarchical dualisms" (2011, 470). Marder suggests plants should be granted vegetal rights based on both intrinsic value and vulnerability, proposing a "Universal Declaration of Plant Rights" that might include acknowledgement of vegetal worth and freedom from genetic sterilization for example (2013b, 50).

Trees, as enduring individuals within the environment, have an arboreal form of place. Casey and Marder have phenomenologically investigated trees and place, endeavoring to "give place back to trees" (2024, 48). They find trees as having and creating unique and individualized place focused on becoming, what they term arborescence. They claim, "the vegetal taking up of space is giving place" (Casey and Marder 2024, 57). This arboreal placialization happens outwardly, inwardly, and laterally, resulting in place as experienced around the central structure of the trunk or grove. While beyond the scope of this paper, Casey's work on edges as phenomenological inquiry also applies to trees both in the way trees have edges themselves and create edges within human experience, which Casey and Marder unpack at length in relation to arboreal implacement.

[1] Trees, like all plants, convert sunlight into sugars via photosynthesis, and participate in the carbon dioxide-oxygen cycle, exchanging carbon dioxide for oxygen, exchanging breath, with humans and non-human animals.

From a parallel perspective, geographers Paul Cloke and Owain Jones have examined trees in relation to place and ethics from a geographical lens. They base their ethical stance on non-human agency and relationality with attention to the specifications of human-tree interactions. They state, "the places which are co-constituted by the relational agency of trees form interesting contexts in which [relational] encounters can be found at work, and where any reconceived forms of ethical mindfulness have to be operationalised critically" (Cloke and Jones 2003, 196). Cloke and Jones propose a place-based ethical mindfulness towards trees and other non-human beings, which is "both place-located and bound into wider relational matrices" (2003, 212). They find place as a locus of moral concern, particularly for trees, and as a prompt toward environmental awareness. This view of arboreal place compliments Casey's implacement.

Both humans and trees make place. While trees are often backgrounded as merely part of the scenery, they are involved in arboreal activities unbeknownst to humans that impact their surroundings. Trees change the light that filters to the ground, provide homes for other plants and animals, and shift the local climate. Arboreal bodies stand in place, enduring throughout time both beneficial and challenging conditions. A tree can provide nutrients to neighbors and kin and ward off unwanted insect attacks (Gorzelak et al. 2015). Trees exist worldwide with, as of 2015, over 400 trees for every human being on the planet (Crowther et al. 2015). Further, at least 33 million humans globally depend on forests for their economic livelihoods, and approximately 95% of people who do not live in cities live nearby a forest (FAO 2022, 5, 11). Hall states that, "close observation of plant life-history demonstrates that plants are communicative, relational beings—beings that influence and are influenced by their environment" (2011, 158). Rather than being a feature of place or merely at a site, living trees are concretely engaged with and create place. The scientific research on how trees and plants participate in place supports the philosophical work reconsidering trees as beings worthy of greater ethical consideration.

15.3 The Science and Philosophy of Plant Behavior

Science showing plants exhibit intelligence, agency, and relationality among other capacities has initiated thought in the humanities about the place of plants and human ethical relationships to them (Ryan 2012). Cutting edge research in the biological field of plant signaling and behavior[2] is showing that plants have memory, learning, communication, relationships, and perceptive capacities (Karban et al. 2013; Gagliano et al. 2014; Yokawa et al. 2017; Simard 2017). These studies give foundation to the idea that plants exhibit intelligence in their ability to assess the environment to choose from potentialities, relationality in their lived connections to

[2] Many terms, including include plant cognition, plant neurobiology, plant consciousness, etc., have been proposed and debated within the field with plant signaling and behavior as the currently accepted terminology. See especially Alpi et al (2007), Taiz et al. (2019).

other beings, and agency in their intentionality of action. While the field is extensive and typically focused on smaller plants rather than trees, I will review just a few papers to illuminate the research findings.

In 2003, a study showed that a pea plant can distinguish between its own roots and the roots of other plants, called "self/non-self discrimination," and will avoid competing with its own roots by reducing the volume and length of roots (Falik et al. 2003). Ecologist Richard Karban has studied this faculty in sagebrush and further showed that sagebrush responds differently to more closely related plants, what he terms kin recognition (Karban et al. 2013). Other studies show learning and memory by testing both reaction and non-reaction in plants. Biologist and interdisciplinary scholar Monica Gagliano and her co-researchers experimented on *Mimosa pudica,* the shy plant, which is so named for its propensity to fold its leaves when disturbed. The researchers dropped the plant several feet numerous times until it no longer closed its leaves in reaction to the drop, and the plant then did not react to the same drop days after the initial training. The plant did, however, react to a different disturbance, shaking, by closing its leaves. This habituation to the drop is a simple form of memory, and Gagliano's experiment demonstrated that plants have both learning and memory (Gagliano et al. 2014). One highly cited study published in 2017 reviewed plants' reaction to anesthetics, which researchers had previously thought required a central nervous system to elicit an effect. A research team lead by bioscientist Ken Yokawa tested several plant species including the shy plant, pea plants, and garden cress plants both by exposing them to anesthetic gases through their leaves as well as liquids through their roots. The shy plant lost its ability to close its leaves when disturbed, and the pea plant ceased its growing, which occurs in a rotational movement, for several hours. The authors even suggest that garden cress plants "represent an ideal in aims to illuminate the elusive mechanisms underlying both anesthetic and the phenomenon of consciousness" and propose this plant could be used to study and test human anesthesia (Yokawa et al. 2017, 8).

Some studies do address trees directly. Beyond intelligence, trees exhibit relationality as shown in their exchange of nutrients and chemicals and agency in their capacity for intentional action. Research led by forest ecologist Suzanne Simard found the mycorrhizal collaborative relationship between trees and fungi that facilitates the exchange of chemicals around a forest, and the hub or Mother Trees that share water, carbon, nitrogen, and other nutrients with their kin trees. The Mother Trees also recognize closely related trees, and these decisions to support some plants over others points towards arboreal agency and relationality (Simard 2017).

Inspired by this scientific research, thinkers in the humanities, including philosophy, have addressed vegetal being anew in the interdisciplinary field of critical plant studies (Ryan 2012, 102). In *Plants as Persons,* an early work in the field, Matthew Hall's primary argument is for the personhood of plants and ethical action towards vegetal persons (2011). He finds treatments of plants as persons in Indigenous and animistic views, surveying several cultures including Aboriginal Australian and North American groups. Using the terminology, following Irving Hallowell, "other-than-human persons," Hall states that trees and plants are autonomous subjects worthy of human ethical treatment and respect, which due to their personhood must be included in moral consideration, although also noting that in many Indigenous cultures this does not preclude taking plant life for human sustenance (2011, 100). Practically, he advocates for reducing unnecessary waste of plant lives and over-consumption, as well as restoration practices that directly impact plant flourishing (Hall 2011, 163–169). His stance comes from a relational concept of personhood as an "overarching category of being" (Hall 2019, 5). Others, including geographer Thomas Puleo, have extended the moral argument for plants based on the potential for vegetal personhood (Puleo 2019). Marder is critical of vegetal personhood, and he claims bestowing autonomy and personhood on trees and plants would, "render more refined the violence human thought has never ceased unleashing against these beings, for instance by forcing plants into the mold of appropriative subjectivity," instead suggesting a rights-based approach to vegetal ethics (2013a, 50, 2013b, 46–50).

We need not develop a wholly novel approach. Here is where Western thought can respectfully and non-appropriatively find insights and advice from cultures with differing relationships to plants and other non-humans, such as Indigenous philosophies. For just one example, Robin Wall Kimmerer, who is a biologist and enrolled member of the Citizen Potawatomi Nation in North America, speaks of Indigenous Potawatomi black ash basket makers. Kimmerer says of master basket maker John Pigeon's thought, "traditional harvesters recognize the individuality of each tree as a person, a nonhuman forest person. Trees are not taken, but requested" (2013, 144). For Potawatomi, the ethical relationship includes both consent and reciprocation, and it is not ridiculous to ask for what is needed and receive an answer. Even the very act of killing itself is done with careful intention to respect the gift of life, which also respects the place from which the tree is taken.

With these views arise challenges for human treatment of trees as mere resources and the ethical question becomes more urgent, especially as the global state of forests continues to worsen with ongoing deforestation and climate change. Net forest acres are lost to fires and deforestation each year (FAO 2022, 5). An ethic that considers arboreal being would require a shift in our consideration of trees.

15.4 Precedents for Arboreal Ethics

Place contains ethical concern and addressing ecological ethics involves place. Casey says, "when it comes to being ethical, there is no escaping the imperative of place" (2009, 265). Because trees endure standing in one place throughout their lives, ethical considerations toward trees must take arboreal existence into account, rather than imposing an anthropocentric view. An ethical stance toward trees may not be based on total lack of harm but instead on taking trees with respect for their lives and their place.

Within Western academia, there have been some radical proposals for arboreal ethics, such as a late nineteenth century article titled "Humanity to Trees" which called for a "society for the prevention of cruelty to trees" (Lowell 1857, 96). Another such article, "Animal Chauvinism, Plant-regarding Ethics and the Torture of Trees," published in 1986 by one J. L. Arbor, clearly written under a pseudonym, calls for moral consideration for trees and plants based on intrinsic value and argues against the view that sentience or consciousness is required for moral inclusion. Arbor states, "trees, like animals, and other plants, but unlike machines, have end-states which are not decided by human beings" (1986, 339). The details of ethical action are wanting in his article, although he claims eating seeds and fruit as well as ornamental plantings would be ethically acceptable while propagation and destruction for human use would not. Most famously, Christopher Stone, a legal scholar, argued that trees should have rights in a 1972 article, "Should Trees Have Standing?" Stone suggests that what he calls "natural objects" should be allowed legal rights (1972, 450–501). This article, along with a book of the same name, has ignited environmental law and attempts to secure rights for non-human plaintiffs. More recently in 2008, the Swiss Federal Ethics Committee on Non-Human Biotechnology published a brochure titled "The Dignity of Living Beings with Regard to Plants: Moral Consideration of Plants for Their Own Sake". The report seeks to articulate what dignity for life, as required by the Swiss government, means in terms of plants, and the document details the struggles the committee had with this question. While all committee members agreed that plants should not be harmed arbitrarily, the committee was split on ownership, patenting, and genetic modification. Further, the committee noted that a majority of their members, "considers any action with or towards plants that serves the self-preservation of humans to be morally justified" (Federal Ethics Committee on Non-Human Biotechnology 2008, 20). However, the report gives few actionable suggestions and was not legally enforceable. Philosopher Robin Attfield offered a biocentric approach to ethics in the forest as a whole to include widespread sustainable management and consideration of the local and global drivers of deforestation (2018). While Attfield's ethic is a step in the right direction, more can be done to address trees themselves and include trees that exist outside the confines of a forest.

Place as a source for ethical relevance appears both in arguments for rights of nature and in ecofeminist place-based ethics. In most societies, trees and plants do not have any protections approximating rights. This lack of rights includes animals and other non-human beings. However, arguments for rights for non-humans are in process in judicial systems across the globe as part of the rights of nature movement. Place is implicit in arguments for the rights of nature. Some rivers, mountains, and ecological regions are now included in the category of legal persons. Several rivers around the world have been granted legal rights, and there are ongoing cases, like that of Happy, the elephant at the Bronx Zoo in New York, for legal rights for animals supported by philosophical reasoning (Comstock et al. 2021). Forests such as the Los Cedros Forest in Ecuador and the Te Urewera Forest in New Zealand have been protected using arguments from the rights of nature movement and are deeply embedded with Indigenous rights movements (Epstein 2022, 11–12). In the United States, researchers at the USDA Forest Service have investigated place-based planning as a way to localize planning and management of the forest as a place rather than as a location of resources (Kruger and Williams 2007). Place has also developed as a site of moral concern in ecofeminist ethical considerations such as place-based ethics of care. Ecofeminist conceptions of place-based ethics proposes that ethical regard emerges both in situ and in relationship and is involved with caring (Goralnik et al. 2014). This view is resonant with the phenomenological understanding of place Casey has developed and includes trees and forests as part of ecological concern.

While such calls towards rights for non-humans form a strong precedent, action has so far been limited, though movements, especially for the rights of nature, are accelerating. These movements, while including trees and plants, often address nature or non-human others broadly. However, an arboreal ethic has not yet been adequately addressed.

15.5 Toward an Ethic of Standing in Place

Building on these attempts as well as those of Marder and Hall, an arboreal ethic takes into account the needs of trees as well as our human interconnection to plants while also justifying our need to depend on them for our livelihoods. This, however, does not mean that eating, building, or heating is unethical. While thinkers on animal ethics can advocate that humans should not use animals, plants and trees are uniquely ethically problematic. Even a fruitarian diet, eating only the fruits and nuts a tree would typically discard, does not address the use of tree products for building and fuel. Not only is human life contingent on taking plant life, there are many ways trees and plants are used instrumentally that do not kill the plant and are sometimes beneficial, such as when we eat fruits intended for long-distance seed distribution. Intuitively, we feel that a clear-cut, which indiscriminately fells living trees, is wrong and sustainable forestry is a good that increases the thriving of the trees and the forest ecosystem. But this is complex. In some cases, burning a forest regularly may avert a greater, more damaging fire later. In some cases, sustainable forestry removes trees that allow neighboring trees to thrive or suggests planting a monocrop tree farm that reduces ecological resilience.

Because of these challenges for an arboreal or vegetal ethic, trees and plants may not have a right to life or bodily integrity. When Heidegger proposes in his *What is Called Thinking?* that we, "for once let [the tree] stand where it stands," (1968, 44) he is suggesting that the tree not be deprived of its place. Here we can "meet a tree in bloom, face-to-face" (Heidegger 1968, 41). In many cases, this means allowing the tree or plant to remain and thrive. However, there are cases where taking a tree or plant is essential for human livelihood. In these cases, an ethic of place may offer an alternative to an ethic based on total lack of harm.

Trees occupy and create the place where they grow. When a tree must be taken for human purposes, the place itself could be allowed to regenerate with or without help from the harvester and with consideration for ecological change over time. An ethic of standing in place considers both individual and surrounding, acknowledging ecological interrelationships. The place encompasses the trees, other plants, fungi, animals, and even humans that depend on one another in a locale. Once a tree or forest has been cut, place irrevocably changes. However, forests depend on change and death for their overall health, which may be an opening for intentioned human taking.

Humans have a long history of removing trees from their place. Forest disruption historically has been ongoing in pre-historic Europe and China for approximately the last 6000 years (Williams 2003, 10). Tropical deforestation continues as one of the main drivers of climate change. An ethic of place could break from this past to allow trees primacy on remaining forested land and other land where there are currently trees. As Casey and Marder state, "no two arboreal places are ever just the same but differ significantly from tree to tree, each tree not only calling for its own place but *creating* that place, *being* its place" (2024, 62). Allowing the place to stand, even if an individual tree does not, creates the opportunity for life to grow again.

It can be argued the language of rights is not appropriate to trees and plants. Philosopher David Wood finds that attending to plants "is an ethical enterprise not in requiring us to recognize the rights of plants, but in reanimating wonder, opening our eyes to the unexpected depth, complexity, and dignity of plants, and the extraordinary ways in which our lives are entangled with them" (2020, 34). Similarly, Cloke and Jones recognize the enchantment, and attending disenchantment, engendered by trees and forests that can serve "as a prompt to personal moral impulses" (2003, 211). Even if trees are not granted a specific right to place, place is still primary to such wonder, awe, and enchantment that humans can experience in the presence of a tree. However, if value and potentially personhood are extended to vegetal beings, considering rights for trees and plants then follows, though these rights will be arbor or vegetal rights specific to these beings. Place is one potential locus from which to determine such rights.

15.6 Conclusion

As Irigaray affirms, "we must learn how to look at a tree, not to perceive its present form in order to re-present it mentally and fix it by naming it. Rather, we must gaze at its being as living and changing" (Irigaray and Marder 2016, 49). The science and thought regarding arboreal being supports reconsidering ethical human treatment of trees. An arboreal ethic has implications beyond the arboreal or even only the vegetal, as humans, especially in the Western lineage, reconsider relations with non-human beings including plants, animals, and even fungi. Leaning on Casey's phenomenology of place, an arborcentric treatment of rights based on place opens possibilities for more respectful human relationships with trees and forests. This initial articulation of an arboreal ethic of place does not address several important considerations, and more could done, for example, on issues of succession, ecosystem-specific concerns, and cross-cultural approaches to human interfaces with arboreal place. If accepted, such an ethic would require reassessing ongoing environmental ills like tropical deforestation. These practices transform the places trees have created, and an arboreal, place-based ethic would allow these places to remain even if the individual trees themselves must be taken for human needs. As trees continue to be reconsidered from within a Western perspective, an arboreal ethic of place may not justify our robbery but may compensate for it.

References

Alpi, Amedeo, Nikolaus Amrhein, Adam Bertl, Michael R. Blatt, Eduardo Blumwald, Felice Cervone, Jack Dainty, et al. 2007. Plant Neurobiology: No Brain, No Gain? *Trends in Plant Science* 12 (4): 135–136. https://doi.org/10.1016/j.tplants.2007.03.002.

Arbor, J.L. 1986. Animal Chauvinism, Plant-regarding Ethics and the Torture of Trees. *Australasian Journal of Philosophy* 64 (3): 339. https://doi.org/10.1080/00048408612342551.

Attfield, Robin. 2018. Forest Ethics. In *Plant Ethics: Concepts and Applications*, ed. Angela Kallhoff, Marcello Di Paola, and Maria Schörgenhumer, 121–130. New York: Routledge.

Casey, Edward S. 2001. Between Geography and Philosophy: What Does It Mean To Be in the Place-world? *Annals of the Association of American Geographers* 91 (4): 683–693. https://doi.org/10.1111/0004-5608.00266.

Casey, Edward S. 2009. *Getting Back into Place: Toward a Renewed Understanding of the Place-world*, 2nd ed. Bloomington, IN: Indiana University Press.

Casey, Edward S., and Michael Marder. 2024. *Plants in Place: A Phenomenology of the Vegetal*. New York: Columbia University Press.

Cloke, Paul, and Owain Jones. 2003. Grounding Ethical Mindfulness For/in Nature: Trees in Their Places. *Ethics, Place, and Environment* 6 (3): 195–214. https://doi.org/10.1080/1366879042000200660.

Comstock, Gary, Sue Donaldson, Andrew Fenton, Tyler M. John, L. Syd M. Johnson, Robert C. Jones, Will Kymlicka, et al. 2021. *The Philosophers' Brief in Support of Happy's Appeal*. In New York State Appellate Court.

Crowther, T.W., H.B. Glick, K.R. Covey, C. Bettigole, D.S. Maynard, S.M. Thomas, J.R. Smith, et al. 2015. Mapping Tree Density at a Global Scale. *Nature* 525: 201–205. https://doi.org/10.1038/nature14967.

Epstein, Seth. 2022. Rights of Nature, Human Species Identity, and Political Thought in the Anthropocene. *The Anthropocene Review* 1–19. https://doi.org/10.1177/20530196221078929

Falik, Omer, Perla Reides, Mordechai Garsani, and Ariel Novoplansky. 2003. Self/non-self Discrimination in Roots. *Journal of Ecology* 91 (4): 525–531. https://doi.org/10.1046/j.1365-2745.2003.00795.x.

Federal Ethics Committee on Non-Human Biotechnology. 2008. *The Dignity of Living Beings with Regard to Plants: Moral Consideration of Plants for Their Own Sake*. Berne, Switzerland: Federal Ethics Committee on Non-Human Biotechnology.

Food and Agricultural Organization of the United Nations (FAO). 2022. *The State of the World's Forests 2022: Forest Pathways for Green Recovery and Building Inclusive, Resilient and Sustainable Economies*. Rome.

Gagliano, Monica, Michael Renton, Martia Depczynski, and Stefano Mancuso. 2014. Experience Teaches Plants to Learn Faster and Forget Slower in Environments Where It Matters. *Oecologia* 175 (1): 63–72. https://doi.org/10.1007/s00442-013-2873-7.

Goralnik, Lissy, Tracy Dobson, and Michael Paul Nelson. 2014. Place-based Care Ethics: A Field Philosophy Pedagogy. *Canadian Journal of Environmental Education* 19: 180–196.

Gorzelak, Monika A., Amanda K. Asay, Brian J. Pickles, and Suzanne W. Simard. 2015. Inter-plant Communication Through Mycorrhizal Networks Mediates Complex Adaptive Behaviour in Plant Communities. *AoB PLANTS* 7: plv050. https://doi.org/10.1093/aobpla/plv050.

Hall, Matthew. 2011. *Plants as Persons: A Philosophical Botany*. Albany, New York: State University of New York.

Hall, Matthew. 2019. In Defence of Plant Personhood. *Religions* 10 (5): 1–12. https://doi.org/10.3390/rel10050317.

Heidegger, Martin. 1968. *What is Called Thinking?*, trans. Fred D. Wieck and J. Glenn Gray. New York: Harper & Row.

Irigaray, Luce, and Michael Marder. 2016. *Through Vegetal Being: Two Philosophical Perspectives*. New York: Columbia University Press.

Karban, Richard, Kaori Shiojiri, Satomi Ishizaki, William C. Wetzel, and Richard Y. Evans. 2013. Kin Recognition Affects Plant Communication and Defence. *Proceedings of the Royal Society* 280. https://doi.org/10.1098/rspb.2012.3062.

Kimmerer, Robin Wall. 2013. *Braiding Sweetgrass: Indigenous Wisdom, Scientific Knowledge, and the Teachings of Plants*. Canada: Milkweed Editions.

Kruger, Linda E. and Daniel R. Williams. 2007. Place and Place-based Planning. In *Proceedings from the National Workshop on Recreation Research and Management General Technical Report PNW-GTR-698*, eds. Linda E. Kruger, Rhonda Mazza, and Kelly Lawrence, 83–88. Portland, OR: U.S. Department of Agriculture, Forest Service, Pacific Northwest Research Station.

Lowell, James Russell. 1857. Humanity to Trees. *The Crayon* 4 (3): 96.

Marder, Michael. 2011. Vegetal Anti-metaphysics: Learning from Plants. *Continental Philosophy Review* 44 (4): 469–489. https://doi.org/10.1007/s11007-011-9201-x.

Marder, Michael. 2012. Plant Intentionality and the Phenomenological Framework of Plant Intelligence. *Plant Signaling & Behavior* 7 (11): 1365–1372. https://doi.org/10.4161/psb.21954.

Marder, Michael. 2013a. *Plant-thinking: A Philosophy of Vegetal Life*. New York: Columbia University Press.

Marder, Michael. 2013b. Should Plants Have Rights? *The Philosopher's Magazine* 62: 46–50. https://doi.org/10.5840/tpm20136293.

Puleo, Thomas J. 2019. Incorporating Nonhuman Subjectivity into World Society: The Case of Extending Personhood to Plants. In *Modern Subjectivities in World Society: Global Structures and Local Practices*, eds. Dietrich Jung and Stephan Stetter, 211–227. Cham, Switzerland: Palgrave Studies in International Relations.

Ryan, John Charles. 2012. Passive Flora? Reconsidering Nature's Agency Through Human-plant Studies (HPS). *Societies* 2: 101–121.

Simard, Suzanne. 2017. The Mother Tree. In *The Word for World is Still Forest*, ed. Anna-Sophie. Springer and Etienne Turpin, 66–72. Berlin: Bettina Warzecha Verlag für Planung und Organisation.

Stone, Christopher D. 1972. Should Trees Have Standing?: Toward Legal Rights for Natural Objects. *Southern California Law Review* 45: 450–501.

Taiz, Lincoln, Daniel Alkon, Andreas Draguhn, Angus Murphy, Michael Blatt, Chris Hawes, Gerhard Thiel, et al. 2019. Plants Neither Possess nor Require Consciousness. *Trends in Plant Science* 24 (8): 677–687. https://doi.org/10.1016/j.tplants.2019.05.008.

Whitehead, Alfred North. 1978. *Process and Reality*, ed. David Ray Griffin and Donald W. Sherburne. New York: The Free Press.

Williams, Michael. 2003. *Deforesting the Earth: From Prehistory to Global Crisis*. Chicago: University of Chicago Press.

Wood, David. 2020. *Thinking Plant Animal Human: Encounters with Communities of Difference*. Minneapolis: University of Minnesota Press.

Yokawa, K., T. Kagenishi, A. Pavlovic, S. Gall, M. Weiland, S. Mancuso, and F. Baluška. 2017. Anaesthetics Stop Diverse Plant Organ Movements, Affect Endocytic Vesicle Recycling and ROS Homeostasis, and Block Action Potentials in Venus Lytraps. *Annals of Botany*. https://doi.org/10.1093/aob/mcx155.

Chapter 16
On the Efficacy of Cultivating Environmental Reverence for Forests

Kimberly M. Dill

Abstract In this piece, I develop a philosophical account of environmental reverence, as induced by more-than-human entities and environments. Utilizing a relational ethical framework, I conceive of environmental reverence as a moral emotion, which—through habituation and cultivation–carries the potential to grow into a fully-fledged environmental virtue. By reference to the empirical, psychological literature, I show that environmental reverence is positively affective (i.e., induces subjective wellbeing), inherently motivating, and promotes efficacious conservation behaviors. So conceived, reverence is constrained by a set of both internal and external success conditions; its proper implementation thus requires on-the-ground relationship building between communities and the local, biodiverse spaces near which they dwell. I focus on the important role that reverence plays in motivating the efficacious conservation of sylvan ecosystems, drawing on eco-anthropological research into the high concentrations of biodiversity found in residentially managed 'sacred groves'. Altogether, I recommend a few first-order, ethical prescriptions: first, mass environmental discourse ought to highlight the motivational efficacy of cultivating cultural narratives that center environmental reverence and the key, carbon-sequestering role that forests play in climate-change mitigation. Finally, it is imperative for environmental ethicists to elevate the epistemic status afforded to traditional ecological and residential land managers.

16.1 Introduction

Imagine being immersed in an old-growth, redwood grove, wherein 800-year-old living giants tower overhead. Those who come into perceptual contact with trees of this magnitude and age often experience a range of emotions—from wonder and respite to awe and reverence. A variety of environmental ethicists—including ecofeminists (Shiva 1988; 2013), some proponents of Traditional Ecological Ways

---

K. M. Dill (✉)
Santa Clara University, California, USA
e-mail: kdill@scu.edu

M. Di Paola (ed.), *The Vegetal Turn*, The International Library of Environmental, Agricultural and Food Ethics 38, https://doi.org/10.1007/978-3-031-68671-9_16

of Knowing (Atleo 2005; Kimmerer 2013), global grassroots movements, and bioregionalists[1]—implicitly or explicitly emphasize that the cultivation of other-centered, *relational* emotions like these over time often effectively motivates and sustains fruitful conservation and restoration efforts at the local level.

This piece serves as a robust analysis of one moral emotion—reverence—, which is emphasized by environmental ethical theories that are either relational (Plumwood 1993; Hourdequin & Wong 2005), virtue ethical (Kawall 2003), or which conceive of moral emotions as integral to their prescriptive framework (Callicott 1989; McShane 2007; Leopold 1949). Environmental reverence warrants a deeper and more unified investigation due to its unique status amongst moral emotions—conferred by its psycho-physiologically restorative effects and its capacity to motivate efficacious conservation behaviors. I frame this discussion of environmental reverence through an interdisciplinary lens, which spans philosophical, ecological, and psychological literatures, respectively. Ultimately, I claim that there is a unifying thread that runs throughout the outlined accounts: namely, the full realization of environmental reverence requires that subjects satisfy some set of internal and external success conditions, measured by reference to changes in motivational states (belief, desire, intention) and behavior (via e.g., their positive, environmental impact). Accounts that centralize its cultivation therefore prescribe on-the-ground relationship building between communities and the local ecosystems near which they dwell.

Narrowing the scope, I then apply this analysis to the conservation and restoration of local, forested ecosystems. More specifically, the ecological and anthropological literature demonstrates that residential communities with cultures that emphasize environmental reverence boast disproportionately high levels of biodiversity in the forested groves that they revere as 'sacred'. At the global scale, this is important to bookmark, for the successful conservation and restoration of plant-, fungi-, and nutrient-rich forests constitutes one viable form of carbon sequestration that additively combats climate change and its detrimental effects. Forests (particularly old growth groves), more specifically,

> [...] continue to accumulate carbon at a much greater rate than was previously thought, making them important carbon sinks which researchers say must be factored into global climate models. (Gilhen-Baker et al. 2022)

Conversely, the degradation of forested ecosystems that are rich in carbon sequestering organisms—e.g., the trees, shrubs, flowering plants, fungi, and bacteria-rich soil that dwell therein—releases large quantities of CO^2 into the atmosphere, further contributing to rising global temperatures (Gilhen-Baker et al. 2022). Altogether, this suggests that the cultivation of cultural narratives that centralize environmental reverence for forests is one efficacious (if surprising) method, which–when fully realized and implemented–sparks and mobilizes conservation, restoration, and climate change mitigation efforts at the local level.

---

[1] e.g., Alianza Internacional de Reforestación and The Greenbelt Movement.

16.2 Environmental Reverence

16.2.1 Norms for Feeling

Following McShane (2007), I maintain that some attitudes are "incompatible with thinking that [an entity or environment's] value is entirely dependent on the satisfaction of our interests" (2007, 175). There are, in other words, certain moral emotions—e.g., love and *reverence*—which are not merely felt, phenomenological excitations. Instead, emotions in this vein maintain a degree of "fittingness" (Anscombe 1957). It is, more specifically, reasonable to feel them, as they are induced by entities and environments that warrant them as an appropriate response. In this sense, they are other-centered (2007, 175), for they track and are induced by an entity or environment's intrinsic (or—as I will later claim—*relational*) value. Though McShane does not focus on reverence in particular, her illustrative analysis of *love* for more-than-human landscapes and beings emphasizes the importance of moral emotions and the norms for feeling that constrain them more generally. It will thus be illuminating to briefly focus on her analysis of ecological love, with an eye toward comparatively elucidating the specific contours of environmental reverence.

To justify the claim that love is genuinely instantiated if and only if a lover attributes intrinsic value to their beloved, McShane considers the (affective and motivating) case of friendship. More precisely, if one (call them *S*) is privy to a subjective feeling of love for another (call them *F*), then this feeling, McShane argues, is not exhaustively contingent on whether *F* can satisfy *S*'s interests. In other words, *S* does not love *F* only in virtue of *F*'s perceived instrumental value—that is, whether *F* satisfies (or fails to satisfy) *S*'s interests. If this *were* the case, then one could properly charge *S* with a sort of relational inauthenticity and, by extension, a failure to act in accordance with moral norms for feeling (McShane 2007, 174). After all, *S*'s emotions, though *resembling* love, respect, or adoration, would be entirely self-interested. For this reason, the set of emotions that *S* feels would be more properly characterized as (something like) narcissistic adoration or codependent attachment.

16.2.2 Relationalism

According to McShane, authentic love requires that *S* properly acknowledges and attributes value to their beloved, which is independent of the mere satisfaction of *S*'s own interests. This, I think, is rather insightful, though warrants the following, theoretical modification: I urge that a *degree* of self-interest is necessary for the cultivation of healthy (e.g., loving or reverential) relationships. After all, relationships that are constituted by *S*'s self-abnegation for the sole sake of their counterpart's wellbeing are inherently self-destructive. With some radical exceptions (e.g., the sacrifices afforded by a parent for the sake of their child under extreme conditions), self-neglect therefore *also* constitutes a kind of relational failure. With this consideration in mind, moral emotions—whether they are induced by and between two members of our species, *Homo sapiens,* or between *Homo sapiens* and the more-than-human world of plants and fungi-rich soil—motivate the cultivation of relationships that are healthful, interdependent, and reciprocal (that is, mutually-beneficial) in nature.

This view—*relationalism*—has been implicitly or explicitly introduced into the environmental ethics literature by a variety of authors, traditions, and theorists, including (but not limited to) Aldo Leopold's Land Ethic (1949), Arne Naess's Deep Ecology (1973), Daoism (Chinn 2013), Confucianism (Hourdequin and Wong 2005), and a variety of indigenous philosophical systems. For example, the ever-insightful Anishinaabe environmental scientist, activist, and writer, Robin Wall Kimmerer (2013) suggests that healthy relationships are best understood and defined by reference to the degree of *reciprocity* that they exhibit. In other words, a relationship qualifies as genuinely loving or reverential if both *relata* mutually contribute in some way to each other's flourishing. When one *relatum* in an overarching pair fails to flourish as a direct, relational result of e.g., emotional or physical depletion, then the relational value—and, by extension, integrity of the overarching relationship—dwindles. As Kimmerer so beautifully writes, "all flourishing is [thus] mutual" (2013, 15).

One key, unifying feature of these diverse worldviews is their common commitment to the ontological intertwinement of *Homo sapiens* with both members of our own species and more-than-human entities, processes, and systems.[2] As a result of this ontological inextricability,

> The most ethically valuable aspect of an identity tied to particular people and to particular places in the world—with their animals, plants, neighborhoods, rivers, playgrounds, and shops—is experiencing oneself as accountable to others and as an agent with an active role in the world. In relationships, one learns that one's actions matter. Relationships allow us to call one another to account, and connections to family and community at the local scale may be the first and best place to gain this understanding (Hourdequin and Wong 2005, 29)

On this view, self-identity, and thus self-interest, are also conceived of relationally; that is, an individual or culture is inherently entangled with the larger-than-human (plant, animal, fungi, and bacteria-rich) world in which they find themselves situated.[3]

By conceiving of self-interest in this broadened way, relationalism circumvents some of the deeply entrenched philosophical debates—so ubiquitous in contemporary environmental ethics—between intrinsic and instrumental value theories. More specifically, it proposes that environmental actions are justified or warranted by reference to the degree of relational value that they either preserve, maximize, or

---

[2] Within the United States, the bioregionalism movement is a precise and practical articulation of this commitment, which makes salient a logical entailment: that is, if one's self-identity and interests are contextualized within one's local environment, then it is in one's self-interest to work in ways that improve the wellbeing of (at a minimum) those local environmental entities that constitute one's immediate sense of extended selfhood (Hourdequin and Wong 2005).

[3] Consider, for example, the deeply-rooted relationship between the Dai people of China's Yunnan Province and *Camellia sinensis*—tea. Indeed, *Camellia sinensis* plays an important cultural-ecological role throughout mainland China, which is evidenced through the ubiquity of ceremonial and social *Gong Fu Cha* (skillful tea serving) practices (d'Abbs 2019).

serve to instantiate; and this relational value is interdependently determined by the degree to which both the agent and the other being-in-question mutually flourish—whether synchronically (at-a-time) or diachronically (over time).[4] By centering *eudaimonistic* flourishing in this way, relational value theories synthetically integrate the insights articulated in both intrinsic and instrumental value frameworks, while simultaneously making sense of the claim that relationships bear a unique kind of value-in-themselves.[5]

16.2.3 Toward a Taxonomy of Moral Emotions

Despite their deep and rich similarities, it is worth noting that love and reverence may be conceived of and characterized differently. More specifically, each has a unique—though admittedly overlapping—phenomenological texture and a diverse set of representational and motivational contents. So, for example, love represents one's beloved as intrinsically (not merely instrumentally) valuable and recommends changes in one's behavior that will contribute to the flourishing of the loving relationship. To return to our earlier case, if *S* genuinely loves *F*, then *S* will be motivated to invest in their relationship with *F* by e.g., communicatively acquiring knowledge about *F*'s likes and dislikes to appropriately satisfy their needs. When reciprocal, this loving process effectively and fruitfully contributes to the flourishing of *both* *S* and *F*, in addition to their overarching, loving relationship. Similarly, reverence attributes intrinsic (not merely instrumental) value to the revered (*R*), is reciprocally beneficial, and recommends modifications in *S*'s behavior that will contribute to *R's* flourishing. Reverence is, however, distinct from love, for its phenomenological character uniquely elicits or facilitates (epistemic or personal) transformation (Paul 2016; Sarkar 2012) by inducing peak experiences (Maslow 1943), psycho-physiological restoration, humility, and a felt expansion of one's sense of self in the presence of the revered.

Admittedly, a precise taxonomy of these diverse moral emotions is difficult. For example, genuine love, like reverence, involves a deep and abiding sense of mutual respect, while at times eliciting humility or a phenomenological sense of ego dissolution in the lover. Nevertheless, the absence of precise taxonomical joints should not be troubling, for the psychological literature indicates that emotions more generally are likewise often nested and vague.[6] In any case, we can pragmatically distinguish

[4] Inspired by Aristotelian *eudaemonia*, flourishing should be understood contextually; more precisely, it is indexed to the kind of being that each *relatum* is. After all, actions that contribute to the flourishing of redwood trees (*Sequoia sempervirens*) will look markedly different from those which contribute to the flourishing of fairy castle cacti (*Acanthocereus tetragonus*).

[5] On this view, relationships are metaphysically emergent and unique: they contain yet supersede their constituent *relata* and, by extension, bear a unique kind of value-in-their-relational-selves.

[6] Even more broadly, a sharply delineated taxonomy of emotions is, at times, difficult. For example, chronic anger may obfuscate a deeper, unprocessed grief (Goleman 2005), creating what Daniel Goleman has termed an 'Anger Iceberg'.

between reverence and love by positing that subjects often attribute a high degree of *transcendence* or *sacredness* to those beings whom they revere—a point to which I will shortly return (Woodruff 2001, 65).

Love (for either land or specific organisms) has typically been prioritized in the environmental ethics literature as the prime moral emotion that motivates care for more-than-human beings and ecosystems (Leopold 1949; McShane 2007; Kimmerer 2013: 122–123). My project here, however, is to recommend a deeper investigation into and sharper focus on the efficacious and unique role that environmental reverence plays in conservation and restoration contexts. Reverence, more precisely, is properly characterized as a relational moral emotion (that, with proper cultivation, carries the potential to blossom into a virtue), for it motivates a set of reciprocally beneficial behaviors in the subjects whom it transforms. Its practical efficacy is, moreover, evidenced by the modified—e.g., caring or conservation-oriented–behaviors that human individuals and communities exhibit in relation to those beings and ecosystems which they revere.

16.2.4 The Psychology and Phenomenology of Reverence

Before diving more deeply into our discussion on the applied efficacy of cultivating environmental reverence for biodiverse forests, it is important to a bit more general philosophical book-keeping. Psychologically, I endorse a cognitivist conception of reverence. This means that reverence has both affective (i.e., felt) and intentional (i.e., representational) components. More specifically, reverence is induced by striking phenomenological experiences (as of, say, grand or very old environmental entities) and is often bolstered by knowledge pertaining to an entity's historical origins or its broader ecological context. So, for example, an individual who beholds an old-growth coast redwood experiences reverence partially due to the redwood's great size (an unmediated perceptual experience), its great age (an experience conditioned by eco-historical knowledge), and knowledge of facts pertaining to the beneficial role that it plays in its broader ecosystem (through, say, water capture and carbon sequestration).

Affectively, reverence is positively valenced in the sense that it contributes to feelings of subjective wellbeing, is aesthetically enjoyable, and is inherently motivating; that is, individuals who experience reverence are thereby motivated to conserve or care for the entities or environments that they revere. This feature distinguishes reverence from experiences of awe or the sublime, which (when induced by e.g., starry night skies or the striking vistas displayed by the Grand Canyon) can also be quite terrifying (Burke 1757; Woodruff 2001) and, by extension, immobilizing in nature. Experiences of awe, the sublime, and reverence do nevertheless share an important commonality.[7] As noted by Elkins et al. (1998), all three involve some sense of *self-transcendence.* Phenomenologically, self-transcendence consists in a felt sense of expansive unity with or connection to something beyond the bounds of this self—including, for example, the more-than-human world.[8] Finally, Haidt (2003) and other positive psychologists emphasize reverence's transformative potential, which is integral to the psychophysiological health, functioning, and wellbeing of members of our species, *Homo sapiens.* The important transformations induced by experiences of reverence include changes in one's felt preferences (Sarkar 2012), self-conception, motivation, and the resulting behaviors that flow from these.

There are a variety of empirical studies that investigate the psychophysiological mechanisms that may best explain reverence's restorative effects. Ai et al. (2009), for example, studied patients who underwent coronary artery bypass graft surgeries. The results of their study indicate that those patients who regularly experienced and exhibited reverence (measured both attitudinally and by reference to their modified behaviors) suffered fewer postoperative complications and enjoyed a shortened hospital stay (Ai et al. 2009). These positive psycho-physiological benefits obtained whether subjects were religious or non-religious, confirming that reverence can be construed as an interfaith, spiritual, or secular feeling and attitude, which is stimulated by a variety of entities, objects, persons, and environments. Similarly, Powell et al. (2003) showed that experiences of reverence produce (or correspond to) a "quieting effect," understood in terms of increased emotional regulation and its biological analogues (e.g., decreased cortisol) (Ai et al. 2009; Powell et al. 2003). Fredrickson (2001; Fredrickson et al. 2003) likewise linked experiences of reverence with quickened recovery rates (measured by reference to cardiovascular activity) after subjects were exposed to anxiety-inducing scenarios. Altogether, these studies justify (and to a high degree) the claim that reverence is psycho-physiologically restorative. It is, in other words, healing, regulative, and is thus conducive to human wellbeing or flourishing.

16.2.5 Reverence as a Moral Emotion and Potential Virtue

Ai et al. (2009) define reverence as a "cross-faith indicator [of] a feeling or attitude of 'deep respect, love, and awe, as for something sacred: veneration' (Guralnick 1978; Woodruff 2001, 8, 117)" (Ai et al. 2009). For the purposes of this piece, I endorse a pluralistic understanding of the 'sacred', granting that reverence can be induced in either secular or religious individuals and contexts. Indeed, articulating

---

[7] The conception of reverence that I here endorse shares some similarities with that proposed by Fromm (1950), who conceives of reverence as being one emotion in an overarching set of "peak experiences," which include awe, the transcendental, and wonder (Elkins et al. 1998).

[8] I do grant that feelings of self-transcendence can and often are achieved with other human individuals, groups of people, and even the species *Homo sapiens,* conceived of as one. Consider, for example, the moments of self-transcendence experienced by concert goers who experience a felt sense of at-oneness with both the crowd and the music (that, I am stipulating, they enjoy).

a precise definition of the sacred is tricky, for 'sacredness' is typically understood as tracking a feature or property that, when perceived, is never wholly articulable in conceptual or linguistic terms. Nevertheless, I posit that entities may be conceived of as sacred if they bear a transformative (e.g., restorative) power, elicit a profound sense of felt connectedness, exhibit features of or induce transcendence (in and of the singular self), bear intrinsic or relational value, and warrant reverence as an appropriate response.

It is worth noting that this definition does not—as Durkheim (1915), by contrast, posits—assume a sharp dichotomy between the sacred and profane and is not contingent on one's subscription to some stripe of theism. In this sense, reverence can be induced in naturalists by more-than-human, biodiverse environments and their constituents, great works of art and other aesthetic objects, and abiotic environmental features (including mountains and starry night skies). Reverence may also, of course, be induced in those who identify as either religious or spiritual. Moreover, the distinction between these two ways of being oriented toward the world—naturalism and spiritualism, respectively—does, at times, blur. As astronomer Carl Sagan so eloquently notes,

> Science [i.e., naturalism] is not only compatible with spirituality; it is a profound source of spirituality. When we recognize our place in an immensity of light-years and in the passage of ages, when we grasp the intricacy, beauty, and subtlety of life, then that soaring feeling, that sense of elation and humility combined, is surely spiritual. (Sagan 1996, 32)

He further writes that,

> In its encounter with Nature, science invariably elicits a sense of reverence and awe. The very act of understanding is a celebration of joining, merging, even if on a very modest scale, with the magnificence of the Cosmos. (Sagan 1996, 32)

Indeed, a direct translation of *anima* from the Latin—soul or spirit, the root of 'spirituality'—yields the English terms, 'breath, air, and life'. That which is spiritual is, in its most basic sense, that which either gives or honors the animate—those beings who live or that which makes life possible.

Drawing more explicitly on Schweitzer's conception of 'reverence for life' is informative here. Inspired by Buddhist and Jain conceptions of non-harm or *ahimsa*, Schweitzer's moving articulation of biotic reverence is broadly pluralistic; it is, in other words, a philosophical and conceptual tool that can be in principle integrated into purely naturalistic, theistic, or agnostic ethical systems. The crux of reverence, importantly, lies in its life- and world-affirming capacity. Reverence, for Schweitzer, is a broadly,

> Spiritual act in which a man begins to live reflectively and begins to give himself to his life with reverence in order to realize its true value. [...] The fundamental fact of human awareness is this: "I am life that wants to live in the midst of other life that wants to live." A thinking man feels compelled to approach all life with the same reverence he has for his own. Thus, all life becomes part of his own experience. From such a point of view, 'good' means to maintain life, to further life, to bring developing life into its highest value. A man is ethical only when life, as such, is sacred to him, that of plants and animals as that of his fellow men, and when he devotes himself helpfully to all life that is in need of help. (James 2005, 30)

Similarly to Schweitzer, I endorse the claim that reverence is best conceived of as a relational moral emotion, which is psycho-physiologically transformative *and* efficaciously motivates ethical behaviors that contribute to the flourishing of the biotic *relatum* whom one reveres. However, and like patience, it is also clear that instances of reverence can be felt only fleetingly (when, say, one briefly visits the gorgeous and spectacular Sequoia National Park for one weekend). Even so, part of reverence's unique power lies its capacity to *transform* those who experience it. Accordingly, and following Woodruff (2001) and Kawall (2003), it thus makes sense to posit that the cultivation of a disposition toward reverence may be appropriately conceived of as a kind of environmental virtue. Like all virtues, reverence thus requires continual practice through habituation (i.e., through the ongoing satisfaction of a set of internal and external success conditions), which ultimately results in the cultivation of a reverential character—and ideally larger culture—over time.

16.3 The Efficacy of Cultivating Environmental Reverence for Forests

16.3.1 Reverence for Sacred Groves

There are plenty of anthropological case studies that demonstrate the reciprocal efficacy of cultivating actual-world, cultural reverence for biotic entities and biodiverse environments. For the sake of brevity, I'll restrict the scope of my applied analysis to reverence for sacred groves—forested areas, consisting of plants, fungi, animals, and bacteria-rich soil, that are preserved due to the important cultural significance that they bear for local populations (Kandari et al. 2014, 1–2; Chandran and Hughes 1997). Broadly speaking, sacred groves,

> [...] around the world are community conserved areas that often have associated limitations on activities within the forest; these traditional rules can serve a conservation role. Sacred groves or forests are conserved by local residents for a variety of reasons, ranging from belief in a forest deity to protection of a spring or as a sacred space where ancestors are buried. (Archarya and Ormsby 2017)

There are a wide variety of residential and local communities that have engaged in the conservation of sacred groves, in locations "including Ethiopia, Japan, Morocco, India, and Ghana" (Kandari et al. 2014: 4). At over 100,000 sacred groves, India boasts the highest number of preserved forests globally (Malhotra et al. 2007; Kandari et al. 2014). This fact should perhaps be unsurprising, for India's history is rich in a Vedic commitment to *Aranyana Samskriti*—that is, a 'Culture of the Forest' (Shiva 1988). Importantly, local populations do not conserve sacred groves solely for the purposes of instrumental resource-management. Instead, these biodiverse spaces are conceived of relationally, blending the boundaries between the human and more-than-human realms. Forests—which encompass entangled networks of the plants, fungi, animals, and bacteria-rich soil that comprise them—are, more specifically, conceived of as material manifestations of divinities or complex webs of emergent intelligence or sentience. Intimate human–more-than-human relationships between revered groves and the people who honor and devote themselves to them are thus forged through a variety of ritualistic, management, and subsistence practices.

For example, the Lepcha people in the Dzongu Valley of North Sikkim[9] forge strong relationships with the forests they revere—rich in *Ficus*, *Rhododendron arboretum*, *Oxalis corniculata*, and *Litsea cubeba*—through a variety of practical, care-taking and knowledge-gathering behaviors. Through cultivating forest-based reverence as a societal virtue and practice, the inseparability between "[c]ulture, politics, and ecology [given their] strong place-based identities" (Arora 2006, 104) has become firmly and enduringly established in Sikkim. To further illustrate, consider the sacred grove which surrounds the Tholung Buddhist Monastery, located within Khangchendzonga National Park. Tholung, "comprises an uninhabited human tract of mountainous forests adjacent to the wish-fulfilling pilgrimage site of the Thong temple […]" (Arora 2006, 65). Revered and protected by Lepcha communities, Tholung is considered inseparable from its people. Indeed, many Lepcha self-identify as "guardians of their sacred grove" (Arora 2006, 65) and follow a variety of rules of propriety, which govern and sharply restrict the kinds of activities that are considered permissible for them to perform therein. More specifically, Arora (2006) reports that, "the chief injunction to all Tholung pilgrims" [conveyed by Lepcha community member, Chumsay Nangpa] is "not to disturb the vegetation and cause harm to the animal life in this sacred forest" (67). These behavioral norms—extensive codes of ecological conduct and accountability—encode and demonstrate the culture of reverence that the people experience, cultivate, participate in, and express for these groves (Acharya and Ormsby 2017). Forests like Tholung, therefore,

[encompass] diverse meanings: a forest, the residence of the guardian deities of the area, a place of Buddhist pilgrimage, a fount of customary wisdom, a repository of Sikkimese national treasures, […] a place affirming ancestral connections with the landscape and expressing indigeneity in land, a forest where society rejuvenates itself, members cement alliances. (Arora 2006, 71–72).

Furthermore, and through their reverential engagement with sacred groves, the Lepcha people have acquired an extensive body of Traditional Ecological Knowledge (TEK) pertaining to the flora and fauna that dwell therein. Reverence, for the Lepcha people, is thereby also epistemically motivating. As, O'Neill et al. (2017) observes,

> […] Sikkim has a rich biocultural heritage that includes knowledge pertaining to over 1100 species of animals, fungi, and plants. Local people not only know about the useful properties of these species, but also the community ecology and life histories of diverse organisms. These aspects of ethnobiological knowledge, which encompass abundance, distribution, and phenology, significantly influence community management practices and can therefore benefit conservation planning in Sikkim. For instance […] Lepcha communities were engaged by government researchers to understand the population status of under-surveyed bird species; local communities were found to provide "data" at the accuracy needed to make

[9] The Dzongu Valley constitutes a reserve for the Lepcha people.

management decisions. Our reviewed records also implied that faith traditions and community taboos sustain many ethnobiological relationships in Sikkim, and cultivate a sense of stewardship toward critical habitat. (O'Neill et al. 2017)

So, the Traditional Ecological Knowledge of the Lepcha people (and other populations who revere and manage forests more generally) includes an in-depth understanding of the interdependent, ecological relationships that obtain between a variety of biodiverse species, including an array of tree populations, medicinal herbs, flowering plants, wild foods, fungi, and more-than-human animals. This extensive body of knowledge is passed down inter-generationally via word-of-mouth (through e.g., storytelling) and is further cultivated via extensive observation and pragmatic engagement (e.g., the responsible harvesting of medicinal foods). There is thus, to quote Whyte, a robust "connection between [a people's, including, for example, the Lepcha's] spirituality and science," which "reveals [...] how spiritually oriented processes of empirical inquiry promote accountability within societies and respect for our interdependence with nonhumans and the environment" (Whyte 2018).

More broadly, Singh et al. (2017) have collected extensive data from two sacred groves (Hariyali Devi and Tungnath) in Uttarakhand (located in the Western Himalayas) to study the role of sacred groves in the conservation of local biodiversity, concluding that, "[sacred] groves, in general, are a valuable tool of biodiversity conservation." (2017, 10). Indeed, and through engaging in a global and comparative meta-analysis of 35 studies across 17 countries, Sullivan et al. (2024) likewise found that higher levels of plant biodiversity are preserved in sacred, as opposed to adjacent, forested groves. However and worryingly, "people's changing attitudes, erosion of traditional beliefs, and human impact have caused [the] degradation of sacred groves over the years" (Singh et al. 2017, 10). Nevertheless, the inlayed spiritual, cultural, and ecological practices of the global peoples who revere and restrict activities within their sacred groves provide empirical evidence in favor of the claim that fully-fledged and habitual reverential practices are both ecologically motivating and generally effective.

Altogether, the intricate, varied, and multiplicitous cultural tapestries of the people who revere and conserve forests—threaded on the warp and weft of reverential codes of conduct—are inextricably interwoven with the flourishing of biodiverse systems. Sacred groves therefore serve as living libraries that preserve both cultural *and* ecological histories. Tragically, however, both such histories are now under threat, as contemporary increases in outside pressures to obtain resources from sacred groves in e.g., Sikkim contribute to sharp declines in biodiversity and the erasure of residential cultural practices and life-ways.

16.3.2 Local Forest Conservation Mitigates Climate Change

Forest conservation and restoration are integral to effectively mitigate the increasing dangers posed by global climate change. Generally, forests stabilize ecosystems and serve as carbon sinks, for they are largely comprised of carbon-sequestering organisms (e.g., flowering plants, fungi, shrubs, and trees), which effectively remove carbon dioxide from the atmosphere. Spectacularly, forest biomass absorbs approximately 7.6 billion metric tonnes of carbon dioxide per year, pulling two-thirds of global CO^2 fossil fuel emissions from the atmosphere (IUCN 2015; Harris and Gibbs 2021). Furthermore, as deforestation—the product of over-logging and preventable wildfires[10]—produces 20% of overall, global greenhouse gas emissions, climate change mitigation pressingly requires the widespread prevention of deforestation (Rizvi et al. 2015, 14). Altogether, the collective conservation of existent forests, restoration of depleted groves, and prevention of further deforestation is imperative to combat rising global temperatures and their deleterious social and ecological effects (e.g., displaced climate refugees and the loss of global biodiversity). Indeed, the restoration and conservation of forests could,

> contribute over one-third of the total climate change mitigation scientists say is required by 2030. [Restoring] 350 million hectares of degraded land in line with the Bonn Challenge could requester up to 1.7 gigatonnes of carbon dioxide equivalent annually. (International Union for the Conservation of Nature)[11]

At this point, it is imperative, I contend, to broadly re-emphasize the efficacy of Traditional Ecological bodies of Knowledge in forest management and restoration contexts. As Vandana Shiva (1988) has rigorously argued, local, residential populations in e.g., India have *already* inherited a rich tradition of effective forest management. Within the current discourse, the traditional ecological practices of the Lepcha People provide an alternative set of climate change mitigation proposals, which centralize reverential practices through the social, cultural, and physical structuring of local communities (by e.g., cordoning off local groves and conditioning the salient relationships that visitors forge when visiting through epistemically informative interpretation). Furthermore, the ecological model recommended by *Aranyana Samskriti* (trans., the "Culture of the Forest") has demonstrably and successfully enacted the preservation of biodiversity, with the added benefit of simultaneously contributing to the sovereignty of local, residential communities, the empowerment of women, and cultural preservation. More specifically, local populations flourish through the reverential cultivation and maintenance of their sacred groves, which reciprocally satisfy their basic needs via "integrated ecosystem" management that involves "multipurpose utilization" and cultural rejuvenation (Shiva 1988, 61). Indeed, traditional ecological knowledge systems of this sort have set the historical precedent for effective environmental movements globally: *Aranyana Samskriti,* exemplified by the Chipko Movement of 1970s in Uttarakhand, has inspired (and continues to inspire) a variety of non-violent 'tree-hugging' forest conservation movements since.

In short, Traditional Ecological Systems that centralize environmental reverence are effective. Unlike models that incentivize conservation via purely economic or

[10] The direct result of e.g., historical fire suppression initiatives in the United States and Australia.

[11] https://www.iucn.org/resources/issues-briefs/forests-and-climate-change.

financial means (which carry a potentially colonial, paternalistic, or overly instrumental tone), reverential conservation strategies promote community sovereignty and adaptive resilience to changing environmental conditions in a way that is culturally sustainable. It would thus behoove our environmental discourse to emphasize the import of cultivating environmental reverence in local communities throughout the Global South *and* North, while simultaneously working to uplift, preserve, and rejuvenate those cultural systems that have come under threat (via e.g., colonial or capitalistic encroachment).[12]

16.3.3 Cultivating Pluralistic Cultures of Reverence

Emphasizing the cultural rejuvenation—or first-time-implementation—of reverential emotions, virtues, and practices in communities where they have no deeply-rooted history is no simple or seamless feat. As noted by e.g., Erin McKenna (2021), our ecological attitudes and proclivities are largely shaped by the cultural narratives that we inherit—via e.g., the literary, religious, filmic, or other mythic representations of human interactions with the more-than-human world that we consume. The above arguments in favor of relationalism and the efficacy of cultivating ecological reverence should then encourage, at a more basic level, the adoption of stories, myths, narratives, and philosophical theories which decenter human beings as sole ecological protagonists or bearers of value. For those in the global North, cultivating and actively engaging in some set of efficacious and reverential practices will require a cultural shift: to mindfully re-craft the tales we tell, the ways in which we build, and how we engage with local, biodiverse environments. In any case, it behooves human communities in general to work toward the cultivation of reverence (*qua* moral emotion and enduring virtue) and reverential practices (via their pragmatic implementation). After all, and as suggested above, the regular stimulation of reverence in *Homo sapiens* is partially valuable insofar as it is psycho-physiologically transformative and restorative. Through inducing a felt, phenomenological sense of self-transcendence and ecological interconnectedness, reverence furthermore inherently motivates care for (say) more-than-human entities and ecosystems—including forests.

---

[12] This proposal shouldn't seem so far-fetched to those dwelling in the United States, for a secular form of this recommended strategy is already implemented in the spectacular Cathedral Grove of Muir Woods National Monument (California, USA). In this "silent preserve [which is] home to the tallest, oldest redwoods at Muir Woods," visitors are explicitly encouraged to "[consider] quiet contemplation as you make your way through this part of the woods" (National Park Service). Walking quietly and attentively through these dappled, golden halls of emerald and rust elicits, quite powerfully, a sense of humility and reverence for the mighty *Sequoia sempervirens* who dwell therein. See https://www.nps.gov/places/000/cathedral-grove.htm.

16.4 Conclusions

Throughout this piece, I have shown that an array of indigenous philosophers, residential populations, and proponents of Traditional Ecological Knowledge hail from and represent traditions historically rich in environmentally reverential narratives. This strongly encourages centering and elevating the epistemic authority and status (Dotson 2018) of Traditional Ecological land managers (including e.g., the Lepcha people who successfully steward Sikkim's sacred groves) in conservation, restoration, and other environmental contexts.

Toward this end, it is worth noting that the legal efficacy of eco-reverential worldviews has already, and most impressively, been demonstrated. Consider, for example, the precedent set by the Maori people of New Zealand, who successfully afforded protections to the Whanganui River in 2017. Revered as sacred by its people, the Whanganui River plays a complex role in Maori self-determination, cosmology, and personal identity. More specifically, the term *Te Awa Tupua* is used to denote the river's personhood, marking its status as an ancestral Maori elder. Reverence is afforded to the Whanganui in reciprocal gratitude for the material, psychological, and spiritual sustenance that the river provides to its people (Whanganui River Claims Settlement Act 2017). In the Crown's Court of New Zealand, *Kaitiakitanga* (trans., River Guardians who are appointed as intermediaries to speak on behalf of the river) were able to afford legal protections to *Te Awa Tupua* by utilizing a characteristically relationalist argument. Their goal was achieved by (for example) attributing a high degree of relational kinship and even *personhood* to the Whanganui River, which warrants its reverential treatment and protected status.

There are deep questions about whether communities more broadly (in, say, the global North) must adopt an animistic worldview, which attributes high degrees of complex consciousness or personhood to forests, rivers, or more-than-human animals, before they can feel genuine reverence for them. Nevertheless, I have suggested that reverence *qua* moral emotion and virtue can be stimulated in *either* secular or spiritual contexts, leaving open the possibility of a compatibility between its cultivation and a multiplicity of metaphysical worldviews (e.g., animism, panpsychism, naturalistic reductionism, etc.). Before concluding, I'd thus like to re-emphasize the psycho-physiological and material efficacy of reverence in conservation and restoration contexts, which (by reference to the empirical data) seems to obtain regardless of variations in deeper philosophical and ontological commitments. Altogether, the historical and contemporary tapestry of relationalism is promisingly rich, varied, diverse, textured, and vast.

Bibliography

Acharya, Amitangshu and Alison Ormsby. 2017. The Cultural Politics of Sacred Groves: A Case Study of Devithans in Sikkim, India. *Conservation and Society* 15 (2): 232. https://doi.org/10.4103/cs.cs_14_29.

Ai, Amy L., Paul Wink, Terrence N. Tice, Steven F. Bolling, and Marshall Shearer. 2009. Prayer and Reverence in Naturalistic, Aesthetic, and Socio-moral Contexts Predicted Fewer Complications Following Coronary Artery Bypass. *Journal of Behavioral Medicine* 32: 570.

Anscombe, G.E.M. 1957. *Intention.* Oxford: Basil Blackwell.

Arora, Vibha. 2006. The Forest of Symbols Embodied in the Tholung Sacred Landscape of North Sikkim. *Conservation & Society* 4 (1): 55–83.

Atleo, Richard. 2005. *Tsawalk: A Nuu-chah-nulth Worldview.* Vancouver: UBC Press.

Burke, Edmund. 1757 [2015]. *A Philosophical Enquiry Into the Origin of Our Ideas of the Sublime and Beautiful.* Oxford: Oxford University Press.

Callicott, J. Baird. 1989. *In Defense of the Land Ethic.* Albany: State University of New York Press.

Chandran, M.D.S., and J.D. Hughes. 1997. The Sacred Groves of South India: Ecology, Traditional Communities and Religious Change. *Social Compass* 44: 413–427. https://doi.org/10.1177/003776897044003008.

Chinn, Meilin. 2013. Sensing the Wind: *The Timely Music of Nature's Memory. Environmental Philosophy* 10 (1): 25–38.

Critical Investigations of Resilience. 2018. A Brief Introduction to Indigenous Environmental Studies and Sciences. *Daedalus* 147 (2): 136–147.

d'Abbs, Peter. 2019. Tea Art as Everyday Practice: Gongfu Tea in Chaoshan, Guangdong, Today. *The Asia Pacific Journal of Anthropology* 20 (3): 213–231. https://doi.org/10.1080/14442213.2019.1611908.

Dill, Kimberly. 2021. In Defense of Wild Night. *Ethics, Policy, and Environment* 25: 153–177.

Dotson, Kristie. 2018. Accumulating Epistemic Power. *Philosophical Topics* 46 (1): 129–154.

Durkheim, Émile. 1915. *The Elementary Forms of the Religious Life.* London: George Allen & Unwin.

Elkins, David N., L. James Hedstrom, Lori L. Hughes, J. Andrew Leaf, and Cheryl Saunders. 1998. Toward a Humanistic-phenomenological Spirituality. *Journal of Humanistic Psychology* 28 (4): 5–18.

Fredrickson, Barbara L. 2001. The Role of Positive Emotions in Positive Psychology. The Broaden and-Build Theory of Positive Emotions. *American Psychologist* 56(3): 218–26. https://doi.org/10.1037//0003-066x.56.3.218. PMID: 11315248; PMCID: PMC3122271.

Fredrickson, Barbara L., Michele M. Tugade, Christian E. Waugh, and Gregory R. Larkin. 2003. What Good are Positive Emotions in Crises? A Prospective Study of Resilience and Emotions Following the Terrorist Attacks on the United States on September 11th, 2001. *Journal of Personality and Social Psychology* 84 (2): 365–376. https://doi.org/10.1037/0022-3514.84.2.365. PMID:12585810; PMCID:PMC2755263.

Fromm, Erich. 1950. *Psychoanalysis and Religion.* New Haven: Yale University Press.

Gilhen-Baker, Melinda, Valentina Roviello, Diana Beresford-Kroeger, and Giovanni N. Roviello. 2022. Old Growth Forests and Large Old Trees as Critical Organisms Connecting Ecosystems and Human Health: A Review. *Environmental Chemistry Letters* 20: 1529–1538. https://doi.org/10.1007/s10311-021-01372-y.

Goleman, Daniel. 2005. *Emotional Intelligence: Why it Can Matter More than IQ.* New York: Random House Publishing Group.

Guralnick, D. B. 1978. In *Webster's New World Dictionary of the American Language* (2nd College Edition), ed. D. B. Guralnick. William Collins + World Publishing Company.

Haidt, Jonathan. 2003. The Moral Emotions. In *Handbook of Affective Sciences*, eds. Richard J. Davidson, H. Hill Goldsmith, and Klaus R. Scherer, 852–870. Oxford: Oxford University Press.

Harris, Nancy and David Gibbs. 2021. *Forests Absorb Twice As Much Carbon As They Emit Each Year. Insights. World Resources Institute.* https://www.wri.org/insights/forests-absorb-twice-much-carbon-they-emit-each-year

Hourdequin, Marion, and David Wong. 2005. A Relational Approach to Environmental Ethics. *Journal of Chinese Philosophy* 32 (1): 19–33.

James Brabazon. 2005. *Albert Schweitzer: Essential Writings.* Orbis Books.

Kandari, Laxman Singh, Vinod Kumar Bisht, Meenakshi Bhardwaj, and Ashok Kumar Thakur. 2014. Conservation and Management of Sacred Groves, Myths and Beliefs of Tribal Communities: A Case Study from North-India. *Environmental Systems Research* 3 (16): 1–10. https://doi.org/10.1186/s40068-014-0016-8.

Kawall, Jason. 2003. Reverence for Life as a Viable Environmental Virtue. *Environmental Ethics* 25 (4): 339–358.

Keltner, Dacher, and Jonathan Haidt. 2003. Approaching Awe, a Moral, Spiritual, and Aesthetic Emotion. *Cognition and Emotion* 17 (2): 297–314. https://doi.org/10.1080/02699930302297.

Kimmerer, Robin Wall. 2013. *Braiding Sweetgrass.* Canada: Milkweed Editions.

Leopold, Aldo. 1949 [1986]. *A Sand County Almanac.* Toronto: Ballantine Books.

Malhotra, K.C., Y. Gokhale, S. Chatterjee, and S. Srivastava. 2007. *Sacred Groves in India.* New Delhi: Aryan Books International.

Maslow, Abraham H. 1943. *A Theory of Human Motivation.* Psychological Review 50 (4): 370–396.

McKenna, Erin. 2021. *Living with Animals: Rights, Responsibilities, and Respect.* Rowman and Littlefield.

McShane, Katie. 2007. Anthropocentrism vs. Nonanthropocentrism: Why Should We Care? *Environmental Values* 16: 169–185.

Naess, Arne. 1973. The Shallow and the Deep, Long-Range Ecology Movement: A Summary. *Inquiry* 16: 95–100.

O'Neill, Alexander R., Hemant K. Badola, Pitamber P. Dhyani, and Santosh K. Rana. 2017. Integrating Ethnobiological Knowledge into Biodiversity Conservation in the Eastern Himalayas. *Journal of Ethnobiology and Ethnomedicine* 13 (21).

Paul, L.A. 2016. *Transformative Experience.* Oxford: Oxford University Press.

Plumwood, Val. 1993. *Feminism and the Mastery of Nature.* London: Routledge.

Powell, Lynda H., Leila Shahabi, and Carl E. Thoresen. 2003. Religion and Spirituality: Linkages to Physical Health. *American Psychologist* 58: 36–52.

Rizvi, A.R., S. Baig, E. Barrow, C. Kumar. 2015. *Synergies Between Climate Mitigation and Adaptation in Forest Landscape Restoration.* Switzerland: IUCN. https://portals.iucn.org/library/sites/library/files/documents/2015-013.pdf

Sagan, Carl. 1996. *The Demon-Haunted World: Science as a Candle in the Dark.* Random House.

Sarkar, Sahotra. 2012. *Environmental Philosophy: From Theory to Practice.* West Sussex: Wiley-Blackwell.

Schulze Ernst Detlef, Dominik Hessenmöller, Alexander Knohl, Sebastiaan Luyssaert, Annett Boerner, and John Grace. 2009. Temperate and Boreal Old-growth Forests: How Do Their Growth Dynamics and Biodiversity Differ From Young Stands and Managed Forests? In *Old Growth Forests. Ecological Studies (Analysis and Synthesis)*, eds. Christian Wirth, Gerd Gleixner, and Martin Heimann, 343–366. Berlin: Springer.

Shiva, Vandana. 1988. *Staying Alive: Women, Ecology, and Development.* London: Zed Books.

Singh, Sushma, Mudasir Youssouf, Zubair A. Malik, Rainer W. Bussmann. 2017. Sacred Groves: Myths, Beliefs, and Biodiversity Conservation—A Case Study from Western Himalaya, India. *International Journal of Ecology*. https://doi.org/10.1155/2017/3828609.

Sullivan, M. K., Browne, L., Penagos Zuluaga, J. C., Liu, J., Surendra, A., & Estrada-Villegas, S. (2024). Sacred forest biodiversity conservation: A meta-analysis. *Conservation Science and Practice*, 6(1), e13055. https://doi.org/10.1111/csp2.13055

Whyte, Kyle Powys. 2017. Food Justice and Collective Food Relations. In *Food, Ethics, and Society: An Introductory Text with Readings*, 122–134. New York: Oxford University Press.

Whyte Kyle. 2018. Critical Investigations of Resilience: A Brief Introduction to Indigenous Environmental Studies & Sciences in Daedalus, 147 (2): 136–147. https://doi.org/10.1162/DAED_a_00497

Woodruff, Paul. 2001. *Reverence: Renewing a Forgotten Virtue.* Oxford: Oxford University Press.

Chapter 17
Is Soil Merely a Means? Reconsidering the Nexus Among Humans, Plants, and Soil

Tsuyoshi Teramoto

Abstract The world will almost certainly remain highly populated for the next 100 years. During this period, it will be necessary to engage in two pursuits that may conflict with each other: increasing food production and addressing soil degradation. This situation may prompt a rethinking of the nexus among humans, plants, and soil. If soil degradation is an urgent problem, one could argue, as eco-modernists do, that we should make maximum use of technology to decouple agricultural production from the natural world. By doing so, we might be able to reduce the land area used for agricultural production, returning as much soil as possible to nature and encouraging soil recovery. This kind of argument, which places great hopes on new technologies and promotes soil substitution, expresses a somewhat limited perspective. Appropriate agricultural practices for soil conservation must be determined not only based on quantitative assessments of environmental impacts, but also by considering aspects such as soil values and technological uncertainties. By pointing this out, this chapter attempts to provide clues for reconsidering the nexus among humans, plants, and soil.

17.1 Introduction: Population, Food, and Soil

The United Nations Department of Economic and Social Affairs estimates that the world's population will increase to about 8.5 billion by 2030, 9.7 billion by 2050, and then peak at about 10.4 billion in the 2080s, remaining at that level until 2100 (UN DESA 2022). On the other hand, a study has forecasted that the world's population will peak at 9.73 billion in 2064, then begin to decline; and, if certain targets included in Sustainable Development Goals (SDGs), such as educational attainment and meeting contraceptive needs, are achieved, then the world's population will decline to 6.3 billion by 2100 (Vollset et al. 2020). Nevertheless, even if the latter projection is correct, the world will almost certainly remain highly populated for

T. Teramoto (✉)
Faculty of Science and Engineering, Chuo University, Tokyo, Japan
e-mail: teramoto.994@g.chuo-u.ac.jp

M. Di Paola (ed.), *The Vegetal Turn*, The International Library of Environmental, Agricultural and Food Ethics 38, https://doi.org/10.1007/978-3-031-68671-9_17

the next 100 years. Even if the world's population were to decline to 6.3 billion by 2100, that number would be roughly equal to the world's population in 2003, and it is not certain that total human activity in 2003 was in harmony with the natural environment. To sustain human society and achieve an appropriate relationship between humans and the natural world, it is essential to make continuous efforts to keep human impacts within the carrying capacity of the Earth over the next 100 years or so, when the population is expected to be large.

These efforts include increasing food production and addressing soil degradation, which may be in conflict with each other. Assuming a population of about 10 billion in 2050, the Food and Agriculture Organization of the United Nations projects that demand for agricultural products will increase by about 50% from 2013 levels, even assuming moderate economic growth (FAO 2017). Meanwhile, "33 percent of land is moderately to highly degraded due to the erosion, salinization, compaction, acidification, and chemical pollution of soils" (FAO 2015). Soil degradation has led to a sharp decline in available agricultural land per capita (0.45 ha in 1961, 0.25 ha in 2000, and a forecast of 0.19 ha in 2050) (FAO 2015). If this trend is allowed to go on, it will be difficult to maintain or increase food production, and global future food needs will not be met. In addition, soil is the keystone of the natural material cycle, providing diverse ecosystem services such as biomass production, storing, filtering, and transforming nutrients, substances, and water, as well as acting as a biodiversity pool and a carbon pool, etc. (FAO 2015). A deterioration of these services can also threaten food production.

This situation may prompt a rethinking of the nexus among humans, plants (especially as food and feed), and soil. One could argue, as eco-modernists do, that agricultural production should be decoupled from the natural world by making maximum use of technology (Asafu-Adjaye et al. 2015). This approach would reduce agricultural land use, return as much soil as possible to nature, and promote soil restoration. For example, increasing yield per land area through intensification (including through genetic engineering of crops), expanding hydroponics in plant factories, and culturing cells to produce meat might ensure sufficient food without using soil or with reduced use of soil. Should we steer in this direction?

I agree that new agricultural methods that reduce soil use have a possibility of contributing to both agricultural production and environmental conservation. I also think that it may be ethically problematic to miss or narrow opportunities to achieve harmonious relations with nature during a phase of population growth by ignoring possible technical solutions. On the other hand, I believe that arguments placing great hopes on new technologies and promoting soil substitution express a limited perspective about the human-plants-soil connection. Soil is not merely a means to an end—it can be an essential means for productive activities, and it may be more than a means. This chapter attempts to present a broader perspective on the connections among humans, plants, and soil, by contrasting soil-based and soil-less farming.

I begin by confirming the existence of two types of soil: soil as a natural ecosystem and soil as a means of production. Next, intensive farming with reduced soil or soil-less farming is characterized by its detachment from the natural ecosystem. Furthermore, I argue that the value of soil is diverse, and that it is not possible to give a univocal answer on future food production only from technical characteristics and quantitative measures of environmental impacts. Finally, I argue that because of the inherent uncertainty of technological solutions, the development of food production technologies should be incremental and diverse.

17.2 Soil as a Natural Ecosystem and as a Means of Production

Soils are formed in a variety of ways by the influence of factors such as climate, organisms, topography, and time on inorganic material (parent material) in the surface layer of the Earth's crust (Jenny 1994). Among these factors, organisms have a unique influence on soil formation. Soil organisms account for 25% of the total number of extant species identified so far on the Earth, with an estimated 360,000 species (Decaëns et al. 2006). This includes a wide range of organisms, such as algae, fungi, mosses, lichens, plant roots, bacteria, unicellular and microscopic animals, invertebrates, insect larvae, worms, arthropods and their larvae, mammals, reptiles, and amphibians. Interactions of these living things in the forms of symbiosis and food chains result in the formation of soil clusters and pore spaces, the cycling of material through organic matter decomposition, and the modification of soil physiochemistry (Lavelle et al. 2006; Blouin et al. 2013). These processes, sustained over a period of 100 to 1,000 years, result in the accumulation of fertile soil, on which plants and other organisms further flourish. Consequently, soil is one of the natural ecosystems and a part of other natural ecosystems formed by interactions among non-living matter, dead organisms, and living organisms.

However, when the human factor is added to soil as a natural ecosystem, soil begins to take on a different aspect. Humans farm for food, which implies the intentional and systematic use of plant growth processes in the soil ecosystem. At this point, soil can be seen as a means of production. This tendency becomes more pronounced when humans "process" soil to increase production. Tilling the land to encourage crop growth, for example, increases the likelihood of soil loss due to erosion caused by wind and rain. Besides, producing crops on the same land for years depletes the soil. If the distance between farms and consumers is great (and it usually is now), most of the nutrients taken out of the soil as food do not return to the soil, so the soil must be fed in some way to sustain farming. To halt soil degradation, active management and intervention are needed, such as designing farms to reduce soil runoff, leaving the land fallow from time to time, and increasing fertility through fertilization. In this sense, humans "make" soil for food production, and soil, at least in agroecosystems, is no longer just a natural entity, but is also an artificial means that is processed for the purpose of growing plants.

In modern industrial agriculture, which supplies vast amounts of food, soil is regenerated on such a large scale as to affect global nutrients cycles (Rockström et al. 2009; Steffen et al. 2011). In the past, natural materials, typically sourced near farms, were added to soil to increase its fertility. However, to continuously produce food in large quantities, much of the nutrition must now come from outside the agroecosystem. In the nineteenth century, guano (coagulated seabird excrement) was brought from the coasts of Peru and Chile to Europe. It replenished nutrients lost from agricultural land. Later, with the establishment of the Haber–Bosch process for fixing atmospheric nitrogen, artificial fertilizers were applied to the soil. As a result, the amount of global nitrogen fixation has approximately doubled in the subsequent 100 years compared to before industrial nitrogen fixation began with the Haber–Bosch process (Galloway and Cowling 2002; Gruber and Galloway 2008).

These developments were also encouraged by the theory of mineral nutrition of plants and the law of the minimum in agricultural chemistry, as articulated by Carl Sprengel and Justus von Liebig (van der Ploeg et al. 1999). The former is the theory that all plants grow on water, carbon dioxide, and various minerals, not organic matter. The latter is the idea that various minerals must be absorbed in certain quantities for plants to grow, and that if one component is deficient, growth will be poor, even if other nutrients are supplied in abundance. According to this, plant growth is constrained by the mineral nutrient that is given the least. These theories provided a guideline for what materials to use as input during crop production, and their amounts. It encouraged the development and active use of chemical fertilizers to improve soil for production. Today, an important technique and task for farmers is to "build the soil", by devising the composition of fertilizers and the timing of fertilizer application. In this sense, the soil has increasingly become an artificially built means for production.

On the other hand, these theories focused on the soil components necessary for growing plants and made it possible to separate soil and nutrition. Around the mid-nineteenth century, against the backdrop of these theories, in an academic study Julius von Sachs grew plants by immersing their roots in aqueous solutions of various inorganic salts. In the 1930s, William F. Gehrig developed a commercial method of soil-less, nutrient-based cultivation in the United States. Since then, the technology of soil-less agriculture has improved (somewhat intermittently) and, today, highly developed soil-less agriculture is practiced successfully in the Netherlands, Israel, and other countries (Fussy and Papenbrock 2022).

It thus turns out that soil is not only a natural ecosystem but is also an artificial means of production. Humans have improved their skills in processing natural soil into a material suitable for food production. If the epoch of the Anthropocene is characterized by the inevitability of human intervention in the natural environment (Steffen et al. 2011), then soil is one of the parts of nature that most strongly embodies the characteristics of the epoch. The artificialization of soil was also a process whereby we refined our recognition of the growing mechanisms of plants and extracted the essential elements for their growth. As a result, soil has come to be positioned as one among various possible "containers" of nutrients; and the link between food production and soil is no longer as inevitable as it once was. In this context, it is no surprise that there are growing expectations for agricultural technologies that minimize soil use to cope with soil degradation.

17.3 Soil-Less Farming as Isolating Farming

Soil conservation is also possible with soil-based farming. We can change the current problematic soil use (including pollution caused by excessive fertilizer inputs and anthropogenic changes in nutrient cycles) and develop and promote conservation-oriented agriculture such as organic farming. In general, such environmentally friendly soil use is expected to reduce yields; and, in such cases, farmland may need to be expanded to provide sufficient food. But that would not be a problem if the agroecosystem were in better harmony with the other ecosystems. If no-till cultivation, use of cover crops, crop rotation, pesticide-free cultivation, and fertilizer-free cultivation, etc. are practiced with proper care, the soil biodiversity may be maintained. Properly managed rice paddies can provide riparian areas that can serve as habitats for a variety of organisms. The expansion of these agroecosystems can be considered as not hindering the coexistence of living organisms and humans.

Thus, soil-based farming is characterized by its connection with the natural environment. In contrast, less-soil or soil-less farming can separate food production from the external environment, at least to some extent. This characteristic enables it to avoid some drawbacks of soil-based farming. For example, if it is possible to reduce land use for agriculture by increasing yield per land unit via intensification through selective breeding, returning more land and soil to the natural ecosystem becomes more viable. The intention here is to separate soil as a means of production from soil as a natural ecosystem, and to use the former efficiently in agroecosystems for reducing negative impacts on the latter.

The above characteristics are more pronounced in closed-type plant factories. For example, a hydroponic cultivation system inside a building can be constructed and operated at places that are unsuitable for soil cultivation—such as arid regions, sandy soil, salt-affected areas, and cities lined with buildings—provided a small supply of water is available. Food production with an indoor system can be carried out year-round, independent of climate conditions. In addition, because cultivation is carried out in a controlled space using liquid fertilizer and water, heavy labor for soil cultivation including soil preparation, fertilizer application, weed removal, and disinfection for pests and diseases, can be avoided. Soil contamination and water pollution caused by excessive applications of pesticides and fertilizers (whether chemical or organic) can also more easily be controlled by soil-less cultivation because the production process is isolated from the outside.

Yet, although it isolates the production process, soil-less farming does not completely sever its ties from the outside. Various equipment and materials must be introduced from outside to initiate and maintain the production inside. To isolate the production process, it is necessary to build a huge infrastructure, which requires significant capital investment and resource consumption. In addition, to create ideal conditions for indoor agricultural production, it is necessary to prepare heating and cooling facilities, equipment for water and liquid fertilizer supply, and information and communication technology equipment for controlling them, as well as the energy required to keep all this running. The superiority of this technology in terms of environmental compatibility will only become known when a rigorous life-cycle analysis reveals the actual effects of the whole process on the external environment.

17.4 What Soils Are to Be Conserved and for What Purpose?

A quantitative comparison of each of the farming methods described above that only looked at their respective environmental impacts is not enough to determine how to cope with soil degradation. There are two types of soil: soil as a natural ecosystem and soil as a means for food production. The former is closely linked to other ecosystems, both globally and locally, and provides a variety of ecosystem services that support the survival and well-being of humans and other living things now and in the future. The latter contributes to essential food production for human survival and well-being. The two are loosely connected but can be separated in terms of function. The soil conservation approach to be taken depends on which of the two soils is to be conserved and for what purpose. There are several possible arguments for this question, as shown below.

17.4.1 The Anthropocentric Argument

When the issue of soil degradation is linked to food security for present and future human beings, soil is to be conserved for realizing anthropocentric goals. In this case, the focus of soil conservation is to conserve soil as a means for food production. This can be done either with soil-based or soil-less farming. In the former case, soil-based farming in an environmentally adaptable manner (such as properly implemented organic farming) would be promoted for soil conservation. If yield reduction is expected from this form of farming, it would be necessary to expand farmland to provide sufficient food. There are no problems with expanding farmland if that does not result in environmental problems such as biodiversity loss and/or climate change and eventually does not threaten the survival and well-being of humans. However, if the reduction in natural soil has profound consequences for ecosystem services, which in turn hinders sustainable food production, then expanding agricultural land is an inappropriate way to realize a sufficient and sustainable food supply. Moreover, even if the decline of natural soil does not hinder food production, if the ecosystem services provided by natural soil play an essential role in human survival and well-being in other respects, their decline will create difficult trade-offs between food supply and other values.

In the latter case, soil-less farming may ensure the possibility of using soil in the future by using no soil or a limited amount of soil now. In this case, farmland becomes fallow or is abandoned. If it is not properly managed in the meantime, soil improvement may be required when using that preserved soil in the future. If knowledge about soil and techniques for using it are lost while abandoning farming, the soil would not be used properly in the future, even though it would have been preserved. In some cases, it may be necessary to continue cultivation with soil on an appropriate scale to maintain both soil quality and the know-how necessary to ensure such quality.

17.4.2 The Biocentric Argument

On the other hand, if we appreciate the intrinsic value of not only the survival and well-being of humans, but also the survival and well-being of other living things, we ought to reduce the use of soil as a means for food production and restore it to its original state in natural ecosystems. If natural ecosystems contribute to the survival and well-being of non-human living things more than agroecosystems (though livestock or pet animals may prefer the latter), then natural ecosystems have a higher instrumental value for them than agroecosystems. In this case, it would be more ethical to promote new less-soil or soil-less farming technologies.

17.4.3 The Ecocentric Argument

The above argument assumes that the survival and well-being of humans and/or other living things are of intrinsic value, and that natural and artificial soils have an instrumental value in the way they contribute to protecting and promoting that intrinsic value. However, it is also possible to find value in soils that goes beyond their instrumental value. If there is an intrinsic value or if we recognize an intrinsic value in the integrity of natural ecosystems themselves or the flourishment of biodiversity, then natural soil ecosystems are valuable in themselves and should be preserved. In this case, it is unethical to eliminate natural soils by expanding agricultural lands, and we should develop and apply soil-less farming technology for food production.

17.4.4 The Social and Cultural Identity Argument

In contrast, if we believe that soil-based agriculture constitutes our social and cultural identity or our equitable and virtuous social being, then soil-based food production itself has intrinsic value (Thompson 2011). Besides, if farmers and consumers find positive value in growing plants with the productive forces of soil, and this is an essential factor for their good life, then soil-based agriculture itself may have intrinsic value. Food sovereignty, broadly defined as the right of people to control their own food systems including markets, ecological resources, food cultures, and production modes (Wittman 2011), may also be understood as the right to realize this kind of intrinsic value. In these cases, substituting soil-based farming with soil-less farming would lead to the loss of important values, and we would then need to rather promote environmentally friendly agriculture that incorporates soil conservation as an essential component of food production.

The above is just an incomplete list of value arguments when considering what soils are to be conserved and for what purpose. Perhaps more diverse and in-depth value arguments could be developed for soil. For example, if not simply feeding people but supplying foods with certain tastes is profoundly important, then we might have to consider a different technological choice. Large-scale production of grains through soil-less farming is still unrealistic due to the huge costs involved in infrastructure construction and procurement of materials. If grains are essential for satisfying our food preferences including taste, or if they are at the core of our food culture and even constitute our identity (Wright et al. 2001; Guerrero et al. 2009), then, at least under present technological conditions, we should give priority to soil-based farming. Conversely, if fulfillment of such preferences can be deemed to be an extravagance under the current population growth, such as luxury emissions in the argument of climate justice (Shue 1993), then we should put up with soil-less farming to reduce agricultural land use.

Consequently, a quantitative analysis, such as a life cycle analysis, will not provide a decisive answer to the question of what farming measure to choose, unless a reasonable agreement on soil values is also achieved. Nor will there be an answer as to what kind of quantitative analysis should be taken into consideration unless it is clear what soils are to be conserved and for what purpose.

17.5 Technological Incrementalism and Diversity

Even if a consensus on the values of soil can be reached, that will not be enough to decide whether soil-based farming should continue or be replaced by soil-less farming. To consider what farming method should be selected in the face of global soil degradation, we must also take account of technological uncertainties.

It is inevitably uncertain whether the introduction of new technologies will succeed in solving problems. Under such conditions, it is desirable to proceed with incremental decision making through trial-and-error learning, when introducing new technologies (Collingridge 1980, 1992). An effective process of trial and error is possible only when the negative impacts of an error are acceptably small. If the cost is fatal, one will lose everything before learning anything from one's mistakes. With that in mind, it seems that a rapid and massive introduction of new technologies can ruin an effective process of trial and error. To implement a new technology appropriately, it is necessary to proceed with room for correction and turning back or changing direction if it fails (Collingridge 1980, 1992). If, in the face of an uncertain future, we operate a wholesale replacement of known technologies that have so far worked to support our societies with alternative, unfamiliar ones, then at least some ways out will be closed off in case of failure, and the cost of failure will be significant.

The same applies to the issue of soil-based agricultural technology and its alternatives. The idea of solving environmental problems by accelerating technological progress, as advocated by eco-modernism, pursues speedy responses to alleged critical situations. However, we must be aware that going too fast has its own risks. As mentioned earlier, once soil as a means is abandoned and the know-how of soil-based farming is lost, it will take a considerable cost and effort to regain them. If we abandon the soil-based food production option just because a new technology has emerged to replace it, we might be taking a risky gamble and entering a situation with no possibility of turning back. Replacement, if ever done, should be incremental and reversible.

This is not to deny the possibility of solving problems with new technologies. If sustainability is enhanced by leaving a wide variety of opportunities for succeeding generations and reducing the conditions that constrain them (Norton 2015), then we need to have a variety of solutions for human food production (from an anthropocentric perspective). Even a society whose means of agricultural production is fully compatible with the environment would not be highly sustainable if it is the only means. Rather, a society that can avail itself of multiple environmentally compatible means for food production is more sustainable. From the perspective of dealing with technological uncertainty, it is necessary to make the means currently in place more sustainable, while developing and introducing a variety of sustainable alternatives. In this respect, it is desirable to continue soil-based farming while minimizing soil degradation, and to diversify agricultural production options by gradually introducing alternative technologies in areas that are now unsuitable for agriculture or are likely to become increasingly less suitable in the foreseeable future. In this case, soil will not be seen and used as just any easily replaceable tool but will remain an essential basis for agriculture.

17.6 Conclusion

When Ebenezer Howard proposed the Garden City concept in 1898, with cities of up to 32,000 people so as to maintain self-sufficiency and manageable waste recycling (Howard 1898), he did not have to think about new production technologies. Soil was an essential part of the local closed-loop food system in a garden city, in which food was produced and consumed and food residues were returned to farms as fertilizer (Pothukuchi and Kaufman 2000). If Howard were alive today, would he incorporate plant factories into the local food loop?

Today, there is a growing awareness of many problems with the global food system and the rise of urban local agriculture, in which agricultural production is carried out inside and around cities (Di Paola 2017). This has ecological benefits, such as adaptive and mitigative functions to climate change, and social benefits, such as community development, education, social inclusion, job creation, and fostering psychological health (Di Paola 2017). Perhaps, some of these benefits do or can also play an important role in constituting our identity and, in this sense, have or come to have intrinsic value. Urban agriculture often uses soil, and in some cases, large amounts of soil are brought to rooftops for cultivation (Appolloni et al. 2021). Why is soil-based farming so often chosen over soil-less farming even in farm-scarce cities? One of the leading reasons may be the start-up costs of soil-less urban plant factories. However, if the cost issue were resolved, could urban plant factories not offer us the benefits typically thought to be bestowed by their soil-based counterparts? Could production in urban plant factories ever come to constitute our identity?

It may be possible, yes. But perhaps the low cost is not the only reason why soil is preferred in urban agriculture. At least for now, we recognize more diverse instrumental and intrinsic value in growing plants in soil, which we believe cannot be achieved in plant factory production. We are still sensitive to these values.

We are facing soil degradation, and we have means of production that can replace certain functions of the soil. However, soils still have essential values for us and should not be given up easily. If agrarian stewardship implies an ethos that encompasses a wide range of values and virtues related to the long-term sustainability of food production (Thompson 1995), then it includes a responsibility to keep asking ourselves about the values of our most important means of food production, the soil.

References

Appolloni, E., F. Orsini, K. Specht, S. Thomaier, E. Sanyé-Mengual, and G. Pennisi. 2021. The Global Rise of Urban Rooftop Agriculture: A Review of Worldwide Cases. *Journal of Cleaner Production* 296: 126556. https://doi.org/10.1016/j.jclepro.2021.126556.

Asafu-Adjaye, J., L. Blomquist, S.B. Brand, B.W. Brook, R. DeFries, E.C. Ellis, C.A. Foreman, D.A. Keith, M.J. Lewis, M. Lynas, T. Nordhaus, R.A. Pielke, R. Pritzker, J.Roy, M. Sagoff, M. Shellenberger, R.W. Stone, and P. Teague. 2015. *An ecomodernist manifesto*. https://static1.squarespace.com/static/5515d9f9e4b04d5c3198b7bb/t/552d37bbe4b07a7dd69fcdbb/1429026747046/An+Ecomodernist+Manifesto.pdf. Accessed 4 May 2024.

Blouin, M., M.E. Hodson, E.A. Delgado, G. Baker, L. Brussaard, K.R. Butt, J. Dai, L. Dendooven, G. Peres, J.E. Tondoh, D. Cluzeau, and J.-J. Brun. 2013. A Review of Earthworm Impact on Soil Function and Ecosystem Services. *European Journal of Soil Science* 64: 161–182. https://doi.org/10.1111/ejss.12025.

Collingridge, D. 1980. *The Social Control of Technology*. New York: St Martin. Palgrave Macmillan.

Collingridge, D. 1992. *The Management of Scale. Big Organizations, Big Decisions, Big Mistakes*. London and New York: Routledge.

Decaëns, T., J.J. Jiménez, C. Gioia, G.J. Measey, and P. Lavelle. 2006. The Values of Soil Animals for Conservation Biology. *European Journal of Soil Biology* 42: 23–38. https://doi.org/10.1016/j.ejsobi.2006.07.001.

Di Paola, M. 2017. *Ethics and Politics of the Built Environment: Gardens of the Anthropocene*. Springer.

Food and Agriculture Organization of United Nations and the Intergovernmental Technical Panel on Soils. 2015. *Status of the World's Soil Resources (SWSR)—Main Report*. https://www.fao.org/3/i5199e/i5199e.pdf. Accessed 4 May 2024.

Food and Agriculture Organization of United Nations. 2017. *The Future of Food and Agriculture—Trends and Challenges*. Rome. https://www.fao.org/3/i6583e/i6583e.pdf. Accessed 4 May 2024.

Fussy, A., and J. Papenbrock. 2022. An Overview of Soil and Soilless Cultivation Techniques—Chances, Challenges and the Neglected Question of Sustainability. *Plants* 11 (9): 1153. https://doi.org/10.3390/plants11091153.

Galloway, J.N., and E.B. Cowling. 2002. Reactive Nitrogen and the World: 200 Years of Change. *Ambio* 31: 64–71. https://doi.org/10.1579/0044-7447-31.2.64.

Gruber, N., and J.N. Galloway. 2008. An Earth-system Perspective of the Global Nitrogen Cycle. *Nature* 451: 293–296. https://doi.org/10.1038/nature06592.

Guerrero, L., M. D. Guardia, J. Xicola, W. Verbeke, F. Vanhonacker, S. Zakowska-Biemans, and M. Hersleth. 2009. Consumer-driven Definition of Traditional Food Products and Innovation in Traditional Foods. A Qualitative Cross-cultural Study. *Appetite* 52(2): 345–354. https://doi.org/10.1016/j.appet.2008.11.008.

Howard, E. 1898. *To-Morrow: A Peaceful Path to Real Reform.* London: Swan Sonnenschein & Co.

Jenny, H.1994 (1941). *Factors of Soil Formation: A System of Quantitative Pedology*. Dover Publication. INC. New York.

Lavelle, P., T. Decaëns, M. Aubert, S. Barot, M. Blouin, F. Bureau, P. Margerie, P. Mora, and J.-P. Rossi. 2006. Soil Invertebrates and Ecosystem Services. *European Journal of Soil Biology* 42 (Supplement 1): 3–15.https://doi.org/10.1016/j.ejsobi.2006.10.002.

Norton, B.G. 2015. *Sustainable Values, Sustainable Change: A Guide to Environmental Decision Making*. University of Chicago Press.

Pothukuchi, K., and J.L. Kaufman. 2000. The Food System: A Stranger to the Planning Field. *Journal of the American Planning Association* 66 (2): 113–124. https://doi.org/10.1080/01944360008976093.

Rockström, J., W. Steffen, K. Noone, Å. Persson, F. S. Chapin, III, E. Lambin, T. M. Lenton, M. Scheffer, C. Folke, H. Schellnhuber, B. Nykvist, C. A. De Wit, T. Hughes, S. van der Leeuw, H. Rodhe, S. Sörlin, P. K. Snyder, R. Costanza, U. Svedin, M. Falkenmark, L. Karlberg, R. W. Corell, V. J. Fabry, J. Hansen, B. Walker, D. Liverman, K. Richardson, P. Crutzen, and J. Foley. 2009. Planetary Boundaries: Exploring the Safe Operating Space for Humanity. *Ecology and Society* 14 (2): 32. http://www.ecologyandsociety.org/vol14/iss2/art32/

Shue, H. 1993. Subsistence Emissions and Luxury Emissions. *Law and Policy* 15 (1): 39–59. https://doi.org/10.1111/j.1467-9930.1993.tb00093.x.

Steffen, W., J. Grinevald, P. Crutzen, and J. McNeill. 2011. The Anthropocene: Conceptual and Historical Perspectives. *Philosophical Transactions of the Royal Society A* 369: 842–867. https://doi.org/10.1098/rsta.2010.0327.

Thompson, P.B. 1995. *The Spirit of the Soil: Agriculture and Environmental Ethics*. London: Routledge.

Thompson, P. B. 2011. The Ethics of Soil: Stewardship, Motivation, and Moral Framing. In *Sustaining Soil Productivity in Response to Global Climate Change: Science, Policy, and Ethics*, eds. T. Sauer, J. Norman, and M.V.K. Sivakumar, 31–42. Wiley.

United Nations Department of Economic and Social Affairs, Population Division. 2022. *World Population Prospects 2022: Summary of Results*. UN DESA/POP/2022/TR/NO. 3. https://www.un.org/development/desa/pd/sites/www.un.org.development.desa.pd/files/wpp2022_summary_of_results.pdf. Accessed 4 May 2024.

van der Ploeg, R., W. Böhm, and M. Kirkham. 1999. On the Origin of the Theory of Mineral Nutrition of Plants and the Law of the Minimum. *Soil Science Society of America Journal* 63: 1055–1062. https://doi.org/10.2136/sssaj1999.6351055x.

Vollset, S.E., E. Goren, C.W. Yuan, J. Cao, A.E. Smith, T. Hsiao, C. Bisignano, G.S. Azhar, E. Castro, J. Chalek, A.J. Dolgert, T. Frank, K. Fukutaki, S.I. Hay, R. Lozano, A.H. Mokdad, V. Nandakumar, M. Pierce, M. Pletcher, T. Robalik, K.M. Steuben, H.Y. Wunrow, B.S. Zlavog, and C.J.L. Murray. 2020. Fertility, Mortality, Migration, and Population Scenarios for 195 Countries and Territories from 2017 to 2100: A Forecasting Analysis for the Global Burden of Disease Study. *Lancet* 396 (10258): 1285–1306. https://doi.org/10.1016/S0140-6736(20)30677-2.

Wittman, H. 2011. Food Sovereignty: A New Rights Framework for Food and Nature? *Environment and Society* 2 (1): 87–105. https://doi.org/10.3167/ares.2011.020106.

Wright, L., C. Nancarrow, and P. Kwok. 2001. Food Taste Preferences and Cultural Influences on Consumption. *British Food Journal* 103: 348–357. https://doi.org/10.1108/00070700110396321.

Chapter 18
Plants and Global Health Ethics

Daniel Elliot Weissglass

Abstract The One Health revolution has recentered global health discourse so as to emphasize the impact of the non-human on human health, but has so far largely neglected the vegetal. In this chapter, I address the impact of vegetal turns on global health and the ethical issues which surround it. I review three ways that a vegetal turn might emerge in global health: emphasis on the instrumental value of the vegetal, recognition that interactions with the vegetal may inculcate virtues and vices in humans, and consideration that plants may themselves have intrinsic value. I review broadly, admittedly sacrificing depth, to suggest the scope and scale of the impacts that each of these would have to global health and related ethical considerations—presenting a sketch of the space of problems rather than offering solutions. Nonetheless, I hope that this chapter may serve as an effective call to action in illustrating a range of open and important questions at the intersection of the vegetal and global health.

18.1 One Health

The 'One Health' framework emphasizes that the health of human and non-human systems are integrally connected. Though admittedly anthropocentric in interest, One Health requires that we center non-human systems even as we pursue human goals, and has produced proposals for significant intervention into the non-human to protect human health—such as vaccinating wild animals to reduce spread of disease to humans (Monath 2013; Zatechka 2016). These include fairly radical proposals, like using transmissible vaccines, which may spread from animal to animal as though contagious, to allow the vaccination of animal populations which may be hard to otherwise vaccinate (Nuismer and Bull 2020). The imposition of these proposals on the non-human raises ethical concerns, and a growing conversation has developed surrounding these concerns, and the adequacy of existing—anthropocentric—ethical frameworks to protect animal, environmental, and other non-human goods

D. E. Weissglass (✉)
Duke Kunshan University, Kunshan, China
e-mail: dew34@duke.edu

M. Di Paola (ed.), *The Vegetal Turn*, The International Library of Environmental, Agricultural and Food Ethics 38, https://doi.org/10.1007/978-3-031-68671-9_18

(e.g., Barrett and Bouley 2015; Degeling et al. 2015; Johnson and Degeling 2019; Lederman and Capps 2020; Van Herten et al. 2019) upon which they would impose.

The vegetal has remain marginalized in the One Health literature, though this is beginning to change as work increasingly challenges the tendency of One Health discussions to overlook the vegetal—chiefly by emphasizing the instrumental importance of plant health in food security (e.g., Andrivon et al. 2022; Fletcher et al. 2009; Hoffmann et al. 2022; Rizzo et al. 2021). This increasing attention to the vegetal has resulted in a growing range of proposals for human-serving interventions into the vegetal, including invasive proposals like 'molecular farming' which uses modified plants as living factories to produce vaccines, pharmaceuticals, or other chemicals for use in health promotion (Fausther-Bovendo and Kobinger 2021; Rybicki 2017). As with the growth of interest in interventions into other non-human systems, the shift of the vegetal from the margin to the center of health discourse—and health interventions—raises ethical questions which may challenge our current thinking, and perhaps even our broader ethical frameworks. This chapter surveys a range of these questions and challenges by review three, increasingly demanding, ways of construing the value of the vegetal—instrumental, instructive, and intrinsic vegetalisms—and sketching the impact that each would have on structing the ethical discourse surrounding global health. Focusing more on breadth than depth, this chapter will introduce rather than solve problems, aiming to be a call to action to consider seriously questions of vegetal value in One Health ethics.

18.2 Instrumental Vegetalism

It is hard to overstate the role of plants in human health. They are the foundation of the global food system, with more than 80% of human food being plant-based and much of the rest derived from plant-fed livestock (Fletcher et al. 2009; Rizzo et al. 2021). Plants, especially in the form of traditional medicines, are critical to the health care system of up to 80% of the worldwide human population (Hao and Xiao 2020), and roughly 25% of medicines sold are plant-derived (Calixto 2019). The importance of plants to food and healthcare places the vegetal at the center of ongoing global health crises. In some cases, these crises are directly linked to human deprivation of vegetal goods. For instance, child and maternal malnutrition remains one of the greatest threats to human health worldwide (GBD 2019 Risk Factor Collaborators 2020) and nearly 2 billion people lack access to essential medications, many of which are exclusively plant-derivatives (Chattu et al. 2023; Raskin et al. 2002). Other crises result from the consumption of harmful plant products. Tobacco and alcohol—predominantly plant derived—are particularly notable for the grievous harms to human health they have caused worldwide (GBD 2019 Risk Factor Collaborators 2020; He et al. 2022). Still other crises have non-vegetal causes, but will likely require turning to the vegetal for solution. As rising antimicrobial resistance threatens to make existing medicines less effective (World Health Organization

2015), plants are perhaps the most promising source of new medicines, with plant-derived compounds feature prominently across a wide range of medical applications (Bergonzi et al. 2022).

Instrumental vegetalism in global health recognizes the unique and extensive impact of the vegetal qua vegetal as a determinant of health outcomes. It does not ask us to look beyond the conventional treatment of plants chiefly as means to ends—but nonetheless centers plants in global health discourse in a way requiring special attention to the nature and context of the vegetal. We will chiefly encounter familiar conflicts balancing safety and freedom, equity and access, etc. colored, but not changed in general outline, by the vegetal turn. When we pay attention to what plants are, and the way they are situated in biological and social environments, aspects of the problem which are otherwise easily overlooked become more obvious. The importance, for instance, of plants to social practices and ways of life demands that we consider more deeply how health directed interventions might impact our relationship with the vegetal. That is, while instrumental vegetalism construes of the vegetal still chiefly as a tool, and therefore does not grant it any special moral regard, it would require recentering global health practice—and our corresponding global health ethics discourse—to consider plants more seriously than we might otherwise.

One of the most vital areas of health policy targeting the vegetal is the protection of the global agricultural system. The impact of agricultural disruption has been tragically demonstrated during the Great Famine of Ireland, caused in large part by a potato fungus, which resulted in estimated one and a half million excess deaths through starvation and disease and a similar number of refugees forced to seek new life abroad, reducing the total population of Ireland by a third (Powderly 2019). It is no exaggeration to say that plant health played a key role in permanently reshaping the experience of a people and the shape of their culture. Unsurprisingly, considerable effort is made to prevent such tragedies from happening again, but the measures used bring with them ethical concerns of their own. The use of agrochemicals to protect against pests and diseases is important to maintaining food security, but may endanger the health of agricultural workers and risk environmental damage (Fears et al. 2014; Hoffmann et al. 2022). Mandated destruction of crops that threaten the larger agricultural systems limits global risk, but places a large burden on individual agriculturalists—and the general lack of recompense for that burden is a notable difference from livestock management practices (Wilkinson et al. 2011). Import/export policies that prevent the spread of plant disease and/or invasive species impose on freedom of trade and travel at the discretion of national governments, resulting in both unjustified barriers and failures to adapt to emerging problems of globalization, climate change, etc. (Fears et al. 2014). Fully assessing these issues will require careful attention to both the nature of the vegetal and the way that it has been embedded in human cultures. The use of agrochemicals, for instance, may be influenced by genetic modification—making these chemicals more effective or less necessary (Fears et al. 2014)—which means that direct interventions into vegetal natures may provide a path to modifying health outcomes for agricultural workers and the environment at large, but not without risks that require careful understanding of the vegetal and its role in the environment and human society to assess. Understanding the impact of agricultural policies, similarly, will require understanding what agriculture means to agriculturalists and their communities, the social systems in which they exist, and the political structures which support them. These issues, that is, cannot be adequately addressed without serious attention given to the vegetal as such.

The same is true with policies which limit the production and distribution of plant products motivated by their health effects. The significance and scale of such policies is perhaps best demonstrated by the fact that one of three World Health Organization documents with legal force is the Framework Convention on Tobacco Control, which is designed specifically to limit the harmful effects of the consumption of (some particular) plant products (World Health Organization 2003). Individual jurisdictions, likewise, often place regulations on the consumption of tobacco, (disproportionately plant based) alcohol, plant-derived narcotics, etc. While many of these restrictions may well be justified by their impact on human health outcomes, we should note that they impose sharp restrictions on various freedoms—both market freedoms relating to free enterprise and personal freedoms regarding home manufacture. Bans or restrictions on cultivation may alter the dynamics of human environment interactions, agricultural communities, and traditional practices; the vegetal—both in general and particular form—and its integration into natural and social ecosystems must be carefully understood to answer these questions. When we consider that plants which may be targeted by such regulation (e.g., opium, coca) have long histories of cultivation and use which have shaped both human and ecological systems, the disruption of which may variously impact those systems, it becomes clear that these regulations must be considered in light of the features of plants in general, and the regulated plants in particular.

Similar concerns arise due to the restrictions imposed by intellectual property practices, already subject to criticism in global health for restricting access to essential medicines (Bhalotra and Pogge 2014). The ownership of plants is likely as old as agriculture, and plant derivatives have long been subject to patent, but patents on genetically modified plants result in the ownership of all possible instances of a kind of plant. This has disrupted practices historically essential to agriculture, like retaining seed for replanting, as large biotechnology firms have successfully sued small farmers to prevent these practices which may threaten their monopoly (Harris 2013; Peschard 2019). As GMO crops have been important bulwarks against global hunger (Roberts 2018), these practices of ownership are not only historically anomalous but may leave many needlessly without. Response to particular egregious forms of this practice, through which firms claim ownership of plants in ways which deny access to variants or derivatives of indigenous plant species to indigenous communities, has required coining the term 'biopiracy' (Then et al. 2012). Biopiracy is perhaps most apparent in global health when we consider the roughly 80% of plant-derived medicines which are used to achieve effects indicated by traditional use (Veeresham 2012), where indigenous knowledge was an essential and uncompensated component of drug discovery. Increasingly complex relationships between human innovation, plant products, and medicine—such as the modification of plants for the express purpose of biomanufacturing pharmaceuticals or their precursors (Fausther-Bovendo and Kobinger 2021; Gupta et al. 2021)—make questions of ownership and its limits even more difficult to navigate. The need to coin a new term, 'biopiracy', to make sense of this phenomena draws attention to the importance of the nature of plants as living things, and surrounding discourse of indigeneity emphasizes that understanding the vegetal as having its own sort of indigeneity—one which stands in relation to indigenous systems of knowing and cultural practices—will be necessary to parse these issues.

The problem of access to the vegetal can be generalized further by considering the number of positive impacts that exposure to plants and plant dominated spaces may have on human health. There is good evidence of instrumental value of spending time around plants and plant-spaces. Gardening appears to be protective of health, and community gardens are associated with a variety of social and personal benefits. Personal and community gardening also seem to encourage healthy eating and activity, improve various health indicators (e.g., cortisol, BMI and body fat, etc.), mental health, quality of life, and overall subjective well-being (Howarth et al. 2020). Home agriculture is also an important tool for achieving food security, especially among low-income populations and/or during periods of instability (e.g., the recent COVID-19 pandemic) (McMahan et al. 2014; Turnšek et al. 2022). In fact, mere exposure to green spaces appears related to better overall health, higher activity levels, lower premature mortality, etc. However, access to gardens and other green spaces is not equitably distributed—with members of marginalized groups often having less access than others(Rigolon et al. 2021; Wolch et al. 2014)—and the time, land, and resources required to engage in substantive gardening is not equally accessible to everyone. Understanding the conflicts that this raises requires us to first seek to understand the relationship between humans and plants, and the role that plant cultivation plays in human wellbeing. It is possible, of course, that this is simply a mechanical benefit due to activity and diet, but the possibility of there being something unique to the relationship of the human and the vegetal will bring us to a deeper view of the value of the vegetal.

18.3 Instructional Vegetalism

Plants are tools, but they may also be partners in important relationships such that through the cultivation of plants we cultivate ourselves and our societies. So-called 'agrarian' views suggest that agricultural practice—especially in traditional and intimate form—evoke a wide range of virtues: self-reliance, independence, integrity, honesty, awe of nature and/or the divine (Thompson 2008). Likewise, some have suggested that caring for plants might serve as a practice which supports the development of virtues essential for personal wellbeing (O'brien 2013), healthy relationships (Schörgenhumer 2018), or environmental stewardship (Paola 2015).

Instructional vegetalism holds that relating properly to plants allows us to develop important virtues, and so will have much to say about what these relationships ought to look like. It is hard to specify precisely the ideal nature of this relationship without taking a particular position, but it is fairly easy to see the sort of concerns that most varieties of instructional vegetalism may have in common about our current practices. The precipitous decline in professional agriculturalists worldwide (from 43% of the world's working population in 1991 to 26% in 2024 (The World Bank 2024)) implies that far fewer of us are connected to agricultural production than in the past. At the same time, the character of that production has changed. Contemporary industrialized agriculture operates on mass scale with a long-run trend of consolidation resulting in agriculture leaving less and less room for the small farmer, increasing specialization promotes the isolated cultivation of particular varieties of particular crops for particular purposes in the place of once varied crops, and technological changes bring about increasing automation and mechanization (Cameron 2006; Janssen 2018; MacDonald 2020; Mehrabi 2023).

The result of these changes may be a profound and historically anomalous alienation from the vegetal, even among those employed in agricultural fields. The modern agriculturalist, certainly more than in the past, may be involved only in one aspect of the cultivation of a particular kind of crop, managed chiefly through large machinery and from which they operate at a relative distance. The comparatively intimate process of working alongside plants in planting, growing, and harvesting—where agriculture would imply building relationships with the plants one grew and the land on which you grew them—has been disrupted by industrial processes, and the relationship of the agriculturalist to the vegetal may have been permanently changed as a result. To the extent that virtues like patience, care, and perseverance are honed through such intimate cultivation, and left behind by the industrial, we might worry that contemporary agricultural practices, for all their virtues, bring out our vices. By changing the relationship between agriculturalists and the vegetal, industrial agriculture may have impaired the development of culturally and personally important virtues. Likewise, if cultivating plants is a key to cultivating ourselves, the increasingly many, especially in urban environments, who have been deprived of significant opportunities to develop vegetal virtues are of great concern.

Instructional vegetalism may sometimes coincide with instrumentally useful practices, but the tendency of instructional vegetalism to refer chiefly to intimate models of vegetal-personal cultivation suggests that conflicts may also be common. We would need to think beyond the safety of contemporary agricultural practices for the physical health of agriculturalists and consumers to consider also their spiritual and moral health. Automation may improve yield, but elicit a sort of mindless treatment of the vegetal as product. Accelerated growth through hydroponics or genetic modification might teach us impatience. Genetic modification might yield hubris. If these concerns bear out, then virtue vegetalism would give us good reasons to be concerned about the practices of industrial agriculture and might warrant a rethinking of those practices so as to reduce alienation and increase virtuous relationships between agriculturalists and their plants.

One path likely to be recommended by instructional vegetalism is the encouragement of as many people as possible to enter into caring relationships with plants: the promotion of personal and community gardens, making public green spaces political priorities, and lionizing the houseplant. In the process, we should note that engagement with the vegetal is not the exclusive product of individual choice. Whether one has access to the circumstances which permit the inculcation of vegetal virtues is determined largely by social forces beyond the individual's control; socioeconomic, climate, and political factors will play significant roles in determining how one is able to engage with the vegetal. If being unable to participate in meaningful care of plants is a hazard to our moral virtue, then systems which distance us from the vegetal are both objectionable for the harms they cause and exculpatory of the less-virtuous behavior it brings about in those distanced. In context of global health, this will be closely related to contemporary debates about individual responsibility for health outcomes. While many health outcomes are the product of lifestyle choices—what we eat, drink, and do—our responsibility for those choices seems to be limited by the impact of the so-called social determinants of health, a set of socially determined influences (e.g., wealth, education, etc.) which have remarkable impact on health outcomes. If we take this limit to be strong enough, it becomes a decisive argument against policies which make access to care or aid conditional on one's compliance with desirable health outcomes—and a similarly decisive argument in favor of addressing limits in the social determinants so as to address the injustices therein (Friesen 2018; Ruger 2004). This will be especially the case if our vegetal virtues include precisely those features—like patience and fortitude—which may be essential in achieving long-term health goals.

18.4 Intrinsic Vegetalism

Instructional vegetalism posits a clear moral significance for plants beyond their practical value, but this value is still rooted in their relationships with humans. A more radical way of valuing the vegetal is one which finds value in the vegetal itself rather than in how they affect other bearers of value. Such a view posits intrinsic value to the vegetal, and calls us to consider the impacts of interventions on plants themselves as of moral significance independent of their impact on others.

Intrinsic vegetalism is most commonly supported by arguments that suggest some property which is generally understood to be a source of value (e.g., life, sentience, intelligence, dignity, goal orientation, etc.) (Attfield 2005; Federal Ethics Committee on Non-Human Biotechnology (ECNH) 2008; Pellegrino 2018) is found in plants, thereby suggesting that plants warrant consideration as value-bearers. These arguments are fundamentally analogical—plants are like something else of paradigmatic value (usually humans) in ways related to the reasons for that value and are therefore (likely) valuable themselves. While I will not be able to dive into these arguments in depth, it is worth noting that they tend to suffer two common problems: over-extension and misdescription. Over-extension occurs when the purportedly value-conferring property cited is present in things which seem to lack intrinsic moral value, suggesting that it is a poor ground for value. Mere life, for instance, as a ground for intrinsic value would dubiously grant intrinsic value to bacteria, and goal-orientation as a ground would grant value to sufficiently complex thermostats. Proponents of such arguments (e.g., Sandler 2018) have reminded us that counter-intuitive conclusions can still be correct, and that we should not mistake absurdity for error. Bacteria and thermostats may simply matter in surprising ways. While this is surely correct, it also misunderstands the problem—which is not the ad hoc rejection of the inclusion of a particular kind of thing, but taking the inclusion of paradigmatically unvaluable (or even disvaluable) things by some standard of value as a good indicator that the standard is a poor one. This is especially plausible when the proposed property—such as life simpliciter or goal-orientation understood this broadly—has no clear moral valence. The over-extension objection is strong, then, except in cases where the extended property is of clear moral value—but these cases seem to run into misdescription problems. Misdescription occurs when the proposed extension relies on properties which it is difficult to demonstrate are appropriately applied to plants—sentience and subjectivity seem good candidates here. The challenge to be addressed here is to demonstrate that plants do, indeed, possess the extended property, which is not easily done.

Many such arguments operate through ambiguity which when clarified in one way results in over-extension and when clarified in another results in misdescription. Such arguments present a property which is under one description present in plants, under another of obvious moral value, but under no description both. Consider plant communication. Plants do in fact send signals to one another in various ways, signals which allow them to coordinate behavior and adaptively respond to environmental conditions. When humans communicate, we often involve various cognitive and phenomenal capacities that are likely of obvious moral value—self-concept, self-interest, experience of (dis)satisfaction, etc. An ambiguously erroneous argument for intrinsic plant value, based on this, would effectively equivocate the two sorts of communication—the one a biological pattern of signaling, the other a product of a sentient and aware cognitive system. However, there is no reason to suppose that all communication of the first kind is of the second, and this is perhaps nowhere better indicated than by the fact that humans engage in the first—through unintentional nonverbal cues, pheromones, etc.—without also engaging in the second. Such arguments, it turns out, will often depend on an ironic anthropocentrism—the presumption that all tasks are completed by all systems in ways similar to their paradigmatic completion in humans.

An alternative approach to establish intrinsic vegetalism is through insistence—possibly brute but not unreasoned—that plants simply have such value. We can see parallels in discussions of moral status with respect to humans, here in the insistence that humans are simply valuable—even if deprived of sentience, agency, life, intelligence, and all other properties which might ground such claims. Something brute of this kind, for instance, seems to be at play in some uses of the concept of human dignity—for instance in explaining the wrongness of using the head of a human corpse to play football (Foster 2015). We might approach a vegetalism of this kind by appropriating Dworkin's conception of 'inviolable value', a sort of value intrinsic to an object or being which is not increased by bringing about more of the same (i.e., in the way that we might wish always to bring about more knowledge, beauty, or pleasure in virtue of some intrinsic value of these things). Grounded as it is in the origin of the valued, inviolable value could be used to imbue value in plants a particular value "[…] as existing, visible and infinitely differentiated outcomes of the investment of nature, of the millennial evolution and adaptation of natural life on Earth." It is the particular identity and history that would make a plant valuable on such a view, and not any possession of general value-conferring properties (Pellegrino 2018, 21).

The fundamental and unsurprising challenge of this approach is precisely that it provides no path to arrive at this value independent of its acceptance, requiring that we apprehend the unique value of the vegetal somehow more directly. We cannot begin in familiar and uncontroversial cases and travel by easy steps along the path, and it is accordingly unclear how likely this is to be successful either in convincing the uncommitted or in actually establishing the moral status of plants.

With acknowledgements made to the general sorts of arguments given for intrinsic vegetalism and the objections they face, it is worth considering what the implications of intrinsic vegetalism to global health practice would be were we to adopt it, for which it will be helpful to consider both a broadly deontological, rights-based approach[1] and a consequentialist, welfare[2] based approach. Both will depend in content on the grounds of the moral status of plants: if plants bear rights due to sentiency, then their rights will be those connected to sentiency; if plants are welfare subjects due to their possession of certain kinds of ends, then their welfare will need to be understood in terms of those ends. However, to avoid getting caught in the theoretical thickets surrounding the moral status of plants, I will adopt a simplifying anthropomorphism that is often at play in the background of such views. That is, I will take plants to be subject to rights and/or welfare considerations in ways that roughly align with our intuitions about human cases, so that we can meaningfully consider possible consequences of intrinsic vegetalism.

However, it is worth belaboring the obvious point that plants are not human, and to acknowledge that this anthropomorphism is in some sense a betrayal of the vegetal turn's commitment to understanding plants as they are. Plants are radically different from us in both capacities and vulnerabilities, and their alienness makes it questionable how well our intuitions—by and large formed with respect to the human case—apply to them. If the capacities, interests, and ends of plants are simply inconceivable to us, if they experience nothing like what we know as pleasure or pain, and if they would demand of the world things which would never occur to us, then we may have little hope of making sense of the demands of intrinsic vegetalism. This is both another challenge for intrinsic vegetalism—which seems to force a choice between anthropomorphism or incomprehensibility—and a reason for anthropomorphisms. The anthropocentrism of our moral intuitions may render vegetal value

[1] I will presume the Hofeldian presumption that the existence of a right implies a correlative duty, but all of what follows should be translatable into systems which make no such assumption.

[2] Plant welfare might be understood—admittedly with objections of undue anthropomorphism (see Višak 2018)—in familiar terms of happiness or preferences, or in ways which make fewer controversial assumptions of plant capacities like biocentric consequentialism's emphasis on the achievement of the peculiar goods of living things (Attfield 2005, 2006).

incomprehensible to us unless we engage in anthropomorphism, which—though it runs roughshod over the uniqueness of the vegetal—will allow us to articulate a plausible account of various moral obligations that we may have to plants on some—though not all—reasonable—though not necessarily correct or consistent—conceptions of their intrinsic value. For the objectives of this chapter, at the very least, this will be useful.

On most reckonings, intrinsic vegetalism raises a number of serious issues for everyday practices, important to global health, which appear to sacrifice the rights or welfare of plants to benefit others. This can take many forms. Our agricultural practices, for instance, tend to be anthropocentric, and it is likely that we will find choices we make violate plausible accounts of the rights or welfare of the vegetal. Clearing local flora to make space for farms, pruning plants to increase their yield, etc. all seem questionable. We'll also need to think about plant-animal interfaces, considering the way that interventions on animal populations can impact plant populations—demonstrated vividly by the impact of reintroducing wolves to Yellowstone National Park in the United States as a cause of restoring willow thickets[3] (Painter and Tercek 2020)—which would complicate animal culls, reintroductions, etc. Of course, there will also be plant-plant conflicts, such as when the growth of one plant threatens another, or a diseased plant might infect others—and these conflicts are made more difficult to assess by the complex interdependence of humans, animals, plants, and other value-bearers. The range of decisions which will be affected by such considerations is likely quite broad: quarantines and travel/work restrictions may leave crops untended, hospital construction may involve tearing up local flora, harvesting medicines may harm the plants from which they are harvested, etc.

While a significant departure in terms of the scale and scope of our moral concerns, the task faced here is not unfamiliar—as balancing values against one another is perhaps the fundamental task of applied ethics. We would, however, have to answer some challenging questions about the relative weight of the values considered. We might default to prioritizing the human or animal over the vegetal, in which case intrinsic vegetalism will amount to relatively little in practice as the values of others will rarely be overridden by those of plants. If, however, we take the vegetal to have more than this nominal value, such that the rights or welfare of plants can be expected to reasonably alter moral conclusions, we will find ourselves in quite unfamiliar territory. Balancing is made especially complicated by the fact that we must often compare costs and benefits of different kinds. Consider regulations preventing the planting of genetically modified crops by those who have not paid for the right to do so, and the resulting lawsuits filled against farmers by large biotechnology conglomerates when seeds have blown into their farmlands by natural process. While these cases already raise a number of questions about intellectual property, corporate ethics, and wealth disparity—the possibility of the growing plant itself possessing rights raises another sort of question: can the owner of genetically modified organism rightfully demand such a plant to be destroyed or relocated, or would doing so violate

[3] Though, of course, we'd need also to consider the Elk, the predation of which by wolves is thought to be responsible for this effect.

the rights or welfare of the plant itself? In such a case, we are not concerned only with life or health, but also with the moral foundations of property, the goods and harms of GMO crops, and a range of other issues.

Intrinsic vegetalism would have effects too far reaching to address in a short space, but we can take some likely conclusions to be the need to take plant health seriously in our global health thinking, to emphasize the need for vegetal medicine which seeks to cure rather than cull ill plants, etc. Beyond this, it is worth drawing a few conclusions which may be unique to either deontological or consequentialist reckonings. Deontologists (at least of certain cultivars) will likely oppose our instrumentalization of the vegetal for failing to recognize plants as ends in themselves, and will demand that we think seriously about the way that the dignity or rights of the vegetal may be offended by common agricultural practices; genetic modification and eugenic selective breeding, hybridization, grafting and budding—which fuse the living tissues of multiple plants—pruning, weeding, and all those practices which subject the vegetal to manipulation, dissection, or death for purposes which are not their own will be highly suspect. Consequentialists will, in contrast, raise concerns about valorized trends—like organic farming—which seem to reject tools and techniques at first glance beneficial to plants, and may ask us to further those practices offensive to 'plant dignity' when they improve the welfare of plants—supporting the very manipulations the deontological view would find offensive. Of course, the classic challenges of each view—the need to identify the boundaries of rights, the stringency of the consequentialist lack of supererogation, etc.—will remain questions here as well. These, however, could only be resolved in context of particular theories of this value, which I will leave to proponents of intrinsic vegetalisms to specify.

18.5 Conclusion

In this chapter, I have sketched various consequences for global health ethics of centering the vegetal as a bearer of instrumental, instructional, and intrinsic value. I have considered these in what I take to be an order from least to most radical—with instrumental vegetalism implying only a shift in focus, instructional vegetalism a somewhat romanticized but neither unfamiliar nor counterintuitive view of plant cultivation, and intrinsic vegetalism a fundamental rethinking of our ethical practices in global health. There is much more to be said about each of these views than can be said here, and exploring these approaches in context of a One Health ethic will be a task for future work, but it is my hope that this chapter succeeds at least in suggesting the scope and scale of the impacts that the vegetal turn may have for global health.

References

Andrivon, D., J. Montarry, and S. Fournet. 2022. Plant Health in a One Health World: Missing Links and Hidden Treasures. *Plant Pathology* 71 (1): 23–29. https://doi.org/10.1111/ppa.13463.

Attfield, R. 2005. Biocentric Consequentialism and Value-Pluralism: A Response to Alan Carter. *Utilitas* 17 (1): 85–92. https://doi.org/10.1017/S0953820804001414.

Attfield, R. 2006. The Shape of a Global Ethic. *Philosophy & Social Criticism* 32 (1): 5–19. https://doi.org/10.1177/0191453706059843.

Barrett, M.A., and T.A. Bouley. 2015. Need for Enhanced Environmental Representation in the Implementation of One Health. *EcoHealth* 12 (2): 212–219. https://doi.org/10.1007/s10393-014-0964-5.

Bergonzi, M.C., C.M. Heard, and J. Garcia-Pardo. 2022. Bioactive Molecules from Plants: Discovery and Pharmaceutical Applications. *Pharmaceutics* 14: 2116–2020. https://doi.org/10.3390/pharmaceutics14102116.

Bhalotra, S., and T. Pogge. 2014. Ethical and Economic Perspectives on Global Health Interventions. In *The Handbook of Global Health Policy*, eds. G.W. Brown, G. Yamey, and S. Wamala, 1st Edition, 209–227. John Wiley & Sons.

Calixto, J.B. 2019. The Role of Natural Products in Modern Drug Discovery. *Anais Da Academia Brasileira De Ciências* 91 (Supplement 3): e20190105. https://doi.org/10.1590/0001-3765201920190105.

Cameron, H.M.G. 2006. Trends in Agriculture and Agribusiness: Knowledge Is Power. *Journal of Business & Finance Librarianship* 12 (1): 3–32. https://doi.org/10.1300/J109v12n01_02.

Chattu, V.K., B. Singh, S. Pattanshetty, and S. Reddy. 2023. Access to Medicines Through Global Health Diplomacy. *Health Promotion Perspectives* 13 (1): 40–46. https://doi.org/10.34172/hpp.2023.05.

Degeling, C., J. Johnson, I. Kerridge, A. Wilson, M. Ward, C. Stewart, and G. Gilbert. 2015. Implementing a One Health Approach to Emerging Infectious Disease: Reflections on the Socio-political, Ethical and Legal Dimensions. *BMC Public Health* 15 (1). https://doi.org/10.1186/s12889-015-2617-1.

Fausther-Bovendo, H., and G. Kobinger. 2021. Plant-made Vaccines and Therapeutics. *Science* 373 (6556): 740–741. https://doi.org/10.1126/science.abf5375.

Fears, R., E.-M. Aro, M.S. Pais, and V. ter Meulen. 2014. How Should We Tackle the Global Risks to Plant Health? *Trends in Plant Science* 19 (4): 206–208. https://doi.org/10.1016/j.tplants.2014.02.010.

Federal Ethics Committee on Non-Human Biotechnology (ECNH). 2008. *The Dignity of Living Beings with Regard to Plants: Moral Consideration of Plants for Their Own Sake.*

Fletcher, J., D. Franz, and J.E. LeClerc. 2009. Healthy Plants: Necessary for a Balanced 'One Health' Concept. *Veterinaria Italiana* 45 (1): 79–95.

Foster, C. 2015. Human Dignity in Bioethics and Law. *Journal of Medical Ethics* 41 (12): 936–937. https://doi.org/10.1136/medethics-2013-101466.

Friesen, P. 2018. Personal Responsibility Within Health Policy: Unethical and Ineffective. *Journal of Medical Ethics* 44 (1): 53–58. https://doi.org/10.1136/medethics-2016-103478.

GBD 2019 Risk Factor Collaborators. 2020. Global Burden of 87 Risk Factors in 204 Countries and Territories, 1990–2019: A Systematic Analysis for the Global Burden of Disease Study 2019—The Lancet. *The Lancet* 396 (10258): 1223–1249.https://doi.org/10.1016/S0140-6736(20)30752-2.

Gupta, D., G. Sharma, P. Saraswat, and R. Ranjan. 2021. Synthetic Biology in Plants, a Boon for Coming Decades. *Molecular Biotechnology* 63 (12): 1138–1154. https://doi.org/10.1007/s12033-021-00386-9.

Hao, D., and P. Xiao. 2020. Pharmaceutical Resource Discovery from Traditional Medicinal Plants: Pharmacophylogeny and Pharmacophylogenomics. *Chinese Herbal Medicines* 12 (2): 104–117. https://doi.org/10.1016/j.chmed.2020.03.002.

Harris, P. 2013, February 12. Monsanto Sued Small Famers to Protect Seed Patents, Report Says. *The Guardian.* https://www.theguardian.com/environment/2013/feb/12/monsanto-sues-farmers-seed-patents.

He, H., Z. Pan, J. Wu, C. Hu, L. Bai, and J. Lyu. 2022. Health Effects of Tobacco at the Global, Regional, and National Levels: Results From the 2019 Global Burden of Disease Study. *Nicotine & Tobacco Research* 24 (6): 864–870. https://doi.org/10.1093/ntr/ntab265.

Hoffmann, V., B. Paul, T. Falade, A. Moodley, N. Ramankutty, J. Olawoye, R. Djouaka, E. Lekei, N. Haan, P. de Ballantyne, and J. Waage. 2022. A One Health Approach to Plant Health. *CABI Agriculture and Bioscience* 3 (1). https://doi.org/10.1186/s43170-022-00118-2.

Howarth, M., A. Brettle, M. Hardman, and M. Maden. 2020. What is the Evidence for the Impact of Gardens and Gardening on Health and Well-being: A Scoping Review and Evidence-based Logic Model to Guide Healthcare Strategy Decision Making on the Use of Gardening Approaches as a Social Prescription. *British Medical Journal Open* 10 (7): e036923. https://doi.org/10.1136/bmjopen-2020-036923.

Janssen, B. 2018. Small Farms, Big Plans: Mechanization and Specialization as Measures of "The Middle." *Culture, Agriculture, Food and Environment* 40 (2): 96–104. https://doi.org/10.1111/cuag.12221.

Johnson, J., and C. Degeling. 2019. Does One Health Require a Novel Ethical Framework? *Journal of Medical Ethics* 45 (4): 239–243. https://doi.org/10.1136/medethics-2018-105043.

Lederman, Z., and B. Capps. 2020. One Health Ethics: A Response to Pragmatism. *Journal of Medical Ethics* 46 (9): 632–633. https://doi.org/10.1136/medethics-2019-105859.

MacDonald, J. M. 2020, February 3. Consolidation in U.S. Agriculture Continues. *USDA Economic Reseaarch Service: Farm Economy.* https://www.ers.usda.gov/amber-waves/2020/february/consolidation-in-u-s-agriculture-continues/.

McMahan, M., E.J.D. Richey, and A. Tagtow. 2014. Cultivating Food Security: Low-Resource Lowans' Perspectives on Food Gardening. *Journal of Hunger & Environmental Nutrition* 9 (3): 299–317. https://doi.org/10.1080/19320248.2014.929548.

Mehrabi, Z. 2023. Likely Decline in the Number of Farms Globally by the Middle of the Century. *Nature Sustainability* 6 (8): 949–954. https://doi.org/10.1038/s41893-023-01110-y.

Monath, T.P. 2013. Vaccines Against Diseases Transmitted from Animals to Humans: A One Health Paradigm. *Vaccine* 31 (46): 5321–5338. https://doi.org/10.1016/j.vaccine.2013.09.029.

Nuismer, S.L., and J.J. Bull. 2020. Self-disseminating Vaccines to Suppress Zoonoses. *Nature Ecology & Evolution* 4 (9): 1168–1173. https://doi.org/10.1038/s41559-020-1254-y.

O'brien, D. 2013. Gardening and Virtue. *Journal of Therapeutic Horticulture* XXIII (II): 19–27.

Painter, L.E., and M.T. Tercek. 2020. Tall Willow Thickets Return to Northern Yellowstone. *Ecosphere* 11 (5): e03115. https://doi.org/10.1002/ecs2.3115.

Paola, M.D. 2015. Virtues for the Anthropocene. *Environmental Values* 24 (02): 183–208. https://doi.org/10.3197/096327114X13947900181310.

Pellegrino, G. (2018). The value of plants: On the axiologies of plants. In A. Kallhoff, M. Di Paolo, & M. Schörgenhumer (Eds.), *Plant Ethics: Concepts and applications.* Routledge.

Peschard, K.E. 2019. *Monsanto Wins $7.7b Lawsuit in Brazil—But Farmers' Fight to Stop its 'Amoral' Royalty System Will Continue.* The Conversation. http://theconversation.com/monsanto-wins-7-7b-lawsuit-in-brazil-but-farmers-fight-to-stop-its-amoral-royalty-system-will-continue-125471

Powderly, W.G. 2019. How Infection Shaped History: Lessons from the Irish Famine. *Transactions of the American Clinical and Climatological Association* 130: 127–135.

Raskin, I., D.M. Ribnicky, S. Komarnytsky, N. Ilic, A. Poulev, N. Borisjuk, A. Brinker, D.A. Moreno, C. Ripoll, N. Yakoby, J.M. O'Neal, T. Cornwell, I. Pastor, and B. Fridlender. 2002. Plants and Human Health in the Twenty-first Century. *Trends in Biotechnology* 20 (12): 522–531. https://doi.org/10.1016/S0167-7799(02)02080-2.

Rigolon, A., M.H.E.M. Browning, O. McAnirlin, H. Yoon, and (Violet). 2021. Green Space and Health Equity: A Systematic Review on the Potential of Green Space to Reduce Health Disparities. *International Journal of Environmental Research and Public Health* 18 (5): 2563. https://doi.org/10.3390/ijerph18052563.

Rizzo, D. M., M. Lichtveld, J.A.K. Mazet, E. Togami, S.A. Miller, and Link to external site, this link will open in a new window. 2021. Plant Health and its Effects on Food Safety and Security in a One Health Framework: Four Case Studies. *One Health Outlook* 3: 1–9.https://doi.org/10.1186/s42522-021-00038-7.

Roberts, R.J. 2018. The Nobel Laureates' Campaign Supporting GMOs. *Journal of Innovation & Knowledge* 3 (2): 61–65. https://doi.org/10.1016/j.jik.2017.12.006.

Ruger, J.P. 2004. Ethics of the Social Determinants of Health. *The Lancet* 364 (9439): 1092–1097. https://doi.org/10.1016/s0140-6736(04)17067-0.

Rybicki, E.P. 2017. Plant-made Vaccines and Reagents for the One Health Initiative. *Human Vaccines & Immunotherapeutics* 13 (12): 2912–2917. https://doi.org/10.1080/21645515.2017.1356497.

Sandler, R. 2018. Is Considering the Interests of Plants Absurd? In *Plant Ethics: Concepts and Applications*, eds. A. Kallhoff, M. Di Paolo, and M. Schörgenhumer. Routledge.

Schörgenhumer, M. 2018. Caring for Plants: Cultivating Relational Virtues. In *Plant Ethics: Concepts and Applications*, eds. A. Kallhoff, M. Di Paolo, and M. Schörgenhumer. Routledge.

The World Bank. 2024. *World Bank Open Data*. https://data.worldbank.org.

Then, C., R. Tippe, Arbeitsgemeinschaft Bäuerliche Landwirtschaft, Bund Naturschutz in Bayern, e.V, Erklärung von Bern, Gesellschaft für Ökologische Forschung, Greenpeace Deutschland, Verband Katholisches Landvolk, Zukunftsstiftung Landwirtschaft, & V.S. Navdanya. 2012. *Opposition to Patent EP 1 962 578* [Patent Opposition]. https://www.no-patents-on-seeds.org/sites/default/files/2018-05/opposition_to_melon_patent.pdf.

Thompson, P.B. 2008. Agrarian Philosophy and Ecological Ethics. *Science and Engineering Ethics* 14 (4): 527–544. https://doi.org/10.1007/s11948-008-9094-1.

Turnšek, M., S.-L.G. Skar, M. Piirman, Ragnheieth, Thorarinsdottir, ur I., M. Bavec, and R. Junge. 2022. Home Gardening and Food Security Concerns during the COVID-19 Pandemic. *Horticulturae* 8 (9). https://doi.org/10.3390/horticulturae8090778.

Van Herten, J., B. Bovenkerk, and M. Verweij. 2019. One Health as a Moral Dilemma: Towards a Socially Responsible Zoonotic Disease Control. *Zoonoses and Public Health* 66 (1): 26–34. https://doi.org/10.1111/zph.12536.

Veeresham, C. 2012. Natural Products Derived from Plants as a Source of Drugs. *Journal of Advanced Pharmaceutical Technology & Research* 3 (4): 200–201. https://doi.org/10.4103/2231-4040.104709.

Višak, T. 2018. Utilitarian Plant Ethics. In *Plant Ethics: Concepts and Applications*, eds. A. Kallhoff, M. Di Paolo, and M. Schörgenhumer. Routledge.

Wilkinson, K., W.P. Grant, L.E. Green, S. Hunter, M.J. Jeger, P. Lowe, G.F. Medley, P. Mills, J. Phillipson, G.M. Poppy, and J. Waage. 2011. Infectious Diseases of Animals and Plants: An Interdisciplinary Approach. *Philosophical Transactions of the Royal Society B: Biological Sciences* 366 (1573): 1933–1942. https://doi.org/10.1098/rstb.2010.0415.

Wolch, J.R., J. Byrne, and J.P. Newell. 2014. Urban Green Space, Public Health, and Environmental Justice: The Challenge of Making Cities 'Just Green Enough.' *Landscape and Urban Planning* 125: 234–244. https://doi.org/10.1016/j.landurbplan.2014.01.017.

World Health Organization. 2003. *WHO Framework Convention on Tobacco Control* (p. 36). World Health Organization. https://iris.who.int/handle/10665/42811

World Health Organization. 2015. *Global Action Plan on Antimicrobial Resistance*. World Health Organization. https://iris.who.int/handle/10665/193736.

Zatechka, S. 2016. Reservoir-Targeted Vaccines as a One Health Path to Prevent Zoonotic Disease. *International Journal of Vaccines & Vaccination* 2 (6). https://doi.org/10.15406/ijvv.2016.02.00049.

Chapter 19
Ethical Issues in Novel Plant Technologies

Lotte Asveld

Abstract Novel plant technologies present increased opportunities for rational plant management, promising more sustainable and efficient agricultural practices. However, questions about their desirability also arise. Many novel plant technologies emanate from a mechanistic, reductionist perspective on living organisms, whereas from a holistic perspective, other solutions to sustainable food production may be more desirable. These diverging perspectives lead to debates on how to assess the safety of novel plant technologies, achieve global economic justice against the backdrop of the increased mechanisation of plant breeding, and how to achieve sustainable agriculture. This latter debate evolves in the context of the clash between agroecological practices and rational, digitalized agriculture. These debates and the most prominent arguments will be discussed in this chapter. As such the chapter provides an agenda of issues that need to be addressed to arrive at a societally robust evaluation of novel plant technologies.

19.1 Novel Plant Technologies Revive a Recurring Debate

Novel plant technologies such as genetic editing and optimized breeding approaches provide unprecedented means to control the genetic make-up, growth and application of plants. These technological interventions raise wider questions about how human beings should relate to plants and which kind of agricultural practices are desirable. Will plant technologies enable us to produce in an efficient and sustainable manner and hence to continue to feed the world in view of a rising world population? Or will they actually disturb agro-ecological practices and a holistic relation to plants and soil, thereby actually undermining sustainability? The societal debate around these questions is infused by perspectives on plants that either portray them as passive material or as valuable members playing their own part in a vibrant ecological community (Rentmeester, this volume, Chapter 13, Brooks, this volume, Chapter 4).

L. Asveld (✉)
Delft University of Technology, Delft, The Netherlands
e-mail: l.asveld@tudelft.nl

M. Di Paola (ed.), *The Vegetal Turn*, The International Library of Environmental, Agricultural and Food Ethics 38, https://doi.org/10.1007/978-3-031-68671-9_19

These perspectives lead to opposing views on the desirability of novel plant technologies, especially in the light of the need for sustainable agricultural practices and the shift to a sustainable bioeconomy (Asveld et al. 2020).

While this debate continues to evolve, novel plant technologies are developing at a fast pace. Relevant technologies include genetic editing techniques such as Crispr Cas, increased data screening opportunities and robotics that allow for smart agriculture as well as optimized plant breeding approaches. They promise efficient, sustainable and climate adaptive agriculture, through improved crop management systems and rational trait selection in plants.

These novel technologies reframe existing concerns about how we should relate to plants within a techno-agricultural system. Salient issues include sustainability, safety, economic justice and naturalness. These issues will need to be addressed to achieve sustainable agriculture in all aspects of the word, i.e. both social, economic and ecological. This chapter will discuss each of these topics with reference to current developments in novel plant technologies.

19.2 Technologies for Rational Plant Management

The above mentioned set of related technological developments currently warrant interest because of its increasing impact on agricultural practice and the ethical and social issues it raises. These technologies interact with each other to create a highly rational plant management approach in which information on the genetic, molecular level is linked with information on ecosystem level, i.e. the design of crops can be adapted to specific local circumstances (Clapp and Ruder 2020).

Crips Cas is a novel gene editing technique that allows for extremely precise engineering of crops (and other organisms), far beyond what is possible with traditional genetic modification techniques (Pickar-Oliver and Gersbach 2019). It has been likened to the cut and paste functionality for text, but instead for genomes (see for instance Wang et al. 2015). There are huge expectations about the possible positive impacts this tool can have on improving crop resilience and productivity for instance through creating drought resistance or disease resistance (Zhu et al. 2020).

The possibilities to apply Crips Cas in a targeted manner to achieve any desired trait is supported by the high-volume genomic data sequencing and assembly that is becoming more and more mainstream. Whole plant genomes are currently being mapped at high volume and with increasing reliability, thereby opening up promising avenues for optimizing plants at the molecular level at low costs (Gladman et al. 2023), especially in combination with phenotype data which measures the prevalence of specific traits. Machine learning allows to sort and analyse these vast amounts of data to enable a sophisticated understanding of plant functions and responses to their environment, leading for instance to more insight into plant stress factors (Singh et al. 2016).

Ubiquitous data collection thus extends abilities to improve the individual plant, at the same time it also offers ways to improve the overall management of plants within their (agricultural) eco-system. This data is derived from actual plants, but also from sensors and visual monitoring systems that assess soil and climate conditions at specific locations. Such data can be used for smart agriculture, i.e. optimizing crop management to increase both productivity and sustainability (Basso and Antle 2020), but it can also be used to assess and manage biodiversity. Environmental DNA (eDNA) is the latest tool by which samples from water or soil are sequenced for traces of DNA from various species to monitor local biodiversity (Stephenson 2020).

These technologies hence offer prospects to achieve sustainable, efficient, and biodiverse agricultural and other plant practices. However, as stated above, they also raise concerns. These concerns can be traced back to diverging views on the relation between plants, humans and the wider ecology. According to one view, a rational approach to sustainability entails a focus on life from a molecular perspective, in which the smallest parts of plants and other organisms can be re-arranged to create the most efficient living production platforms (Vivien et al. 2019). Another view, which is also expressed in the vegetal turn, is that living organisms, and especially plants, should be understood and valued as participants within their wider ecological, geographical and cultural context (Rentmeester, this volume, Chapter 13). These diverging views have implications for the evaluation of novel plant technologies, as will be discussed below.

19.3 Safety

Safety is always a top concern of anyone dealing with novel technologies and this is especially true for genetic modification of plants. Although genetically modified crops have been used for decades now, there are still widely diverging perspectives on their safety (PewResearch 2020). The EU has traditionally held a precautionary approach to this technology, while the US has not considered genetic modification to be potentially more harmful than other breeding technologies (Asveld 2007). Many European countries still completely ban the cultivation of genetically modified crops within their borders (Habets et al. 2019).

From a reductionist, mechanistic perspective on plant DNA, genetic modification does not need to be considered as very risky. According to this view, control over the plant can be effectively exercised at the molecular level. Redesign of DNA presents great opportunities to control the plant and adapt it to changing circumstances. Also, a modified plant can be tested for any toxicity or pathogenity, thereby eliminating harmful effects in early stages.

However, from a more holistic perspective, changes at the molecular level can have unwanted effects in the larger complexity of the surrounding ecosystem. These effects may not be directly perceptible at the level of the individual plant. They may manifest only in the long-term when it is too late to revert them. Such effects can include the demise of a previously healthy ecosystem or the collapse of a vital insect population.

Novel gene editing technologies bring a new dimension to the existing divergence in views. With technologies such as Crispr Cas, a breeder does not necessarily need to insert foreign DNA to make changes to a plants genetic make-up. The insertion of foreign DNA has always been a major source of concern for opponents of genetic techniques. Proponents argue that these concerns can be put aside with the advent of Crispr Cas technologies.

However, Crispr based editing technologies are still treated with caution by the EU, to the dismay of many researchers and companies who argue that these technologies are safer than other breeding techniques because they are more precise and often produce traits that could have also emerged naturally (Macnaghten and Habets 2020). Additionally, the expectation is that Crispr Cas could produce great benefits in terms of food security (Wada et al. 2020). Therefore, the argument goes, this technology should be treated as classic mutagenesis breeding techniques instead and hence be subject to less strict safety regulations than traditional GMOs.

Opponents however question the proposed safety of Crispr Cas technologies. These technologies are still relatively new and their safety hasn't been proven yet. There is evidence of unwanted off-target effects in organisms after they have been edited. Also, there are many modification steps required at the cellular level, each of these steps potentially brings about risks. Lastly, there could be wider ecological effects if the use of gene-editing techniques leads to the development of monocultures of super plants, similar to what happened with genetically modified crops (Habets et al. 2019).

So far, the European Court of Justice has ruled that gene editing technologies should be treated as GMOs and hence are subject to strict safety regulations. There is an ongoing public debate about whether the European Commission should uphold this position or change to a more lenient regime for gene-editing. At the time of writing (April 2024) the European parliament has accepted a proposal by the European Committee to regulate non-transgenic crispr cas technologies similar to non-genetically modified crops. This was met with great enthusiasm from academics in crop engineering and the agro-industry but disappointed many environmental groups and ecological and environmental scientists. This proposal still has to be accepted by the member states.

A possible approach to deal with the issue of safety of novel plant technologies is Safe-by-Design, which is rapidly gaining ground in policy and research circles (Robaey 2018). Because it is difficult for regulators to address potential risks of new technologies in a timely manner, the idea is that the individuals who are developing novel technologies, also identify potential risks and ways to mitigate those risks. After all they know the most about such novel technologies and hence can be expected to be able to design ways to guarantee safety (Bouchaut and Asveld 2021). Potential Safe-by-Design strategies for plant biotechnology are physical containment or controlling evolution by making hybrids of modified plants so that they are not able to produce offspring (Robaey 2018).

However it can be questioned whether such strategies will convince groups or individuals who are concerned about the safety of engineered plants, because these concerns often arise from a wider set of beliefs and values in which safety is only one element, or because the safety concerns emanate from the belief that plants and their wider ecological surroundings cannot be adequately controlled at the molecular level (Asveld et al. 2020). At the same time, a Safe-by-Design approach could provide a middle ground between precaution and innovation, and in that respect does seem promising for policymakers and researchers.

19.4 Economic Justice

Another contentious issue driven by novel plant technologies is that of ownership, access and benefit sharing of genetic information derived from plants. There is a long-standing international debate around the question of who should receive benefits resulting from the use of genetic, biological resources, whether they are derived from animals, plants or micro-organisms. This may for instance be relevant if a cosmetics company uses plants because of properties that are part of the traditional knowledge of indigenous communities. An example of this is the use of Argan oil by l'Oreal, which was traditionally used by local communities in Morocco. These communities reached an agreement on receiving both monetary and non-monetary benefits (such as health care and literacy programmes) with l'Oreal in exchange for producing pressed Argan oil cakes (Robinson 2014).

The Argan oil case presents an example of how companies might share benefits from biological resources and associated knowledge in a responsible way. However, such sharing does not always go well. There are many cases of so-called bio-piracy in which powerful multinationals exploited the knowledge of local communities without sufficient compensation (Robinson 2010). Also, it may not always be easy to identify with whom the benefits of the use of a genetic resource should be shared as there may be multiple holders of the same traditional knowledge or it is difficult to establish what kind of benefits these holders should get (Lucas et al. 2013).

Several UN frameworks have been formulated which stipulate that for any commercial use of genetic information the donor country or community should be rewarded (Morgera 2016). One of the most prominent of these agreements is the Nagoya Protocol. This Protocol issued in 2010 is built on the premise that protection of biodiversity can be served by sharing access and benefits of genetic resources. Countries rich in biodiversity will be incentivized to protect biodiversity if they can reap benefits from this. Although this protocol has received wide-spread support from more than 60 countries world-wide and has been implemented in many European countries, its execution knows many difficulties and there are questions whether it reaches its goal (Wynberg et al. 2021).

19.5 Digital Sequence Information and Precision Fermentation

The development of ever faster genetic sequence technologies in combination with DNA synthesis and gene editing techniques such as CRISPR have complicate matters even further, as they have rapidly increased the use of Digital Sequence Information (DSI). DSI consists of DNA information that can be spread via digital channels. A researcher can download genetic information from anywhere in the world and use it to modify or construct any organism. This movement towards increased digitalisation in biotechnology has also been termed 'dematerialisation'. Although such technologies offer increased opportunities for researchers and developers, they risk undermining current access and benefit regimes. Digital sequences are usually much harder to trace compared to physical specimen, both upstream and downstream of a biotech innovation trajectory.

The use of DSI opens up the potential for novel microbial production platforms that can replace the use of plants. Specific compounds that are usually derived from plants, such as aromas or pharmaceutical ingredients, can also be produced through precision fermentation. This technology uses modified micro-organisms to process feedstock such as sugar or CO2 into desirable end-products. The micro-organisms are able to produce these end products because they have the relevant DNA sequences built into their metabolic pathways. These sequences have originally been derived from the plants that traditionally produce these compounds.

Precision fermentation has for instance already been applied to artemisinin and vanillin. Artemisinin is an ingredient for anti-malaria drugs that is traditionally derived from Chinese sweet wormwood plant, while vanillin is traditionally derived from the vanilla orchid. The fermentation production pathways have on the one side been hailed as more sustainable, cheaper and reliable alternatives because they require less land and can produce a more consistent quality than traditional plant-based production processes. On the other hand, such fermentation has been criticised as undermining traditional livelihoods and agroecological practices (Asveld et al. 2020). Traditionally many smallholders take care of crops such as the vanilla orchid and the Chinese sweet wormwood plant. Because the synthetic alternative can be produced more efficiently, at lower costs and with greater purity, critics fear that these alternatives will undermine the livelihoods of smallholders. A current pressing problem is also how to ensure fair access and benefit sharing in relation to these production platforms that rely on the use of DSI.

A question underlying many of these issues, is whether the origin of genetic information holds any specific relevance to the actual genetic code. The perspective of many bioengineers is that genetic code is comparable to computer code and that a cell is a programmable entity (Calvert 2012), ready to express any substance as built into its metabolic circuitry. In this view, the origin of the genetic code is completely irrelevant, all that matters is its function (Roosth 2017). However, others may argue that genetic information is actually only relevant within a specific biological context. That is where it has an impact and that is also how it was shaped in the first place. Such a perspective likely aligns more with the holistic vision of nature and environment as often found with indigenous people (Right to food and nutrition Watch 2018).

These diverging perspectives lead to a different assessment of the importance of the origin of genetic information. If genetic information is considered as intimately connected to its environment, this environment should at least be acknowledged and recognized in any further uses of the genetic information. Relevant aspects of this environment could be both ecological, geographical and cultural. A more mechanistic, reductionist approach to genetic information gives little incentive to value the environment from which the genetic information was derived.

During the meeting of the UN Convention on Biological Diversity in 2022 in Montreal it was decided that DSI would be managed under a, yet to be developed, multilateral framework, instead of under more complicated bilateral agreements for access and benefit sharing. In such a framework, all users of DSI would contribute to a common fund which would then be used to provide benefits to donors of DSI. However, open questions remain about how to arrange and formulate these benefits. In the case of vanilla for instance, which is already manufactured through precision fermentation by engineered microbes, it is unclear who should be the recipient of benefits of the use of DSI. The vanilla orchid is now mainly grown in Madagascar, but originally originated from Mexico (Scott and Berry 2017). Furthermore, existing frameworks for access and benefit sharing such as developed under the Nagoya Protocol are being criticised for functioning as a procedural safeguard instead of an incentive for 'potentially transformative collaborations in light of indigenous understandings.' (Morgera 2019).

It has been argued that access and benefit sharing should move away from being a compensation and instead offer something that is structurally beneficial to communities (Laird et al. 2020). In this regard it is to be applauded that the UN has decoupled benefit sharing from actual access to specific DSI. By doing this the focus is moved away from individual actors and their bilateral agreements and instead attention is given to the wider economic system and its problems. The fundamental issues underlying this debate are about global inequality and the reparation of historical (colonial) injustices. What is needed a comprehensive approach that addresses these issues, while at the same time allowing room for sustainable innovation and production. However, how this can be done and who should take responsibility for this remains an open question.

19.6 Food Security and Sustainability

The combination of gene-editing techniques with fine-grained data management promises a potentially game-changing approach to agriculture which could help feed the world under increasing ecological and environmental stress (Klerkx and Rose 2020). Digital agriculture comprises technologies such as artificial intelligence, robots, internet of things and involves huge data processing (Wolfert et al. 2017). Crop-sensors can send real time date to farmers in order to improve their agricultural management, or be used to guide robotic systems in harvesting cycles. This set of technologies is expected to improve efficiency within food production systems as well as post farm monitoring and hence improve food security, safety and sustainability (Wolfert et al. 2017) and improving animal health (Rotz et al. 2019).

However, questions have been raised about the impact of such technologies on existing agricultural practices. Some argue that large integrated systems will come at the expense of the autonomy of farmers, who are increasingly reduced to small elements in a digital agricultural production machine (Clapp and Ruder 2020). Other concerns involve the alienation between farmers and their animals (Blok and Gremmen 2018) and an increased industrialisation of agriculture at the expense of agro-ecological practices (Rotz et al. 2019).

The agritech giant John Deere has developed a smart tractor that collects data from crops and soil through sensors. The software as well as the data are not openly accessible, instead farmers can subscribe and pay for access to this data. While such developments may indeed increase productivity and sustainability because of targeted pest control and fertilizers input, it can be questioned what they imply for the self-determinacy of the farmer. Also, questions can be asked about who this data collection ultimately serves, is it broader societal goals or private corporate interests (Bronson 2018)? Companies like Bayer and Syngenta can use this data to improve their products instead of striving to reduce reliance on chemical inputs.

Whether the sustainable promise of digital agriculture is deemed feasible hence depends very much on the definition of the root problem. If efficiency is considered to be the main problem of current agricultural practice, then digital agriculture could be a solution. However, there are many voices who point out that the grip of the agri tech sector on agriculture is actually the main problem and any technology that allows this group to tighten their grip can only affect food security and sustainability in a negative way. According to this perspective, the real solution lies in diversified, local, agro-ecological approaches (IPES-Food and ETC Group 2021). In this view, digital agriculture may lead us to more efficiency in the short term, but not to sustainability in the long term.

19.7 Naturalness

The issue of sustainability and efficiency is closely linked to that of naturalness. How natural is it for human beings to exert extensive control over plants? And if it is not natural, how problematic is that? Is natural inherently good? As such, the concept of naturalness raises two distinct issues. Firstly, there is the more fundamental question of how human beings should relate to nature. Secondly there is the question of what natural is exactly and which products can be termed natural and which cannot.

The first issue is the more fundamental one that originates from differences in values, beliefs and convictions. For some people (such as environmental activists) nature is something fragile that should be treated with the utmost care, so as not to upset vulnerable ecological balances. For other people (such as some working in bioengineering or in high-risk investment), nature is essentially a resource, which has many wonderful things to offer and can provide viable solutions to pressing problems (Asveld and Stemerding 2018).

These opposing worldviews lead to widely divergent evaluations on the desirability of novel technologies. For some stakeholders, any human intervention in living organisms such as plants is unwanted, because these organisms should be understood as essential elements within a wider fragile ecosystem (Shiva 1993; De Witt et al. 2017). Others may see human interventions in living organisms and especially plants as necessary conditions to achieve and maintain a liveable planet (Gladman et al. 2023; De Witt et al. 2017).

These clashing perspectives have been most notable around the use of genetically modified seeds, but also emerge around novel developments such as the production of aromas through microbial cellular platforms (Asveld et al. 2020). If vanillin is derived from industrial fermentation, can it be called natural? Fermentation is a traditional technique which is embraced by many in the natural food and health movement, yet production of ingredients through industrial fermentation would likely not be considered natural by this group.

Crispr Cas based gene editing techniques could shift the traditional dichotomy between natural and artificial, because edits made by Crispr Cas could potentially also occur naturally. This raises the question whether consumers for whom 'naturalness' is an important quality in their food, will adjust their evaluation of edited plants, if this is based on Crispr Cas. There is indeed some evidence in this direction (Delwaide et al. 2015). However, for many consumers, naturalness is not just related to the process of breeding a new plant species, but also relates to the agricultural system in which this plant in used (Bartkowski et al. 2018). Innovations that are likely to lead to monocultures will be considered less natural by many people compared to innovations that support organic, agroecological practices (Asveld et al. 2020).

19.8 Conclusion

The broad societal debate on the desirability of highly rational, technologically mediated interventions in plants continues to be unresolved. Novel technologies such as Crispr based gene editing and digital agriculture have reiterated this debate and brought some new angles to recurrent issues. Will novel gene editing techniques bring more or less safety? How can we ensure fair access and benefit sharing when genetic resources become increasingly digitized? Will the various technologies framed as digital agriculture help us to achieve sustainable agricultural practices? And how should we evaluate these new technologies in terms of naturalness? This chapter does not provide conclusive answers to these questions, but by indicating the societal issues that novel plant technologies raise, it might help to find ways to formulate robust and convincing answers.

References

Asveld, L., and D. Stemerding. 2018. Social Learning in the Bioeconomy: The Case of Ecover. In *Experimentation Beyond the Laboratory: New Perspectives on Technology in Society*, ed. I. Van de Poel, L. Asveld, and D. Mehos, 103–124. London p: Routledge.

Asveld, L. 2007. Autonomy and Risk: Criteria for International Trade Regimes. *Journal of Global Ethics* 3 (1): 21–38.

Asveld, L., Osseweijer, P., & Posada, J. A. (2020). Societal and ethical issues in industrial biotechnology. *Sustainability and Life Cycle Assessment in Industrial Biotechnology*, 121–141.

Bartkowski, B., I. Theesfeld, F. Pirscher, and J. Timaeus. 2018. Snipping Around for Food: Economic, Ethical and Policy Implications of CRISPR/Cas Genome Editing. *Geoforum* 96: 172–180.

Basso, B., and J. Antle. 2020. Digital Agriculture to Design Sustainable Agricultural Systems. *Nature Sustainability* 3 (4): 254–256.

Blok, V., and B. Gremmen. 2018. Agricultural Technologies as Living Machines: Toward a Biomimetic Conceptualization of Smart Farming Technologies. *Ethics, Policy & Environment* 21 (2): 246–263.

Bouchaut, B., and L. Asveld. 2021. Responsible Learning About Risks Arising from Emerging Biotechnologies. *Science and Engineering Ethics* 27: 1–20.

Bronson, K. 2018. Smart Farming: Including Rights Holders for Responsible Agricultural Innovation. *Technology Innovation Management Review* 8 (2): 7–14.

Calvert, J. 2012. Ownership and Sharing in Synthetic Biology: A 'Diverse Ecology' of the Open and the Proprietary? *BioSocieties* 7 (2): 169–187.

Clapp, J., and S. L. Ruder. 2020. Precision technologies for agriculture: Digital farming, gene-edited crops, and the politics of sustainability. *Global Environmental Politics* 20 (3): 49–69.

De Witt, A., P. Osseweijer, and R. Pierce. 2017. Understanding Public Perceptions of Biotechnology Through the "Integrative Worldview Framework." *Public Understanding of Science* 26 (1): 70–88.

Delwaide, A.-C., L.L. Nalley, B.L. Dixon, D.M. Danforth, R.M. Nayga Jr., E.J. Van Loo, et al. 2015. Revisiting GMOs: Are There Differences in European Consumers' Acceptance and Valuation for Cisgenically vs Transgenically Bred Rice? *PLoS ONE* 10 (5): e0126060. https://doi.org/10.1371/journal.pone.0126060.

Gladman, N., S. Goodwin, K. Chougule, W.R. McCombie, and D. Ware. 2023. Era of Gapless Plant Genomes: Innovations in Sequencing and Mapping Technologies Revolutionize Genomics and Breeding. *Current Opinion in Biotechnology* 79: 102886.

Habets, M., L. van Hove, and Q.C. van Est. 2019. *Genome Editing in Plants and Crops: Towards a Modern Biotechnology Policy Focused on Differences in Risks and Broader Considerations.* Rathenau Instituut.

IPES-Food & ETC Group. 2021. A Long Food Movement: Transforming Food Systems by 2045.

Klerkx, L., and D. Rose. 2020. Dealing with the Game-changing Technologies of Agriculture 4.0: How Do We Manage Diversity and Responsibility in Food System Transition Pathways? *Global Food Security* 24: 100347.

Laird, S., R. Wynberg, M. Rourke, F. Humphries, M.R. Muller, and C. Lawson. 2020. Rethink the Expansion of Access and Benefit Sharing. *Science* 367 (6483): 1200–1202.

Lucas, J.C., D. Schroeder, R. Chennells, S. Chaturvedi, and D. Feinholz. 2013. Sharing Traditional Knowledge: Who Benefits? Cases from India, Nigeria, Mexico and South Africa. *Benefit Sharing: From Biodiversity to Human Genetics* 65–93.

Macnaghten, P., and M.G. Habets. 2020. Breaking the Impasse: Towards a Forward-looking Governance Framework for Gene Editing with Plants. *Plants, People, Planet* 2 (4): 353–365.

Morgera, E. 2016. The Need for an International Legal Concept of Fair and Equitable Benefit Sharing. *European Journal of International Law* 27 (2): 353–383.

Morgera, E. 2019. Under the Radar: The Emergence of Fair and Equitable Benefit-sharing as an Inherent Component of Human Rights Connected to Natural Resources. *International Journal of Human Rights* 23 (7):1098–1139.

PewResearch. 2020. *Many Publics Around the World Doubt Safety of Genetically Modified Food.* https://www.pewresearch.org/fact-tank/2020/11/11/many-publics-around-world-doubt-safety-of-genetically-modified-foods/. Last accessed March 31 2023.

Pickar-Oliver, A., and C.A. Gersbach. 2019. The Next Generation of CRISPR–Cas Technologies and Applications. *Nature Reviews Molecular Cell Biology* 20 (8): 490–507.

Robaey, Z. 2018. Dealing with Risks of Biotechnology: Understanding the Potential of Safe-by-Design. *Report Commissioned by the Dutch Ministry of Infrastructure and Water Management.*

Robinson, D. 2010. *Confronting Biopiracy: Challenges, Cases and International Debates.* Routledge.

Robinson, D. 2014. *Biodiversity, Access and Benefit-sharing: Global Case Studies.* Routledge.

Roosth, S. 2017. *How Life Got Made.* Chicago: University of Chicago Press.

Rotz, S., E. Duncan, M. Small, J. Botschner, R. Dara, . Mosby, and E.D. Fraser. 2019. The Politics of Digital Agricultural Technologies: A Preliminary Review. *Sociologia Ruralis* 59 (2): 203–229.

Scott, D., and D. Berry. 2017. Genetic Resources in the Age of the Nagoya Protocol and Gene/genome Synthesis. Workshop report, Edinburgh University.

Shiva, V. 1993. *Monocultures of the Mind: Perspectives on Biodiversity and Biotechnology.* Palgrave Macmillan.

Singh, A., B. Ganapathysubramanian, A.K. Singh, and S. Sarkar. 2016. Machine Learning for High-throughput Stress Phenotyping in Plants. *Trends in Plant Science* 21 (2): 110–124.

Stephenson, P.J. 2020. Technological Advances in Biodiversity Monitoring: Applicability, Opportunities and Challenges. *Current Opinion in Environmental Sustainability* 45: 36–41.

The Right to Food and Nutrition Watch. 2018. *When Food Becomes Immaterial, Confronting the Digital Age* (10).

Vivien, F.D., M. Nieddu, N. Befort, R. Debref, and M. Giampietro. 2019. The Hijacking of the Bioeconomy. *Ecological Economics* 159: 189–197.

Wada, N., R. Ueta, Y. Osakabe, and K. Osakabe. 2020. Precision Genome Editing in Plants: State-of-the-art in CRISPR/Cas9-based Genome Engineering. *BMC Plant Biology* 20: 1–12.

Wang, J., J. Li, H. Zhao, G. Sheng, M. Wang, M. Yin, and Y. Wang. 2015. Structural and Mechanistic Basis of PAM-dependent Spacer Acquisition in CRISPR-Cas Systems. *Cell* 163 (4): 840–853.

Wolfert, S., Ge, L., Verdouw, C., and M. J. Bogaardt. 2017. Big data in smart farming—A review. *Agricultural Systems*, 153: 69–80.

Wynberg, R., R. Andersen, S. Laird, K. Kusena, C. Prip, and O.T. Westengen. 2021. Farmers' Rights and Digital Sequence Information: Crisis or Opportunity to Reclaim Stewardship Over Agrobiodiversity? *Frontiers in Plant Science*: 1608.

Zhu, H., C. Li, and C. Gao. 2020. Applications of CRISPR–Cas in Agriculture and Plant Biotechnology. *Nature Reviews Molecular Cell Biology* 21 (11): 661–677.

Chapter 20
Herbaria and Dark Botany

Prudence Gibson

Abstract The institution of the Herbarium has played a vital role in the imperialist activity of collecting, identifying, classifying and naming plants. In Australia, herbaria follow a European model of collection, research, conservation and restoration including the latest in genetic plant research. This vital work is critically important for understanding changes in climate, loss of habitats, extinctions in the wild and how to redress issues of endangered plants. Australia's herbaria also use the Linnaean system of naming—Latin and common names—and are sites of exquisite beauty and hidden narratives, where sheet after sheet of specimens reveal the animism of their plants' lives. The colonial pursuits of plant collecting, although they have been effective in establishing herbaria plant collections, can also be seen as forms of excessive dominion. As Eurocentric, androcentric and anthropocentric activities, colonial plant-collecting has a legacy of human and non-human casualties. The collecting and classifying of plants have been achieved at great and violent cost to First Nations peoples, their land, and related marginalised groups. This chapter follows a Dark Botany methodology—the acknowledgement of darker histories and a focus on redressing past injuries. For instance, there are now calls to tri-name, that is, to add an Indigenous name to the database of Latin and common names. There are calls to decolonise herbaria archive collections, by revealing the 'true' stories of the plants and collectors. There are calls to acknowledge that First Nations peoples already knew the names of plants and always understood their utilitarian and medicinal applications. This chapter focuses on the methodology of decolonising plants within the context of the Sydney herbarium collection. It attends to a method of decolonising plants by decolonising humans first. This is Dark Botany, a process of telling alternative botanical stories. The chapter will interrogate the human-plant relations within the Sydney Herbarium, to define an environmental humanities methodology of Dark Botany.

Keywords Decolonising Plants · Herbaria · Dark Botany Methodology · Archives

---

P. Gibson (✉)
University of New South Wales, Sydney, Australia
e-mail: p.gibson@unsw.edu.au

M. Di Paola (ed.), *The Vegetal Turn*, The International Library of Environmental, Agricultural and Food Ethics 38, https://doi.org/10.1007/978-3-031-68671-9_20

20.1 A Methodology for Herbaria

This chapter focuses on the institution of the herbarium and an Australian decolonial critique within the 'vegetal turn.' The 'vegetal turn,' in the context of Australia's ongoing colonial culture, refers to new plant perceptions and values as modes of engaging with the natural world, but which require significant redress. This redress requires a process of Dark Botany, which is 'a methodology of counter-narratives that acknowledges the systematic instrumentalization of plants in botany, in philosophical thinking and its effects on aesthetics. The methodology is one which notes monoculture and extinction threats, food security and biodiversity problems. It tells true stories of colonial plant oppressions and historical extraction for profit, known as 'green imperialism'' (Jottkandt and Gibson 2024).

Plants can no longer be perceived as flat and lifeless, despite this being a common misconception with respect to herbarium specimens. Plants should not be instrumentalised and diminished by masterful colonial approaches. They should not be considered the backdrop to human (white, male, privileged) action. Within the Sydney Herbarium collection, there has been research from 2020–2024 undertaken to *re-value* plants and to critically analyse the role of art, design and narrative in the remediation, conservation and decolonisation of plants' environments (Gibson 2023). This herbarium research serves as an example of creating counter-narratives and forming dark botanical methodologies.

Plant aesthetics is intrinsic to the Sydney Herbarium and its recent art, design and narrative projects. Plant aesthetics sits at the nexus of the vegetal world and visual art. It has been informed by changes and discoveries in plant science, albeit via western (or modern) scientific methodologies. This has altered the way humans perceive and value plant life in the twenty-first century. These changes have contributed to the 'vegetal turn's' shift towards a more equal register of human and non-human being: humans now know that plants can learn, remember, communicate, and evolve quickly (Parise et al. 2020; Calvo et al. 2020; Marder and Gagliano 2019). This has triggered fundamental questions in the humanities: If plants can no longer be considered inert and unmoving, nor a mere background for human action, and are capable of vast and distributed information networks, how should art and narrative relate to the vegetal world? Is there even an appropriate place for the humanities, if humanism has proven itself to be the obstacle to good plant-human relations? (Plumwood 1993). These questions have been taken up by a wave of philosophical and ethical inquiries into human-plant ontologies and ecofeminist understandings of human-plant relations (Vieira et al. 2016; Van Dooren et al. 2016 etc.).

Aesthetic inquiries into herbarium plants have been uundertaken at the Sydney herbarium, in part as a response to the ecological work of Monica Gagliano (Gagliano 2020) particularly her laboratory research proving plants can learn, memorise and make decisions. It also acknowledges the plant philosophical work of Michael Marder (2016, 2019) for his work re-assessing our human interactions with plants on a more equitable philosophical plane. Dark Botany, as a form of critical decolonial research into the institution of the herbarium, includes an investigation of how Australian

Indigenous peoples' cultural knowledge was often not acknowledged or recorded in herbaria documents (*The Herbarium Tales* 2024). The tools of Critical Plant Studies (CPS), a space where plant science, philosophy, literature, and the arts converge, have provided the critical framework for a deep evaluation of how plant collections, including those archives in herbaria, affect our human relationships with plants and reflect our inter-human relations, and the accompanying structures of historical human-plant domination. This critique helps clarify what the problems are, such as ongoing disputes about name changes, who gets to name plants, which material is made accessible to the public or not, which plants attract the most genetic research testing and why—and where cultural and political absences, and even violence (such as erasure of Indigenous cultural plant knowledge) have occurred. This chapter attends to these issues via the Dark Botany methodology.

20.2 Colonial Erasures and New Decolonial Art

Herbaria around the world are sites where plants are collected, dried, mounted on a page and then stored in archival boxes. Specimen sheets present the dried plant, taped or glued to the paper page, which is then annotated with information about where the plant was collected and by whom, along with any superfluous fieldnotes. This forms a large dataset of vegetal knowledge and also carries additional rich cultural knowledge. In the Sydney herbarium, for example, one sheet reads, 'beside a low rocky hill with nepheline deposits, not somewhere to be readily accessed. I don't expect to return to it.' This note was hand-written next to a lost paenula daisy specimen and introduces the personal and anecdotal.

The paenula information reflects the circumstances in which the plants were collected: it reveals more than how the plant was collected, and suggests why it may never happen again. The wider data, at least in Sydney's herbarium, also reveals some discrimination, even racism, due to the erasure of information about the women and Indigenous peoples who were involved in the collecting of many plants. This is the history of Dark Botany. The associated erasure brings to mind one of the deepest influences on a critical examination of herbaria, that is, the work of Australian eco-feminist Val Plumwood. As Plumwood explains, one of the most common forms of denial of women and nature is backgrounding and instrumentalising (Plumwood 1993). This refers to the removal of women and the natural world to the distance, as other, as inert but useful, so that male action can be performed. Women and nature are reduced to assets that can be utilised by and for a male patriarchy.

A solution is to break those historic associations of women with plants, in order to liberate women from an inferior ontological register (Gibson and Gagliano 2017). However, Plumwood warns against rejecting the close association of women with nature, or to see women-nature connections as the failure of humanism, because to remove women from nature results in further human isolation from nature. It perpetuates a violent and continued progression of exclusion, and a further devaluing of nature. In describing who undertakes these kinds of exclusions and devaluing, Plumwood refers to the 'white, largely male elite.' She calls it the master model (Plumwood 1993). The master model is alive and well in the herbaria of the world, including the Sydney Herbarium. There are missing stories, missing acknowledgements of who did the real collecting, who the guides with specialist skills were, and how the plants were often gathered by Indigenous guides. These missing stories are the historical status quo, against which Dark Botany works to disrupt the darker tales of our botanical histories.

One of the best resources for information on these missing acknowledgements, particularly of who collected plants during the colonial era in Australia between 1780–1900, is Philip A. Clarke's 2008 book, *Aboriginal Plant Collectors.* This book fills in the herbarium's gaps by naming and acknowledging such collectors as the Aboriginal guides of the Governor of the New South Wales colony Arthur Phillip, Balloderee and Colebee (Clarke 2008). Female Aboriginal women also worked as plant collectors: Matora, wife of Boongaree, collected specimens for Russian explorer Bellingshausen (Clarke 2008); Abaroo—who was a midwife and helped famous Australian marine William Dawes collect Aboriginal words in the 1790's in Sydney —collected plants for British navigator Captain Flinders (Clarke 2008). It is unknown if these Indigenous collectors were paid or recompensed for their work. They are not acknowledged on herbaria collection data, except for one small phrase on one sheet in the Sydney Herbarium which reads 'Got by Dan'- which refers to Aboriginal botanical guide, Daniel Moowattin.

It is not only information about the lives and roles of Indigenous Australian plant collectors that is absent. Indigenous names for plants are also difficult to find in the Sydney Herbarium because it uses, as most other herbaria also do, the Linnaean system of plant classification, which relies on a European set of criteria and categories that do not represent Australian Indigenous structures of naming. The colonial methods of taxonomy that still exist today do not adequately account for or support multiple Aboriginal names and knowledge. Although she did not directly refer to herbaria, Plumwood best understood these systems and structures of male, European, white, privileged systemic activity and conveyed their contribution to a continuation of colonial masteries, inequalities, and lack of truthful representation.

There is a cultural move to address these missing/erased stories in plant institutions. There is also a move among herbaria around the world to engage artists to help interpret and re-present botanical life in a more sincere and kin-like way than Linnaeus' system allows. The Sydney Herbarium project, mentioned above is titled The HerbariumTales. It continues the work of Plumwood and others to critique the plant institution systems and model new ways of redressing colonial masteries. Art and design are a way of mediating difficult colonial legacies and unresolved tensions, because the effect is not overt. It is subtle and diffuse. The question scholars in the visual arts working with plant institutions such as herbaria ask is this: what can the arts do to re-value plants and the different ways of knowing them, thus filling the gaps in the way we categorize and ultimately think of them? How can the arts direct audience attention to plant conservation, restoration, and future propagation? This recent Sydney Herbarium exhibition, as part of The Herbarium Tales, included presenting artists' work as they responded to the 1.4 million specimens in the collection, all of which can be seen on the Herbarium Tales web site (Jottkandt 2022).

The first commissioned artist was Fabri Blacklock, an academic at the University of New South Wales and a Nucoorilma/Ngarabal/Biripi woman. For our Herbarium project, Blacklock worked with Aboriginal Elder Aunty Zona Wilkinson to hand dye eight lengths of silk and satin silk. These were dyed using multiple species of Eucalyptus leaves which were researched in the Herbarium collection. Blacklock disrupted the colonial elements of the collection by both referring to its wealth of information but also simultaneously turning to her community for cultural knowledge about the Eucalyptus trees, their medicinal and cultural use, and the respect and kinship that exists between plants and peoples. For the haunting sound component of her work, she collaborated with Dr Chels Marshall to record bush sounds and an Aboriginal male reciting the Indigenous names of Eucalypts. These words, so absent from the herbaria data, resounded through the exhibitions space as a reminder that Indigenous culture is not lost, nor forgotten.

Blacklock, by virtue of creating these hanging silks and a soundscape, and by hanging them in an unlikely Herbarium collection space, avoided the aestheticizing of plants. They were not diminished by a Western installation design process, because they were hung in a haphazard fashion. Her works also resisted illustrative reductivity, because they presented the remnant marks of living leaves, rather than representations. They resisted cultural spectacle because the silks still smelled of Eucalyptus leaves. This created a presence, a dynamism and an earthy liveliness that much plant art loses.

The second artist responding to the herbarium collection was Anna Raupach, who is an academic at the Australian National University. Her work made use of a dataset from the Royal Botanic Gardens digital collections database, whereby she geolocated 4000 trees based on the herbarium living collection and then created a visual/audio immersive effect app for phones. After downloading an app, visitors could walk through the bushland, and experience the altered visuals of trees and plants from her dataset. They could also experience a digital soundscape and access herbarium data information about the plants they walked past. This opened up botanical information to a wider public, creating access that is often not available to general publics.

The two artworks drew attention away from conventional naming, documenting and taxonomizing plants. They resisted conventional illustration. They instead worked to re-connect visitors with plants and trees in alternative ways. The third artist Rebecca Mayo (Wolifson 2022) worked with local Indigenous Elder Aunty Deidre Martin to create a large, hand-operated dying machine and dyed fabric with sarsaparilla, and the fourth artist Erica Seccombe also created a video following Aunty Deirdre Martin who guided the audience through the cultural and medicinal value of plants at artist residency and gallery property, Bundanon, in New South Wales, Australia.

All four artists responded to the concept of 'revaluing plants in the herbarium collection' and all four artists resisted the conventions of contemporary plant art by focussing on shared information and open access to botanical knowledge in vivified ways, rather than creating static or inaccessible experiences. Plant art is a function of aesthetic thinking and acting; consequently, it can shave close to age-old dangers of aestheticizing nature. These dangers include flattening the plants visually or diminishing their agency by framing or white-boxing them. There is also the danger of re-presenting plants *for humans* rather than alongside, as co-species. The original cabinets of presented plants and other objects that had been collected from around the world—Wunderkammer—were implicated in this diminishment of plants because they were collected as part of imperial races and nationalistic fervour. Many plants were historically collected in the context of pillaging, stealing and poaching (Gibson 2010). The task for these four artists was to resist those structures and to awaken visitors to the subtlety and complexity of plant life.

20.3 Plant Aesthetics

The aesthetics challenges, that these commissioned artists attended to, as part of the herbarium project, are not new. Through the history of nature painting, aesthetic relations with plants have changed. This can be briefly followed, through the development of nature-as-construct, ranging from seventeenth century French classical formulae of beauty and order, where a distant view is scaled by an abandoned or ruined building, and which created an aesthetic of how a visually pleasing landscape should look. That is, the trees and plants are assets and contribute to this vision of rural beauty. Land has been often presented as part of a masculine gaze upon the unconquered or semi-conquered world, as entity to be predated. Some landscape painting processes incorporated Romantic framing devices such as the fallen tree or the ancient column… all to give the effect of beauty and truth (Burke 2008, 64): these structures of aesthetic framing influenced the processes for establishing the colonial herbaria.

These aesthetic considerations have had an impact on Australian sensibilities too, as there is a history of artists presenting drought-ridden and often scrubby land in the form of a European bucolic bounty, such as the paintings of artists Von Guerard and John Glover (Smith 1945). For a long time, this false aesthetic did a disservice to the multiplicity and bounty of Australian coastal scrublands which are richly diverse and not at all similar to European nature scenes. Unfortunately, this tradition of misrepresenting nature created a division between white Australian settlers and the rough bush they inhabited.

If aesthetics is the critical judgement of beauty and truth (Burke 2008), then perhaps our more urbanised lives (in cities, at least) have also moved humans away from the material power of plants, and to the position of a distant viewer. The perceived nature/culture division and its corollaries, discussed in detail in Plumwood's writing on the mastery of nature have contributed to a deep ontological separation of humans from their plant partners, mistaken as a freedom from nature (Miller 2002, 57). These separations have affected the way we understand the natural world as something that needs to go in a frame, on the wall, on a page or in a book.

Previous human relationships with plants have been undermined by unequal distributions of power (Singh 2018) and the Sydney Herbarium plant art project worked to undermine these masteries further, by disrupting conventional plant representation.

In recent years, Critical Plant Studies, as part of 'the vegetal turn,' has impacted the contemporary art world, with thanks to such advocates as writer and curator Giovanni Aloi, who worked as editor of *Antennae Journal* for many years and promoted and commented on the work of many artists working with plants, their philosophy, intelligence, movement, and subjugation (Aloi 2019). In *The Plant Contract* (2018), I addressed these issues by investigating the areas of wastelands, wildlings, human-plant hybrids, bio-rights, ecofeminism, rhizomes, and death, by applying a critical lens to major international artists' works such as Eduardo Kac who engineered a petunia flower mixed with his own DNA, or Pierre Huyge whose wasteland scenes of weeds disrupted normative art experiences. The investigation included Kris Verdonck's *Exote* where he created a greenhouse that went feral and Maria Fernanda Cardoso whose ongoing connection with Australian plants melds with her memories of violence and colonial fervour in her homeland of Colombia (Gibson 2018).

The artists chosen for discussion in *The Plant Contract* were engaging with aesthetic ways to better connect humans and plants. The book argued that the aesthetic act—visual and performance art and design—can mediate difficult post-Edenic relations with plants, that art can encourage less hubristic relationships with the vegetal, as well as change our perceptions of plants as "other" and instead return humans to the vegetal world (Gibson 2018). By post-Edenic, I refer to the judeo-christian narrative where Adam and Eve were evicted from the Garden of Eden. From that moment, humankind forever wished for an escape from the wilderness, back to the innocent and abundant perfection of the garden. This discourse has affected our contemporary human-nature relationships by mythologising plants and nature, instead of fostering close kinship.

For plant aesthetics, the interest in plants expressed by contemporary artists, evident over the last decade and rising, suggests that artists are engaging with plants as intellectual drivers of new forms of critical thinking and art-making (Aloi 2019). Plants are no longer the inert subject of an artist's creativity, shaving too close to dominion or mastery over nature. Plant aesthetics over the last decade has been far more concerned with performative co-collaborations *with* plants and *for* plants, rather than with art *about* plants. This is a reaction against instrumentalization, as it is a methodology that seeks to avoid human use of plants for human gain, without care or consideration for the independent agency and rights of plants (Gibson and Sandilands 2021).

20.4 Mastery and Nomenclature

Herbaria plants, by virtue of the way they are collected, pressed on a paper page and stored, suffer the above aesthetic challenges of flattening and framing which results in an absence of vivification. Plant specimens are exquisite archival objects that often leave a scent of their former lives, especially the seaweed. Herbaria specimen sheets carry aesthetic complexities which become implicated in the way they are represented in herbaria—that is, for human viewing.

As part of the 'vegetal turn,' interest in how to redress mastery issues of plant aesthetics has widened. Here, mastery refers to the colonial practices of naming and classifying plants without consideration for Indigenous knowledges, nor for the need for equitable access (Gibson and Willoughby 2021). Decolonising plants is a process that responds to imperial power and historical national dominion over the lands from which many plants were collected, often without permission. Decolonising plants refers to a circular model process of learning true stories (the Indigenous stories that have been silenced or erased); decolonising non-Indigenous human selves (understanding the starting-point privilege of white humans); creating equitable access to data (not all people have access to non-Indigenous institutions equally); seeking Indigenous knowledge (seeking expert information from Indigenous Elders); and understanding the dovetailed relationship of plants and aesthetics (especially in the herbaria). Ecofeminism notes that these oppressions usually run together (Warren 1997). Facing the dark truths about these kinds of plant-human relations, in historical and contemporary contexts, reveals insights into colonial empire-building, colonial slave-trading for plantation labour (Gibson 2022), and nationalistic fervour that resulted from colonial collection and the articulation of Western knowledge of plants (Mastnak 2014).

A casualty of colonial histories is the damage enacted upon Indigenous peoples' botanical knowledge. With the rise of colonial botanical collecting, classifying, naming and keeping, there has been an equal and equivalent fall in Indigenous knowledge (TallBear 2011). In Australia, a lack of recognition of Indigenous agriculture and botany created a corollary of violence and racism (Pascoe 2014). At the same time, white settler managers of institutions, including herbaria, did not earn the trust of Indigenous peoples. Part of the reason why there is little information about Indigenous plant knowledge and uses in public records is that Indigenous knowledge-holders did not feel safe sharing their expertise with non-Indigenous institutions. In contrast, among Indigenous teachings and culture, there is a history of talking to and asking permission from plants in terms of use and kinship. This permission-asking is not generally practised in dominant white cultures (Clarke 2008).

The 1.4 million plant specimens in the Sydney herbarium collection have Linnaean names. This nomenclature includes a Latin name of family/genus/species, plus there is often a common name added to the database. As mentioned, there are almost no records among those 1.4 million records that also name an Aboriginal word for the plant. This is further problematised by the fact that there exist multiple Indigenous names for plant parts, as well as different names for different seasons of those plant parts, and different names for the same plant when found in a different Country (Gibson and Willoughby 2021). These problems of naming plants are ongoing colonial issues that are connected with the history of white settlement. Never before have Indigenous names of plants been included on the specimen sheets (except for very rare instances). Hence, naming is part of the decolonising challenge for Australian herbaria.

Mastery scholar Julietta Singh zeroes in on mastery and white nomenclature as thus: 'The power to (re)name, then, comes to signal a mode of masterful relation in which the one who names is also the one who can bestow, classify, and possess' (Singh 2018). The nomenclature processes within colonised countries also reinforce the error of using European systems of classification for trees that always already had Indigenous names. The problems of coloniality in the herbaria are deep and demand the decolonising plant model approach. The first step to elaborate is to tell true stories.

20.5 Dark Botany

Dark Botany is a concept and methodology that acknowledges past violence done to plants and Indigenous peoples of Australia. It 'approaches and actively respects the independent agency of plants. As a counter-narrative approach, telling stories outside the conventions of traditional botanical discourse, it focuses on narratives that are not mainstream and may not have been intended for public knowledge. Counter-narratives are needed to redress the preconceived notion that plants are inert, passive and only make sense within a taxonomic system of naming and classifying' (Jottkandt and Gibson 2024).

In Dark Botany, the arts have the capacity to communicate difficult political and social issues. Even for the arts, lively and dynamic plant aesthetics is a challenge to navigate, because humans don't know what plants want. However, this methodology suggests improved human/plant relationship, whereby plants are respected, their capacity and capabilities are celebrated, their permission asked. Aesthetics can function without constantly requiring human profit, and can be undertaken for the sheer joy of being kin with the botanical world.

The challenges for herbaria as discussed in this chapter suggest a need for Dark Botany, the new decolonial methodology that incorporates improved access to botanical information, increased requests for expert advice, multiple naming praxes, and an end to Indigenous erasure. These are suggestions for better humans-plants relations and a better understanding of their extensive and distributed value.

Critical Plant Studies within the 'vegetal turn' is a crucial field and Dark Botany is a crucial methodology because it incorporates analytical critical approaches to vegetal challenges and works to redress inequities and past violence towards plants. It is a field of study immersed in aesthetics and arts praxes (Aloi 2019; Gibson 2018).

Decolonising plants is the function of Dark Botany. Like all human-plant entanglements, it requires better work from humans, more empathic understanding and respect for plants as kin. This is not as easy as it might sound. Herbaria, like most colonial institutions, are established to allow sharing of plants across the world, to store, propagate, and defend plants; and to enact restoration projects and genetic research across different countries. This work is acutely relevant and necessary. Plant issues are a challenge for all humans but a strong vegetal methodology is needed.

References

Aloi, G., ed. 2019. *Why Look at Plants*. Boston: Brill Rodopi.

Becher, C. 2020. Vegetal Poetics: Narrating Plants in Culture and History. *Kult* 61. https://journals.ub.uni-giessen.de/kult-online/article/view/1015/1204.

Bennett, J. 2020. *Influx and Efflux*. Cambridge: Duke University Press.

Brill series (Critical Plant Studies) . https://brill.com/view/serial/.

Burke, E. 2008. *A Philosophical Enquiry into the Sublime and the Beautiful*. Routledge, London. Originally, 1757.

Calvo, P., M. Gagliano, G.M. Souza, and Trewavas, eds. 2020. A Plants are Intelligent, and Here is How. *Annals of Botany* 125: 11–28. https://doi.org/10.1093/aob/mcz155.

Clarke, P. 2008. *Aboriginal Plant Collectors*. Sydney: Rosenberg.

Gagliano, M., A.G. Parise, and G.M. Souza. 2020. Extended Cognition in Plants: Is it Possible? *Plant Signalling and Behaviour* 1710661.

Gibson, P. 2010. *The Rapture of Death*. Sydney: Boccalatte.

Gibson, P. 2018. *The Plant Contract*. Boston: Brill Rodopi.

Gibson, P. 2021. Earth Voice: Plant Blindness, Magic and Art. *Antennae: The Journal of Art and Nature* 53: 4.

Gibson, P. 2022, June 6. How the Art of Daniel Boyd Turns Over the Apple Cart of Accepted White Australian History. The Conversation.

Gibson, P. 2023. The Plant Thieves. New South Publishing, Sydney. https://unsw.press/books/the-plant-thieves/.

Gibson, P., and M. Gagliano. 2017. The Feminist Plant: Changing Relations with the Water Lily. *Ethics and the Environment* 22: 2.

Gibson, P., and C. Sandilands, eds. 2021. *Performing Plants issue, Performance Philosophy*.

Gibson, P., and S. Willoughby. 2021. Mastery and the Banksia Tree. Europe Now. https://www.europenowjournal.org/2021/11/07/mastery-and-the-banksia-tree/.

Jottkandt, S. (2022–2024). Herbarium tales web site: http://theherbariumtales.org/.

Jottkandt, S., and P. Gibson, eds. 2024. *Dark Botany*: *The Herbarium Tales*. Sydney: Open Humanities Press, Sydney.

Lawrence, A. 2021. Listening to Plants: Conversations between Critical Plant Studies and Vegetal Geography. *Progress in Human Geography*.

Lindon, H.L., L.M. Gardiner, A. Brady, and M.S. Vorontsova. 2015. Fewer than Three Percent of Land Plant Species Named by Women: Author Gender over 260 Years. *Taxon* 64 (1). https://doi.org/10.12705/642.4.

Marder, M. 2016. *The Chernobyl Herbarium*. Open Humanities Press.

Marder, M., and M. Gagliano. 2019. El curioso caso de la Mimosa pudica: cosas que aprende una planta. La Maleta de Portbou. https://lamaletadeportbou.com/articulos/el-curioso-caso-de-la-mimosa-pudica/.

Mastnak, T., J. Elyachar, and T. Boellstorff. 2014. *Botanical Decolonization: Rethinking Native Plants*, 32. Environment and Planning D: Society and Space.

McPherson, H. 2020. Face to Face Interview Between Prudence Gibson and Hannah McPherson at Sydney Herbarium February.

Miller, E. 2002. *The Vegetative Soul: From Philosophy of Nature to Subjectivity in the Feminine*. New York: SUNY.

Morton, T. 2018. *Dark Ecology*. New York: Columbia University Press.

Pascoe, B. 2014. *Dark Emu*. Broome: Magabala Books.

Plumwood, V. 1993. *Feminism and the Mastery of Nature*. Abingdon: Routledge.

Singh, J. 2018. *Unthinking Mastery: Dehumanism and Decolonial Entanglements*. London: Duke University Press.

Smith, B. 1945. *Place Taste and Tradition: A Study of Australian Art since 1788*. Sydney: Ure Smith.

Sumner, R. 1993. Amalie Dietrich and the Aborigines: Her Contribution to Australian Anthropology and Ethnography. *Australian Aboriginal Studies* 2.

TallBear, K. 2011. Why Interspecies Thinking Needs Indigenous Standpoints. Cultural Anthropology. https://culanth.org/fieldsights/260-whyinterspecies-thinking-needs.

Van Dooren, T., E. Kirksey, and P. Münster. 2016. Multispecies Studies: Cultivating Arts of Attentiveness. *Environmental Humanities* 8 (1): 1–23.

Vieira, P., M. Gagliano, and J. Ryan. 2016. Introduction. In *The Green Thread: Dialogues with the Vegetal World*, eds. P. Vieira, M. Gagliano, and J. Ryan. Lanham: Lexington Books.

Warren, K., ed. 1997. *Ecofeminism: Women. Culture, Nature.* Bloomington: Indiana University Press.

Wolifson, C. 2022, December 27. *Weather Based Art a Sign of the Climes*. Sydney Morning Herald.

Chapter 21
Between Jungle and *Jangala*–Vegetal Matters in an Aqueous World

Deepta Sateesh

Abstract Knowledge of plants in the Southwestern Ghats of India, prior to colonial interventions, was enmeshed in human practices and perceptions of the vegetal mediated through everyday encounters. Experience and sensorial aspects of human relatedness to plants was transmitted through mediums of oral stories, songs, and material making. Inhabitants of this monsoonal world roamed diverse terrains, walking through a matrix of sacred groves, grasslands, swamps, rocky outcrops, and cultivation lands that gave them an understanding of shifting seasons, other-than-human worlds. The rigorous colonial recordings and interpretations of vegetal life in the Ghats, using drawings and textual descriptions of local plant knowledge in this region, began in the sixteenth and seventeenth centuries when colonial travellers and surgeons were fascinated with *materia medica*. The most well-known documentation, the seventeenth Century 12-volume *Hortus Malabaricus*, found its way to experimental gardens in Europe and influenced norms of classifying, documenting, and codifying indigenous understandings in the vocabulary of a then emergent science of Botany. This chapter articulates the nuances of transformations that occurred with the recordings of plant life as the ontic register shifted from the encountered multidimensional knowledge towards static text and images; and the possibility of renewed relational understandings framed by a situated ontology.

Keywords Situated Knowledge · Vegetal World · Everyday Practices · Decoloniality · Multiple Ontologies · Translation · Vocabulary · Representation

21.1 Introduction

The Western Ghats of India, a UNESCO World Heritage Site, is known for its rich plant life found as dense, deciduous, and mixed vegetation in this hilly coastal terrain. These 'forests' and 'jungles' are one of the landscape elements used to define the

D. Sateesh (✉)
Srishti Manipal Institute of Art, Design and Technology, Manipal Academy of Higher Education, Manipal, India
e-mail: deepta.sateesh@manipal.edu

M. Di Paola (ed.), *The Vegetal Turn*, The International Library of Environmental, Agricultural and Food Ethics 38, https://doi.org/10.1007/978-3-031-68671-9_21

boundaries of the Western Ghats (Gadgil et al. 2011, 6). Characterised by their unique diversity of cloud- and rainforests, grasslands and swamps, sacred groves and private plantations, the Western Ghats have a past embedded in botanical language and imagery.

In India, the contemporary understanding of "forest", as an entity, can easily be traced to the first Forest Act of 1865 and its making, but the word and its associations have a rich and complex past of image and text, as well as from diverse practices and knowledge of people in the Southwestern Ghats. The official botanical understanding enshrined in the Forest Act came from meticulous studies by colonial surgeons, naturalists, administrators, and botanists who brought with them first an 'eye' for art and aesthetics, and later Linnaeus' classification system. Plants and trees were documented and described by their physical attributes, and simultaneously represented in elaborate drawings as objects with lines, textures, and colours.

Kumar (2008) describes the forest as "protected and productive",[1] taking for granted the idea of the forest as constituted of plants and trees as objects. In both the Western Ghats Ecology Expert Panel (WGEEP) and High Level Working Group (HLWG) Reports,[2] the forest is referred to as a collection of trees carrying the same functions. The 'productive' (Gadgil et al. 2011, 11) subcategories include plantation forests, orchards, agroforestry "dependent on rainwater", historically contributing to the economic growth of the region with the extraction of timber[3] (Kasturirangan et al. 2013; Mann 2014, 184–209; Kumar 2012) and, today, other forest produce such as honey and medicinal plants. The 'protected' subcategories include dense rainforest in inaccessible terrain, dry thorn forest, swamps, grasslands, and sacred groves. Both functions ('protected' and 'productive') call attention to the image of a particular density of trees or collection of plants, an imaging enabled by the discipline of botany, then spatialised into a map. With both WGEEP (Gadgil et al. 2011) and HLWG (Kasturirangan et al. 2013) reports using "forest" as a category to demarcate the boundaries of the Ghats, the intention seems to be for the Ghats, its vegetation and inhabitants, to become managed spaces, through various methods including joint forest management (Ghokale 2004). The categorisation, control and management of these places, and new proposals for protection or production, were/are all dependent on a worldview that relied on and attempted/attempts to establish a "universal" visuality and botanical language.

In *Green Imperialism*, Richard Grove (1995) traces the development of an environmentalist movement in Europe, from the fifteenth and sixteenth centuries through the late Renaissance period and into the Enlightenment. This was a period of exploration of the natural world across the globe, particularly of tropical environments, in search of resources as well as a (Biblical) 'return to nature' (Grove 1995). The environmental

[1] The earliest reference to forest is in Vedic literature, of imagery (described in text) of plants and trees of some value (Kumar 2008).

[2] The WGEEP and HLWG Reports were instrumental in designating the Western Ghats as a UNESCO World Heritage Site in 2012.

[3] Timber was needed for shipbuilding and the railways, and the dense forests of the Western Ghats was seen as valuable for its hardwood such as teak, blackwood, sandalwood (particularly found in the South-Western Ghats) (Cleghorn 1861).

concepts and recording methods that emerged then can be seen in nineteenth Century colonial representations of the Southern Peninsula of India, as illustrations that use certain kinds of lines (Mathur and da Cunha 2006) to see the world objectively, observe it from afar, and re-present it visually. In this way, nature and culture were divided, and both turned into resources (natural and human, respectively). These were reductions of a rich dynamic matrix of vegetal-watery-inhabitant-material relational and situated understandings into static, spatialised universal objects. The nature-culture divide is particularly evident in the contemporary forest-related conflicts in the Ghats, related to extraction of soil and mineral resources, displacement of forest-dwellers, human-animal conflicts, encroachments, and more recently disasters that appear to be 'natural'. In recent years, the Western Ghats have gained much attention from policymakers and scientists, as dramatic shifts in the terrain have been occurring during the monsoon season—impacting lives, livelihoods, food production, and access to water. Many of the areas that were devastated were productive, intensely cultivated lands, and protected forests, inherited from the time of British occupation of the southern Indian peninsula.

These contemporary conflicts in the Ghats stem, at least in part, from an inherited colonial framework of botanical representations, or imagery, and continue due to the associated vocabulary of individual objects and landscapes that is embedded in environmental policies and laws, such as the Forest Act of India (1927), and the Biodiversity Act of India (2002). What now transpires is a worldview that embraces a separation of water from land and air. This "representation" is ontologically rooted in the way nature is *seen*, *known*, and subjected to attempts of control (Castree 2014, 37–68). And yet the inhabitants of the Ghats continue to dwell in a gradient of wetness, and live with no separation of the elements (Fig. 21.1), their practices entwined and corresponding with a dynamic weather-world of emergent phenomena, intercepting wetness to appropriate conditions in everyday ordinary life.

In order to uncover pasts that we have little access to through local written accounts of the colonial era in the Southwestern Ghats, I triangulate amongst archival records, drawings and visual representations, and stories from the field. The narratives, shared in regional languages of everyday practices offer new avenues for thinking through vocabularies that hold knowledge and understandings of change, or *translation as method* (Sarukkai 2016), a way to imagine and access possibilities that have not been considered thus far.

From records of the past, the forest has been politicised and polarised. This fixed image (of forest as a collection of trees that hold a particular visual understanding) perpetuates through environmental laws, from the historical Indian Forest Act, 1865 (protecting the forest), right up to the Forest Rights Act, 2006 (protecting rights of the dwellers), as has its spatiality, but its value changed over time and the structures that govern the forest have shifted and diversified. Despite these laws, ground conflicts continue as the lives of the inhabitants and their relationship with the forest were never considered on their own terms. These protectionist laws allow the inhabitant to stay in their habitat only if they take responsibility to protect the forests and its produce, limiting their rich lives to a singular role of conservators/stewards (Abrell et al. 2009) conforming to the language of the outsider, and to the line drawn to demarcate the

Fig. 21.1 Midday mist in the monsoon of 2014, Agumbe, Southwestern Ghats, India (Image courtesy: Author)

forest. The inhabitants of the Western Ghats, displaced by these activities, did not necessarily see plants as objects separate from their environment, nor the forest as a space to destroy or exploit. There is the possibility that they carry the meaning of '*jangala*'[4] in their daily lives, which differs from the 'cultivated', 'structured', 'civilisation', collection of 'trees' in which one needs to understand the ordering principles to negotiate; while in the *jangala* one is free[5] to occupy and temporally appropriate space, negotiating with complexity and uncertainty (da Cunha 2018). In this chapter, I speculate that these negotiations are based on ordinary everyday practices corresponding with rain.

[4] Jungle comes from a pre-colonial (and perhaps non-Sanskrit) term *jangala*, which means uncultivated open land, later included "uncultivated, tropical trees, dense unordered landscapes" through Rudyard Kipling's representations in *The Jungle Book* (1894).

[5] "Free" here does not imply free will, but free from objectifying or spatialising things; "free" implies relief from fixed systems and structures.

21.2 Tracing Botanical Representations

Travellers, doctors, naturalists, and colonial administrators were all fascinated with the Southwestern Ghats, the hills and valleys, dense vegetation and cultivation, seasons and local knowledge. Their attempts at reading the region included a range of worldviews, varying from curiosity and speculation that left room for the unknown, to prescriptive designs that were adapted from their prior knowledge (of European landscapes and respective knowledge from the Enlightenment period) and contemporary information. Although these travellers, medical practitioners, traders, and rulers documented this region for centuries in various ways, it was the discipline of Botany that initiated our contemporary understanding of the Ghats as a sensitive forested region (as described in the WGEEP and HLWG Reports).

> I have seen a great many trees encircled by and entangled with other plants, while the latter, fixed in the soil with their roots, had formed such a confused mass with the tree with which they had become entangled that it could hardly be distinguished. (Draakenstein 1682a)

The overwhelming plant material that encompassed/engulfed everything in the Ghats, and the deep everyday knowledge of them and their entangled jungles fascinated the seventeenth Century Governor of Malabar, Henrik Adriene von Rheede tot Draakestein (Rheede). He put together a mammoth project of collecting, illustrating, and describing medicinal plants of this rich region. The twelve-volume *Hortus Botanicus Malabaricus*, or 'Garden of Malabar', is known as one of the most significant contributions to botany—a field that, at that time, had not yet been formalised into a *discipline*.

Rheede was able to employ almost two hundred gatherers, mostly local toddy tappers (*Thiyas* and *Billavas)*, who traversed the region's "lofty forests" bringing back leaves, barks, branches, fruits, flowers, plants. In *Hortus Malabaricus*, plants have been visualised and described as objects, along with their medicinal values and other uses (health and trade-related). His efforts in creating *Hortus Malabaricus* intended to blur the line between local knowledge and European understandings, with the books, or textual gardens, themselves evidence of collaborations amongst Itty Achuden (a local healer, or *vaidya,* who was the primary consultant for this project), Rheede, the Raja of Cochin (Ravi Kerala Varma), toddy tappers and other *vaidyas*, locals, as well as four artists, and Konkani Brahmins that formed a committee for vetting the *information* and *vocabulary* (Grove 1995, 85–87). With the intention of recording *materia medica*, Rheede created a "contact zone" (Pratt 1992, 4)[6]—a space where the plants were collected, arranged, understood, debated over, and organised. That space was on Rheede's war ships (Manilal 2012, 54), offshore in the Arabian Sea. It was there that the plant matter was gathered, information was filtered and extracted from a larger body of practical local knowledge (knowledge that could not

[6] Pratt describes the "contact zone" as "social spaces where disparate cultures meet, clash, and grapple with each other, often in highly asymmetrical relations of domination and subordination" (Pratt 1992, 6). They are also spaces of "transculturation", where the dominant and the marginalised are culturally transformed.

be represented in text or image), and re-organised and curated in ways different from those of the inhabitants.

In his writings, Rheede suggested:

> Moreover, it would involve great profit for the Illustrious East India Company, Which indeed would be able to save those expenses it spends on transporting medicaments to India. Indeed, it would be possible to make use, at less expense and with greater profit, of Indian medicaments, either the same or at any rate with the same, if not superior curative virtues, for the above-mentioned reasons. (Draakenstein 1682b, xiii)

This indicated that the plants were commercially valuable[7] (Grove 1995, 91). Trade in this region across the Indian Ocean was indeed common; but visualising the plants as abstract objects in space allowed Rheede to easily imagine the commercial interests beyond simple on-the-spot use of these vegetal matters as medicines. The sheer ability to extract and abstract the living plants from their dynamic environments and study them on a ship made it possible to imagine that these plants, these *things*, could be utilised in novel ways outside their contexts.

Rheede did not complete his study of the plants of Malabar, nor could he complete any of the volumes of *Hortus Malabaricus* in Cochin. Due to political issues between him and his former patron, Van Goens, he left Malabar in 1677 to Batavia and then to Utrecht, where he continued to work on *Hortus Malabaricus* remotely (Heniger 1986).

21.2.1 Tracing Botanical Representations: Plant as Object

Nonetheless, *Hortus Malabaricus* contributed greatly to the formulation and articulation of the discipline of botany. Rheede, while gathering data for *Hortus Malabaricus*, had European (often Dutch) artists create the hand-drawn illustrations (Manilal 2012). What Rheede appreciated more was drawing from live real subjects for the artists that were isolated (by the eye) and extracted and abstracted from their environment. Never did the artists situate themselves amidst the vegetation to draw them live, in-situ, in the environment.

Rheede had also expressed an intent not to draw "similarities and comparisons of these plants with those of Europe"—attepmting to describe the plants of Malabar on their own terms (Manilal 2012). But due to the nature and media of recording and representation, namely visual drawing, comparisons were invited—and indeed it was through comparisons of such kind that Linnaeus often proceeded in his classifications.

Plants were recorded as single, unitary, static objects, separated from their environments both conceptually and physically, as they were brought to Rheede by collectors

[7] The intention was for this new wealth to be taken back to their home countries and their colonies for further (academic) use "albeit frequently turned to more mercantile uses" (Grove 1995, 91).

on his instruction. This unitary view of a plant as an object that is singular[8] and has its own singular functions was distinctly European, infused with notions of *essence* and *intrinsic qualities*, and mingled easily with the equally European interpretation of the tropical as exotic, rare, special, extraordinary—and with the appendix that such *object*, this singular plant, is in various ways (including commercially) valuable. This tendency to isolate functions, deduce uses, and extract such usages from the "object" then enabling of such usages, leaves the larger ecology and natural phenomena in a dark, non-participative background. The plant drawings in *Hortus Malabaricus*, particularly the outlines, are of the singular plant, highlighting them as nouns, its leaves, fruits, seeds, stalk, and reproductive parts. The information provided includes lifecycle/lifetime and growth patterns (again through static drawings). As an object or individual entity, it was now possible to know the plant from reading text and viewing an image or collection of drawings of the plant and its parts.

The imagery in *Hortus Malabaricus* worked to erase the *vitality* (Bennett 2010) of plants as it is expressed through their larger ecological porosities, affordances, and capabilities in context[9]—precisely that which was known by ordinary local inhabitants of the Ghats through their everyday practices. These porosities, affordances, and capabilities, and their materialities and temporalities, remain unseen, unrecorded and unknown in Rheede's monumental works. Erased, thus absent.

Hortus Malabaricus was revered not just at Leiden Gardens and the University, but by all European travellers and scholars of that time, and until today, influencing the design of public gardens all the way from Cape Town, South Africa, to Calcutta, India (Spudich 2008; Manilal 2012; Grove 1995, 13). *Hortus Malabaricus* became, not only the main link for botanists between the southwest coast and the Ghat mountains of India and Leiden Gardens in the Netherlands, but also a key model of the way plants were imaged, named, described and grouped (Grove 1995, 78) (see Fig. 21.2).

All such information was a filtering of particular 'things' of potential commercial interest from a larger body of knowledge that, on the other hand, was excluded from *Hortus Malabaricus* entirely.[10] *Hortus'* visual descriptions extract the plant from its context and place it elsewhere–as an illustration in a book, and then in a library as reference for use ex-situ: a resource, now made available for consumption and distribution. Neither the environment nor the season, time, transformations, in which each plant was found was included, nor was there mention of the other entities that it engaged with and engaged with it. The complex contexts and multiplicities within which these vegetal matters were inhabitants, and the rich relational understandings

[8] "Surveyors were trained in the skills and principles of picturesque art as a way of documenting [..] less an aesthetic uniqueness and more an ability to isolate, categorise, and, above all, transplant things" (Mathur and da Cunha 2006, 123).

[9] "Context" stands for the totality of transformations that the plant is a part of, that which "carries on" and within which the plant is a moment in time of continuities and forces that are dynamic (see Ingold 2011).

[10] Rheede's recordings in *Hortus Malabaricus* are effectively extractions of knowledge from local practitioners, including Itty Achuden and the toddy tappers (Rheede 1682a, Preface to Volume I), and remodulations of it as information. *Information* is a processed representation of something, whereas *knowledge* is familiarity, often tacit, acquired through observation, inquiry and/or experience.

Fig. 21.2 Theka (Tectona grandis). Hortus Malabaricus, Volume IV (*Source* Draakenstein 1683, 94). The imaging of plants found in the Western Ghats allows one to understand the different parts, their functions within the plant body and a record of their size, shape, colour and texture. These measures all rely on visuality (the ability to read/see the drawings). The drawings are devoid of surroundings of biota that depend on or inhabit the plants and vice versa, as well as the conditions (atmospheric and ground conditions), enabling the viewer/reader/learner of botany, to see clearly the drawings of the plants in detail, as objects in space, participating individually.

of these by the local human inhabitants is missing; and the names of places and Linnaeus' botanical taxonomy of the plants (Linnaeus 1750) are inadequate to offer such contexts, being themselves just useful information.

21.2.2 Tracing Botanical Representations: Vegetal Vocabularies

The coastal plains and hills situated along the escarpment of the Southwestern Ghats form a very undulating terrain, of high grounds and swales (Fig. 21.3) characterised by the rich red lateritic rock and clayey soil. Although this region has been cultivated for millennia, there are few descriptions of the situated vegetal practices here in relation to rain and wetness. The region is warm and humid all year round, with the least wet moments around December and January, and the most wet during the

Fig. 21.3 Surrounded by teak plantations, *vayal* (swale), affording multiple practices of cultivation, grazing, water gathering. Muthanga, Wayanad 2018. Image courtesy: Author

monsoon between June and September. This "aqueous territory" (Gattegno 2014, 31) includes mists, fogs, moisture, vapor, steam, and other waters, and the rains themselves are of multiple types, and transform the environment in different ways across the seasons (Viju 2019).

The inhabitant's understanding of the environment, of how plants grow, their possible synchronous relationships with seasons, terrain, other biota, their participation in not just food but rituals, stories, poems and songs, how they transform through the environment and with the atmosphere, have all been excluded from the twelve volumes of *Hortus Malabaricus*. These situated, diverse processes and modes of seeing, learning and knowing that are not written or physically recorded, are the everyday ways that practical knowledge was/is acquired and transmitted through the communities and families of the ordinary human inhabitants through practice. Such knowledge was/is in itself also seasonal, enacted at different times, in different ways, and in different realms or aspects of daily life; and it contains clues to knowledge beyond a discipline (botany) or singular practices (such as medicine or trade).[11] One example in South Canara is an animist practice, *bhuta kola* (an ancient ritual dance practiced by a seer), performed during the least wet season when land is left fallow. The medium, usually from a family who is intimate with the land and its processes, advises the landlord on near future possibilities related to the land and its stewards.

[11] Such as recorded in Sangam literature (Srinivasan 2016) or found in the field where itinerant and marginalised communities carry out multiple and seasonal practices.

Other everyday practices unearth myths and stories, such as *paddanas,* that inhabitants sing into existence rhythms of movements of practices[12] , for example during the transplanting of paddy (Rai 1996). Embedded in these practices and accompanying narratives are relations beyond the social, connecting beyond the human, with the more-than-human—possibly of how to respond to certain environmental conditions. Baindur suggests that these narratives emerge from situated-ness in the environment, and are not universal, but tied to particular contexts and conditions on the ground (Baindur 2009, 45).

In a wet imagination of the Ghats, the language of vegetal matter was and continues to be related to and entwined in daily life and ordinary practices in a diversity of ways—as sentient beings, food, clothing, ritual practices, and more. When moving through and dwelling in this aqueous world, the vocabulary of vegetal matter emerges as an intimate, sensorial, experienced understanding, of place, taste, touch and texture, smell, and diverse concoctions for nourishing and/or healing the body and earth—with words such as *pulli* (sour), *nati* (local or of a particular place or conditions) and *pullu* (that which dwells near the surface). These sounds are dynamically *descriptive* of operations in the everyday, rather than statically definitory of their "objects" while they also offer a way of thinking through everyday practices.

At another scale, the forests of the Southwestern Ghats are/were visualised as collections of plant matter, groves, forests, plantations, fields, and so on. These collective words represent and encourage views from a distance; their referents are *spatial* elements of the landscape. In contradistinction, situated knowledge of inhabitants is simultaneously experienced and expressed through a vocabulary that intricately weaves intimate vegetal matter with understandings of terrain and human practice in a tactical relational manner, of knowing different 'grounds' (Ingold 2020) and changing gradients of wetness. When immersed and dwelling in the environment, plants and their habitats are part of one's experience—particularly when walking. Multiple grounds gather their own gradients of wetness, each presenting its own sets of opportunities for diverse and temporal vegetal engagements (or designs) that respond to these multiple moments of precipitation, condensation, evaporation, transpiration, and sedimentation (Bremner 2020). The verb-like vocabulary that expresses these moments of operational possibilities in the Ghats include *bailu* (low lying area), *kan* (sacred grove), *thota* (kitchen garden), *kadu* (forest), *kumri* (shifting cultivation), and *bena* (grazing ground) (Sateesh 2020). Today, such fluid terrains also host tensions of various development initiatives, including plantations for cash crops and related wildlife monitoring and management, and dams for irrigation, water supply and power. These permanent designs are based on a stark wet-dry ontology, and demand a far more static world to function.

---

[12] In his book on Australian aborigines in *Songlines*, Bruce Chatwin writes of oral stories that provide knowledge of navigating places and practices through song. In this narrative, he describes the landscape coming into existence through the oral practice of song that contains a certain vocabulary and rhythm (1998).

21.3 Conclusion

The continual extraction and abstraction from local contexts and understandings of plants into images and vocabularies of plants as objects, as treated in the discipline of botany, reduced the breadth and depth of the practical knowledge possessed and enacted by the inhabitants of the Southwestern Ghats, reducing it to information about resources.

The inhabitants of the Ghats do not see plants as objects separate from their environment, nor the forest as a space to exploit. Rather than "jungle", they possibly used the word '*jangala*'[13] (uncultivated open ground)—in their daily lives, to signify a place unencumbered by fixed structures, that one is free to occupy and temporally appropriate, negotiating with complexity and uncertainty (da Cunha 2018).

The monsoon, rain, or wetness were never recorded by colonial travellers and surveyors as active everyday participants in the Ghats. The weather and other environmental conditions were placed in the background, an operation only possible through a binary wet-dry colonial lens (Sateesh 2017) never privileging the lifeworld of vegetal matter in the Ghats and of their inhabitants. There is much to understand now, in times of climate instability, in their ways of living and considering weather as an inhabitant of the terrain and its plant life.

References

Abrell, E., I. Kohler-Pollefson, B. Lassen, G. Martin, O. Rukundo, J. von Braun, and P. Wood. 2009. *Bio-Cultural Community Protocols*, eds. K. Bavikatte, and H. Jonas. Natural Justice and United Nations Environment Programme (UNEP). https://wedocs.unep.org/bitstream/handle/20.500.11822/9819/-Bio-Cultural_Community_Protocols_A_Community_Approach_to_Ensuring_the_Integrity_of_Environmental_Law_and_Policy-2009bio-cultral-community-protocols.p.pdf?sequence=3&isAllowed=y

Baindur, M. 2009. Nature as Non-Terrestrial: Sacred Natural Landscapes and Place in Vedic and Puranic Thought. *Environmental Philosophy* 6 (2): 43–58.

Bennett, J. 2010. *Vibrant Matter—A Political Ecology of Things*. Durham and London: Duke University Press.

Bremner, L. 2020. On Sediment as Method. In: Monsoon [+ Other] Grounds, edited by Lindsay Bremner and John Cook, 13–22. London: Monsoon Assemblages. http://www.monass.org/wp-content/uploads/2020/02/Monsoon-other-Grounds_Online.pdf.

Castree, N. 2014. *Making Sense of Nature—Representation, Politics and Democracy*. Oxon: Routledge.

Chatwin, B. 1998. *The Songlines*. London: Vintage.

Cleghorn, H. 1861. *The Forests and Gardens of South India*. London: W. H. Allen & Co.

Da Cunha, D. 2018. The Jungle's Call. Harvard Design Magazine 45 (Into the Woods). http://www.harvarddesignmagazine.org/issues/45/the-jungles-call.

Draakenstein, H.R. 1682a. *Hortus Indicus Malabaricus*, vol. I. Amsterdam.

[13] Jungle comes from a pre-colonial (and perhaps non-Sanskrit) term *jangala*, later to include "uncultivated, tropical trees, dense unordered landscapes" through Rudyard Kipling's representations in *The Jungle Book* (1894).

Draakenstein, H.R. 1682b. *Hortus Indicus Malabaricus*, vol. III. Amsterdam.
Draakenstein, H.R. 1683. *Hortus Indicus Malabaricus*, Vol. III. Amsterdam.
Gadgil, M., B.J. Krishnan, K.N. Ganeshaiah, V.S. Vijayan, R. Borges, R. Sukumar, L. Noronha, V. Nayak, D.K. Subramaniam, S.P. Gautam, R.R. Navalgund, and G.V. Subrahmanyam. 2011. Report of the Western Ghats Ecology Expert Panel, India.
Gattegno, N. 2014. Aqueous Territories. In *Design in the Terrain of Water*, eds. A. Mathur, and D. da Cunha, 31–39. Applied Research + Design Publishing.
Ghokale, Y. 2004. Reviving Traditional Forest Management in Western Ghats: Study in Karnataka. *Economic & Political Weekly* 39 (31): 3556–3559.
Grove, R.H. 1995. Green Imperialism—Colonial Expansion, Tropical Island Edens and the Origins of Environmentalism, 1600–1860, eds. D. Worster, and A.W. Crosby. Cambridge: Cambridge University Press.
Heniger, J. 1986. *Hendrik Adriaan Van Reede Tot Drakestein (1636–1691) and Hortus Malabaricus—A Contribution to the History of Dutch Colonial Botany*. A. A: Balkema Publishers, Rotterdam and Boston.
Ingold, T. 2011. *Being Alive: Essays on Movement, Knowledge and Description*. Routledge.
Ingold, T. 2020. What on Earth Is the Ground? In *Monsoon [+ Other] Grounds*, ed. Lindsay Bremner and John Cook, 71–88. London: Monsoon Assemblages.
Kasturirangan, K., C.R. Babu, J.M. Mauskar, K. Chopra, J. Kishwan, D. Shankar, S. Narain, P.S. Roy, A. Tyagi, and I. Chandrasekharan. 2013. *Report of the High Level Working Group on Western Ghats*, vol. I.
Kipling, R. 1894. *The Jungle Book*. London and New York: Macmillan and Co.
Kumar, B.M. 2008. *Forestry in Ancient India. Asian Agri-History* 12 (4): 299–306.
Kumar, V.M.R. 2012. Colonialism and Green Science: History of Colonial Scientific Forestry in South India, 1820–1920. *Indian Journal of History of Science* 2 (47): 241–259.
Linnaeus, C. 1750. Botanica. Ex Caroll Linnaei, Philosophia Botanica. Stockholm.
Manilal, K.S. 2012. A Study on the Role of Itty Achuden in the Compilation of Hortus Malabaricus. In *Hortus Malabaricus and the Socio-Cultural Heritage of India*, 41–75. Indian Association for Angiosperm Taxonomy: University of Calicut, Calicut.
Mann, M. 2014. *South Asia's Modern History—Thematic Perspectives*, 184–209. London and New York: Routledge.
Mathur, A., and D. da Cunha. 2006. *Deccan Traverses—The Making of Bangalore's Terrain*. New Delhi: Rupa Publications.
Pratt, M.L. 1992. *Imperial Eyes: Travel Writing and Transculturation*. London and New York: Routledge.
Rai, V.B.A. 1996. Epics in the Oral Genre System of Tulunadu. *Oral Tradition* 11 (1): 163–172.
Sarukkai, S. 2016. Translation as Method: Implications for History of Science. *Indian Journal of History of Science* 51 (1): 105–117.
Sateesh, D. 2017. Tile Colonialism. *India in Transition. Center for the Advanced Studies on India*. Philadelphia: University of Pennsylvania.
Sateesh, D. 2020. Movement and Place-Making in a Monsoon Terrain. *Dearq* (26): 16–27. https://doi.org/10.18389/dearq26.2010.02.
Spudich, A. 2008. Such Treasure & Rich Merchandize. National Centre for Biological Sciences, Tata Institute of Fundamental Research, Bangalore.
Srinivasan, T.M. 2016. Agricultural Practices as Gleaned from the Tamil Literature of the Sangam Age. *Indian Journal of History of Science* 51 (2): 167–189.
Viju, B. 2019, August 15. Is Wayanad Facing the Brunt of Climate Change? In *The News Minute*. https://www.thenewsminute.com/article/wayanad-facing-brunt-climate-change-107285. Accessed on 24 April 2021.

Chapter 22
Circumnutating Thought: What's Ahead in the Vegetal Turn?

Michael Marder

Abstract In the brief remarks that follow, I propose to take stock, provisional though such a task necessarily is, of "the vegetal turn" in philosophy and the humanities and to reflect on its possible futures. I will do so by extending myself in time and space, reaching out to what or who is not there yet, elongating my writing, inclining it toward you, the readers, with all the oscillations and turns of a revolving nutation, implied in growth.

In the brief remarks that follow, I propose to take stock, provisional though such a task necessarily is, of "the vegetal turn" in philosophy and the humanities and to reflect on its possible futures. I will do so by extending myself in time and space, reaching out to what or who is not there yet, elongating my writing, inclining it toward you, the readers, with all the oscillations and turns of a revolving nutation, implied in growth.

After all, that is how plant organs grow: not stretching out in a straight line between point A and point B, but rotating around themselves along a spiraling track that, while it is most often elliptical, may also be circular or quite irregular.[1] Figures of turning, tropes of tropisms, are crucial to plant life, to its vivacity, health, and lush development. This volume has proceeded in the same spiraling way, each chapter taking its own spin as it shot off the conceptual nodes of contemporary reflection on plants and built a branching network of nonlinear, often irregular, sometimes elliptical evocations and connections with the others. And I, too, in saluting the readers with this Afterword, will now attempt to write circumnutationally.

You will be tempted to jump to the conclusion that you are in for an elaborate metaphor, or even an allegory. For, what have human writing and thinking to do with vegetal growth? Yet, these are not metaphors or allegories. Already in my first book on plants and philosophy, *Plant-Thinking* (2013),[2] I emphasized that the connections between plant and human processes are neither literal nor metaphorical. The relation

M. Marder (✉)
University of the Basque Country, Vitoria-Gasteiz, Spain
e-mail: michael.marder@gmail.com

Ikerbasque: The Basque Foundation for Science, Bilbao, Spain

M. Di Paola (ed.), *The Vegetal Turn*, The International Library of Environmental, Agricultural and Food Ethics 38, https://doi.org/10.1007/978-3-031-68671-9_22

is always mediated, in the worst of cases by powerful mechanisms of psychic and cultural repression and, in the best, by the dynamics of sublimation: that is why the shapes that vegetal tendencies assume outside the kingdom Plantae, and in us humans, are hardly recognizable as such. Since then, plant-thinking has kept turning around these points, orbiting them elliptically, branching out in many directions including the arts and the sciences, but returning to them each time anew.

Like circumnutational movement, the ties between human thinking and plant growth are neither straightforward (literal, directly translatable into each other's terms) nor purely metaphorical (situated on another plane of meaning altogether). The typically elliptical nature of circumnutation gives us a clue as to the complexities of these ties: an ellipse has two focal points, rather than the one (center) of a circle—a thought crucial to the late writings of Jacques Derrida. Gathering plants and humans around and in itself, plant-thinking is also both and neither literal and/nor metaphorical. Internally decentered, it depends on the noncoincidence, the fecund tension, and the chiasmatic cobelonging of two foci and of the bustling, unfocused middle between them. Stereoscopic theoretical vision, affording viewers the perspective of volume and depth, corresponds to the bifocal nature of physical ocularity.

The vegetal turn, to which plant-thinking has made its contribution, needs to be understood precisely in terms of turning, a turning that is physical as much as psychic and cultural, *toward* and *with* plants. The first kind of movement (turning toward) is tropic. Plants and, in particular, flowers that track the position of the sun in the sky turn heliotropically, a behavior that has been described as early as the works of Marcus Terentius Varro dating back to the second century BCE.[3] The second kind of movement (rotating around and with) is properly nutational, or circumnutational—the movement discussed in depth in the 1880 volume *The Power of Movement in Plants,* coauthored by Charles Darwin and his son Francis.[4]

The vegetal turn in philosophy and the humanities is, similarly, tropic— or better phyto-tropic, a turning *toward* plants, which puts them in the spotlight; but it is also a circumnutational turning *with* plants, which attempts to accompany their capacities, perspectives, manners of dwelling, and the reverberations of their lives and deaths in and outside the immediate environmental contexts of their growth. The combination of turning *toward* and turning *with* is just as indispensable to a creative, ever surprising, and lively engagement with the vegetal as two focal points are to an ellipse.

At the same time, the two turnings are, to some extent, antithetical. Turning *toward* plants entails thematization, drawing attention to all the obscure, sidelined, or previously overlooked references to and representations of plants in philosophy, art, literary studies, theology, and anthropology, among many other fields, practices and disciplines. Turning *with* plants implies the opposite—namely dethematization, treating plants not as intellectual objects but as subjects with their own sort of intelligence, rooted in their sessile, trans-elemental, symbiotic, temporally and behaviorally diverse existence.

If I privilege this latter, circumnutational movement of turning, that is because I observe a disproportionate emphasis on the former, phyto-tropic movement in the academic milieu, with a vast proliferation of studies of "Plants in X [author, work, historical period, etc.]" and often no more than lip service paid to the fine textures of plant subjectivities. To be sure, a cookie-cutter application of "theory" to materials under study triggers lightning-fast growth of apparently new fields of research, but it is growth by accretion, which has little to do with plants. Every emergent field of study freezes the very intellectual and practical turn that has given rise to it in mid-movement—and plant humanities or critical plant studies are far from exceptional here.

My point is not to insist on the unalloyed dynamism of thinking, reluctant to commit to a body of writing, of work, of thought. Stabilization is unavoidable and, strange as it may seem, vital, if only for the sake of effectiveness, that is, of producing actual effects in actuality. But stabilization is acceptable only so long as it also elliptically nourishes and is nourished by destabilization. Once you consider the nexus of life and the living, you will readily discern the shape of this ellipse: life is the indomitable destabilizing force that, in its pure state, is indistinguishable from death; it is temporarily stabilized in the living, whose bodies are substantive, as the biophysical supports for life and the future lifeless bodies, or corpses; only in a constant back-and-forth of stabilization and destabilization, forms and forming, organs and the organizing/disorganizing activity is biology actual. So, the movement of turning *with* plants, which amounts to a thinking circumnutation, runs the risk of getting lost, buried beneath a pile of correct applications of the theory, in research articles and volumes made possible by the vegetal turn. The field of plant humanities would thus fall victim to its own success.

If a crass notion of applicability undermines the phytotropic movement of thought, circumnutational thinking is also plagued by significant problems. On the one hand, there is a sense that the plants' modes of being, temporal rhythms, metabolic and physiological processes, bodily articulations, and interactions with the world are drastically different from their animal counterparts. How can we turn with plants across the ontological gap between us and them, which is a priori unbridgeable? On the other hand (as a development mirroring the first objection and constituting together with it the bifocal structure of a negative ellipse), there are those who assert that plants are persons, like us, or that we are persons like them, with every patent trait of metaphysical subjectivity deployed to support this assertion. In addition, exceedingly literal scenarios of becoming-plant are entertained, emphasizing the underlying vegetal-human sameness.

Many chapters in this volume have shown in different ways and from different angles how both these positions, of sameness and otherness, hold a grain of truth. The timescales of many plant behaviors are indeed often so divergent from animal and human actions as to render them imperceptible to us, causing a disproportionate increase in ontological distance. And yet the metamorphoses of plants that take place in human organisms are of a piece with the very basic necessities of human life, such as food or breathable air, with which plant parts and exudates permeate our bodies through and through. And, vice versa, human bodies can stimulate plant growth with the by-products of their life activity and, ultimately, with their remains.

As Marcello Di Paola noted in the opening chapter to this volume, plants are far away and yet so close to us in so many ways. But these poles, of difference and sameness, radical as each of these qualities may be, must both be kept within the elliptical structure of the plant-human interface, for them to attain philosophical and practical validity. Otherwise, each of them taken separately from the other would ultimately prompt thinking to rotate around a newly constituted center, thus leaving the orbit of vegetal life and starting to revolve around itself, in a return to a narcissistic type of cognition.

Whether dealing with sheer difference or with absolute sameness, thought thinks only itself, even in the guise of thinking its other. A circumnutating thought, for its part, turns around itself and the other—the two focal points of its elliptical movement—in an internal decentering that gives up neither the act of human thinking nor the vibrations of plant intelligence. This will be my rejoinder to Aristotle's "unmoved mover"—the thought that thinks itself in a perfectly circular, unicentric manner. Although circular patterns of vegetal circumnutation are possible, they are exceptional and should be kept so, in life and in thought.

Various chapters in this volume and the volume as a whole show the powers of both the tropic and the circumnutating tendencies in the vegetal turn. I welcome this elliptical ambivalence and its unresolved tensions—and, with that, the very vegetality of this book. And then I look further ahead. When someone asks me about my hopes for the future of plant-thinking and the vegetal turn that it participates in, my answer is this: that these ambivalences and tensions never dissolve, but rather sink even deeper, and that plant-thinking and the vegetal turn rot well—decomposing thoroughly into the subsoil of culture, to nourish from below the development of the thinking to-come.

The vegetal turn cannot, in any circumstances, come to resemble the model of accumulation (of knowledges, sources, texts, works, materials, even of kinds of subjectivity or intelligence, and especially of value) that characterizes capital and that elsewhere I have called "the dump."[5] The phase of exponential growth in our reflections on plants will become suffocating unless it gives way to decay. Growth and decay are the two elliptical foci of plant life: turning not only towards but also with plants entails learning the art of rotting well.

Notes

1. Fernando, Migliaccio, Paola, Tassone, and Alessio Fortunati. 2013. Circumnutation as an Autonomous Root Movement in Plants. *American Journal of Botany* 100 (1):4–13.
2. Michael Marder. 2013. *Plant-Thinking: A Philosophy of Vegetal Life*. New York: Columbia UP.
3. "No less remarkable is the behavior of the flowers which are called 'heliotropes' from the fact that they face the rising sun in the morning and follow its course until the setting, facing it the whole time [Nec minus admirandum quod fit in floribus, quos vocant heliotropia ab eo, quod ad solis ortum mane spectant et eius iter ita secuntur ad occasum, ut ad eum semper spectent]" (*De re rustica* I.46).
4. Charles, Darwin, and Francis, Darwin. 1898. *The Power of Movement in Plants*. New York: D. Appleton & Co.
5. Michael Marder. 2020. *Dump Philosophy: A Phenomenology of Devastation*. London & New York: Bloomsbury.

GPSR Compliance

The European Union's (EU) General Product Safety Regulation (GPSR) is a set of rules that requires consumer products to be safe and our obligations to ensure this.

If you have any concerns about our products, you can contact us on ProductSafety@springernature.com

In case Publisher is established outside the EU, the EU authorized representative is:

Springer Nature Customer Service Center GmbH
Europaplatz 3
69115 Heidelberg, Germany

The manufacturer's authorised representative in the EU is Springer
Nature Customer Service Centre GmbH, Europaplatz 3, 69115 Heidelberg,
Germany. If you have any concerns regarding our products, please
contact ProductSafety@springernature.com

Printed and bound by CPI Group (UK) Ltd, Croydon, CR0 4YY
01/10/2026
02225907-0008